Maintaining & Repairing VCRs

4th Edition

Maintaining & Repairing VCRs

4th Edition

Robert L. Goodman

TAB Books
An imprint of McGraw-Hill

New York   San Francisco   Washington, D.C.   Auckland   Bogotá   Caracas   Lisbon   London
Madrid   Mexico City   Milan   Montreal   New Delhi   San Juan   Singapore   Sydney   Tokyo   Toronto

McGraw-Hill

A Division of The **McGraw·Hill** *Companies*

pbk   1 2 3 4 5 6 7 8 9 FGR/FGR 9 0 0 9 8 7 6 5

Library of Congress Cataloging-in-Publication Data
Goodman, Robert L.
   Maintaining and Repairing VCRs / by Robert L. Goodman.—4th ed.
     p.   cm.
   Includes index.
   ISBN 0-07-024200-3 (p)
   1. Videocassette recorders—Maintenance and repair. I. Title.
  TK6655.V5G66 1995
  621.388'337—dc20   95-22902
                                                 CIP

Acquisitions editor: Roland S. Phelps
Editorial team: Joanne Slike, Executive Editor
              Andrew Yoder, Managing Editor
              Melanie Holscher, Book Editor
Production team: Katherine G. Brown, Director
Design team: Jaclyn J. Boone, Designer   EL3
              Katherine Stefanski, Associate Designer   0242003

Contents

Acknowledgments

Much of the information in this book was provided by VCR manufacturers, their technical personnel, and VCR electronic service technicians. I am most grateful to the following companies and individuals:

General Electric Co. (Mr. R.J. Collins)
JVC (US JVC Corp.) (Mr. Paul E. Hurst)
Matsushita Electric Corp.
Quasar (Mr. Charlie Howard)
Magnavox Consumer Electronics Co. (Mr. Ray Cuichard)
RCA/Consumer Electronics Co. (Mr. J.W. Phipps)
Sony Corp. of America (Ms. Paula V. Duffy)
Zenith Video Tech Corp. (Mr. Terry L. Hupp)
The 3M Company (Mr. Dick Skare)
Gemstar Development Corp.
VCR PLUS+ (Mr. Roy Mankovitz)

And a big thanks to Mr. Larry Schnabel Sencore News, Brian Phelps Electronics Marketing, George Gonos Sales Manager, Bob Eiesland Graphic Artist and Photos, Don Multerer Technical Training, and Al Bowden President of Sencore, Inc. in Sioux Falls, SD.

Introduction

The first chapter of this book introduces you to video cassette recording techniques, then provides a brief history of videotape recording, early home VCE systems, and how the systems developed and improved over the years. The later part of this chapter gives a brief overview of VCR test equipment.

Chapter 2 contains troubleshooting tips and service techniques. Chapter 3 delves into the Sencore All-Format VCR analyzer and how to make up a work tape.

Chapter 4 covers servo troubleshooting while chapter 5 covers routine VCR maintenance procedures, interference tips, and diagrams for VCR/TV connections.

Chapter 6 discusses VCR tuners and IF circuits, and chapter 7 includes special circuits and accessories information.

Chapter 8 shows you how stereo and digital audio circuits work. Chapter 9 shows you the many features of troubleshooting with the Sencore VC93 VCR Analyzer, plus VCR troubleshooting and performance testing.

Chapter 10 discusses ways to analyze video camcorder problems using the Sencore VCA94 "Video Tracker".

Chapter 11 covers features of various brands of camcorders that can help you choose a camcorder. The last chapter offers actual case history VCR problems and solutions to aid you in VCR troubleshooting. And you will also find a technical VCR and camera glossary.

Information on the BETA machine format is included because many of these are still being used, and the Beta tapes are getting hard to find; however, Snow White has just been released in Beta format.

Only Sony sells Beta machines, one stereo model for $799 and a mono unit for $499. You will generally find the Beta VCRs from mail-order firms. Absolute Beta general store, located in Remington, VA, sells new and used Beta models and antiques. I have two Beta machines and one is the first Zenith machine built by Sony. At this time, I think the Beta format gives the best color picture recording and playback viewing.

Video cassette recording

SPACE-AGE ELECTRONIC TECHNOLOGY HAS CREATED A new era of the videotape machine for home use. These video cassette recorders (VCRs) provide an acceptable color picture at a fairly low cost for the machine and the videotape.

This chapter details some basic video recording considerations, a comparison of Beta, VHS systems, and 8-mm format, a brief history of videotape recordings, and the types of test equipment to consider for servicing.

In a broad sense, videotape recording uses the same principles as audio tape recording. It is a magnetic method to record a signal onto tape. However, audio and video signals are very different. Audio has a frequency range of approximately 20 Hz to 20 kHz. Video has a much greater and higher frequency range, from about 30 Hz to 4.5 MHz. These high frequencies are required for picture quality, and the lowest one is for the sync pulses. The sources of the video signals can be from a TV camera, another tape, VCR machine, a prerecorded tape, or an "off the air" TV station signal.

Frequency limitations

The gap effect of the heads is the most important limitation on the range of video frequencies that can be recorded and played back. Figure 1-1 shows an illustration of this effect. It is a curve of the playback frequency response. As the signal to be recorded rises in frequency, a steady increase in the playback signal occurs. This rise is at the rate of 6 dB per octave, an octave being a doubling of the frequency. This rise in output continues until a maximum is obtained. The frequency where this occurs is indicated in Fig. 1-1 as f_m, the maximum frequency. Beyond this frequency value, the output will rapidly drop, reaching zero at twice the frequency of maximum output. The points f_m and $2f_m$ are determined by the size of the head gap and tape speed. The more narrow the gap is, the higher the frequency will be for maximum output. A very narrow gap restricts the output at the low frequencies.

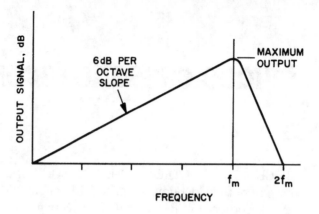

■ **1-1** *Playback frequency response.*

The speed of the tape movement past the head also affects the frequency response of the record-playback system. The greater this speed is, the higher the frequency of maximum output but the lower the signal output at the lower frequencies.

The audio frequency range of about 10 octaves (20 Hz to 20 kHz) applied to the output curve of Fig. 1-1 would show the 20-kHz limit equal to $2f_m$ and the low end, 20 Hz, at a very low level of output. The frequency of maximum output would be 10 kHz. Frequency compensation to raise the low and high ends of the curve and reduce the middle could produce a more constant output with frequency. Several different methods have been used in audio devices for this compensation. A frequency range of 10 octaves, with compensation, is about the maximum range that can be practically recorded and played back.

The approximate range of video frequencies to be recorded and played back for TV video is 30 Hz to 4.5 MHz, or 18 octaves. The signal output curve for a head designed to give maximum output at 4.5 MHz would show the 30 Hz output to be down about 110 dB. These extreme differences of signal level would make compensation and equalization of output at all necessary frequencies impossible. Thus, some other method must be used.

FM video recording/playback

If the range of frequencies for video recording and playback, with an overall difference of approximately 4.5 MHz, is converted to a higher frequency spectrum, the ratio of the high and low video frequencies can be greatly reduced. For example, if the frequency spread were changed to a variation between 5 MHz and 10 MHz (the difference being about equal to the video frequency range used) the frequency spread becomes 1 octave rather than 18. If the head gap is designed for maximum system output at 7.5 MHz, the outputs at 5 MHz and 10 MHz would be down only a small amount, and equalization would be easy. The frequency variations in the video signal could occur within this altered frequency spectrum, that is, by *frequency modulation* (FM).

Frequency modulation of the video signal to be recorded and played back is then the answer to the video frequency response problem. For example, a high frequency sine wave strong enough to saturate the tape becomes the carrier for the FM process. This carrier is modulated by the video frequencies. The lowest frequency corresponds to the sync pulse tips, with the highest corresponding to peak white. See Fig. 1-2. The actual frequencies selected for this method of FM vary from recorder to recorder. FM systems also strictly control the signal amplitude levels, thus reducing noise problems. This is another advantage of the use of FM in the recording/playback process.

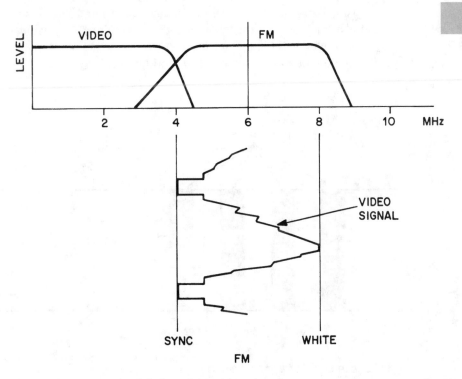

■ **1-2** *Frequency modulation of the video signal.*

The high carrier frequency used for the FM signal process requires a small head gap and high tape speed. The high tape speed, of course, means fast consumption of the tape and therefore larger tape reels. In most cases, this is an impractical situation. Also, the high tape speeds are difficult to control, complicating the drive mechanism. This method of recording by moving the tape past a stationary head is called *longitudinal recording*. This is the method still used successfully for audio recording, but not for video recording.

Transverse recording

Longitudinal recording results in a recorded, magnetic track that is parallel to the length of the tape. Another method of recording that uses much less tape is *transverse recording*. This method produces tracks *across* the tape at right angles to the length of the tape, as illustrated in Fig. 1-3. The head is not stationary in this process; it moves across the tape at a relatively high speed, as the tape moves rather slowly past the head mount. This results in a high head-to-tape speed, also called *writing speed*. This idea of a moving head led to the development of a rotating head mechanism. Two heads are required for this operation. Head A is recording a track on the tape while head B is retracing or returning to the top edge of the tape for its next track. When that position is reached, head A is switched off and head B is switched on. Thus,

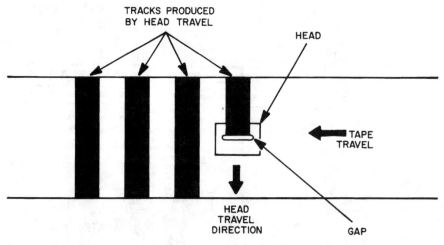

■ **1-3** *Transverse recording technique.*

properly timed switching is an important requirement of the rotating head and transverse recording process.

During the playback mode, proper switching and positioning are very important to ensure that the heads retrace the exact paths made by the heads during recording. Servo systems have been developed to control the head positions—another requirement of videotaping.

Small irregularities such as lumps or holes in the tape coating that are not much of a problem in audio recording become serious in video recording. They can cause loss of contact between the tape and the head resulting in incomplete tracks on the tape and temporary loss of the signal on playback. This signal loss is called *dropout*. It might appear as horizontal flashing on the screen—a portion of a line or up to several lines in duration. Surface bumps on the tape, dirt or residue on the heads, or nonregular tape motion can cause the dropout condition. Video recording requires the maintenance of precise contact between the head and the tape at all times. To ensure this requirement, a slight penetration of the tape by the tip of the head is employed.

The best example of transverse recording is the quad-head recorder, a large and complex machine used much by the broadcast industry, but it is not suitable for the general video recording consumer market. The speed of the tape and the tension on the moving tape must be very steady to guarantee reliable signal recording and especially accurate, steady playback. Hence, tape tension control is another requirement of video recording/playback.

Helical video recording

A videotape recorder that is simpler, smaller, and less expensive than the quad-head machine is one that employs the helical scan method of recording and playback of the video signals. With helical scan, the tape wraps around a large drum that contains one of two heads that rotate in a plane parallel to the base of the machine. The tape leaves the drum at a different level than it approached it. The movement of the tape is along a spiral, or helical path. The track that is recorded on the tape by the rotating head is "slanted" and longer than the path for transverse recording. The head drum diameter and the tape width can be designed to make the recorded track long enough to include a

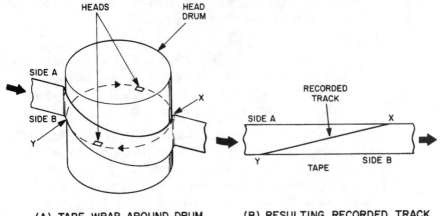

(A) TAPE WRAP AROUND DRUM   (B) RESULTING RECORDED TRACK

■ **1-4** *How tape wraps around the drum for helical scan.*

complete TV field for head control and switching is much simpler than for the quad-head recorder. Looking at Fig. 1-4, notice the development of the slanted tracks on the tape with this helical scan method. These tracks contain the video information. The control and audio tracks are recorded in a longitudinal manner, along the two edges of the tape. As shown in Fig. 1-5, the recorded tracks slant to the right on the tape, as viewed from the side away from the drum. The same tracks, viewed from the other (oxide) side of the tape that contacts the drum and heads, appear slanted to the left as shown in Fig. 1-6. The spaces be-

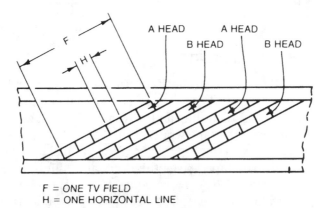

F = ONE TV FIELD
H = ONE HORIZONTAL LINE

NOTE: NOT TO SCALE

■ **1-5** *Helical scan recording tracks.*

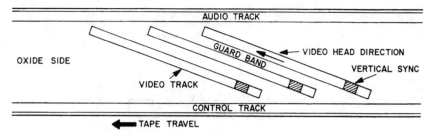

AUDIO TRACK

OXIDE SIDE

GUARD BAND ← VIDEO HEAD DIRECTION

VERTICAL SYNC

VIDEO TRACK

CONTROL TRACK

← TAPE TRAVEL

■ **1-6** *Oxide side of tape for helical scan.*

tween the recorded tracks are the *guard bands*. Their purpose is to eliminate "crosstalk" between tracks.

Helical scan videotape recorders have become dominant in the nonbroadcasting applications of video recording. Although many different models and formats have been developed and used, the basic principles are the same for all helical machines. The tape leaves a supply reel, passes over a tension arm, past an erase head, around the head drum, past a control head and audio head, and through a capstan and pressure roller to the take-up reel. Refer to Fig. 1-7 for this tape path. Figure 1-8 shows the relationship, timewise, between the tracks on the tape and the video information recorded on the tracks. Switching between the heads occurs during the few lines just before the vertical sync pulse.

7

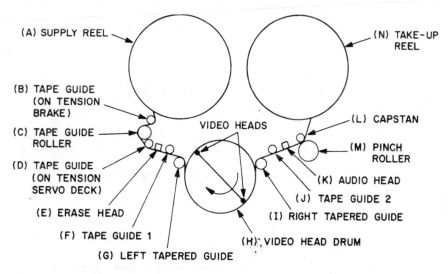

(A) SUPPLY REEL

(N) TAKE-UP REEL

(B) TAPE GUIDE (ON TENSION BRAKE)

VIDEO HEADS

(C) TAPE GUIDE ROLLER

(L) CAPSTAN

(M) PINCH ROLLER

(D) TAPE GUIDE (ON TENSION SERVO DECK)

(K) AUDIO HEAD

(J) TAPE GUIDE 2

(E) ERASE HEAD

(I) RIGHT TAPERED GUIDE

(F) TAPE GUIDE 1

(H) VIDEO HEAD DRUM

(G) LEFT TAPERED GUIDE

■ **1-7** *Tape path for reel-to-reel recorder.*

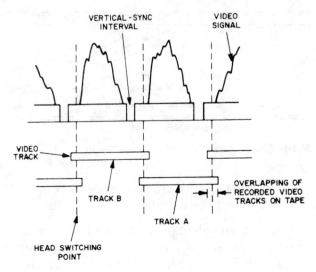

■ 1-8 *Video signal and recorded tracks.*

Color recording

We have looked at the general form and requirements and methods for recording and playing back the luminance video signals. The first videotape recorders were for black-and-white signals only. Later, methods were developed that permitted the recording of color signals. Figure 1-9 shows the National Television Systems Committee (NTSC) signal spectrum.

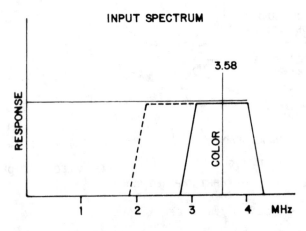

■ 1-9 *The NTSC signal spectrum.*

The basic requirements for a system that records and plays back color information would include the following requirements:

☐ Compatibility with black-and-white television.

☐ Processing of the NTSC signal within the same bandwidth as used in black-and-white recording, without changing the tint of the color signal.

Two methods have been used in the recording of color signals onto tapes. One method is called the *direct method*, and the other is called the *color under method*. In the direct method, the NTSC signal is coupled to the FM modulator, just as is done with the black-and-white signal. The color signal consists of an ac 3.58-MHz signal on a dc level. The dc portion of the signal determines, through the FM modulation action, the carrier frequency for the video and color signal modulation. The ac portion of the signal produces sidebands of this FM carrier. The color signal can be easily demodulated as with the black-and-white signal. One problem with this direct method is that the sidebands on demodulation can cause interference beats in the picture.

The color under method separates the color and luminance signals from the incoming signal. Each portion of the total signal is processed individually. The color is heterodyned down from 3.58 MHz to a lower carrier frequency and recorded directly onto tape. The FM carrier is used as a bias that is amplitude-modulated. During playback in this method, the color carrier is recovered and heterodyned back up to 3.58 MHz. See Fig. 1-10.

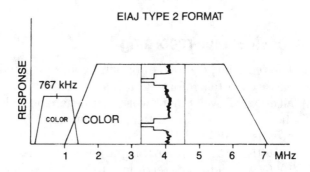

■ **1-10** *Color under method.*

Color playback

The successful playback of color signals recorded onto the tape is a more critical operation than that for black-and-white recording and playback. The major problems of color playback are as follows:

- ☐ Tension changes in the tape movement.
- ☐ Tape stretch.
- ☐ Wow and flutter in the tape movement.
- ☐ Servo instabilities and corrections.
- ☐ Noise problems in the signal.
- ☐ Beats in the picture.

If the problems occur, tint can change, color can be lost, or inter-ference bands appear in the picture. Thus, some means of cor-recting for possible color problems is a definite requirement for color recording and playback machines.

Videotape color requirements

Among the basic requirements of a reliable, successful color video-tape recorder machine are the following:

- ☐ Properly timed switching of the heads.
- ☐ Properly positioned heads.
- ☐ Tape tension control.
- ☐ Continuous head-to-tape contact.
- ☐ Crosstalk reduction.
- ☐ Correction or avoidance of color recording or playback problems.
- ☐ Steady head rotation speed.
- ☐ Steady tape movement speed.

History of videotape recording

The recording of video information onto tape and the successful playback of this information is a great stride forward in commu-nications. Development work that began in the late 1940s led to the first application of this technique in TV broadcasting. Am-prex is credited with inventing a videotape recorder (VTR) in 1956 that was used for this purpose. The machine was a model 1000, a quad-head recorder, using the transverse recording method and a 2-inch-wide videotape. Improvements have been made consistently since then, in the original product and in other versions of video recording. The VTR has definitely become an essential piece of equipment for all TV broadcasters. Tape recordings eliminate the live programs and can be used for de-layed and repeated programs.

The development of helical scan video recorders in the early 1960s made possible the extension of the video recording into other fields, such as education, training, and industry. The first helical scan recorder was developed by Amprex, the model 660, in the early 1960s. It used a 2-inch-wide tape. Then 1-inch machines superseded the 2-inch version. Panasonic, as early as 1964, developed a VTR using a 1-inch-wide tape that was used for medical purposes and also by industrial markets.

The education field has been a major user of VTR playback units, connected into extensive closed-circuit TV (CCTV) systems. Pre-taped programs and information have become great aids for the teacher and trainer. Combined with TV cameras for on the spot, "live" input to the TV monitors, or recorded for later use, the VTR has given added dimension for many instructional activities. Among others, Sony has been particularly successful in this application of VTRs.

The early VTRs featured the reel-to-reel format, with manual tape threading from a supply reel, past the heads, to the take-up reel. This design is still used on some older video production systems. Automatic tape threading and smaller tape package are now possible with the VCR cassette tape units. VCRs were originally used by the teaching profession, but they are now enjoyed by the viewing public for home video recording. The compact design and easy operation of video cassettes opened the field for sales to the consumer market.

An example of early model VTRs using the helical scan technique was marketed by Sony in 1964. The unit was a reel-to-reel machine that recorded black-and-white video on 1/2-inch-wide tape. Figure 1-11 shows the principles of this Sony recorder. The tape is wrapped around one-half or 180 degrees of the drum, which contains two rotating video heads. The top portion of the picture details the angular path of the tape across the drum head. The tape enters the drum area from the supply reel on the left at a higher level than at its exit to the take-up reel. This tape movement past the rotating heads produces the magnetic patterns on the tape as shown in Fig. 1-12. The video tracks are slanted, hence the term *slant track recording*. The audio track and the control track (on opposite edges of the tape) are produced from a stationary head, located in the tape path from the video head drum to the take-up reel.

The control track is part of an automatic servo system that ensures the correct positioning of the video heads relative to the

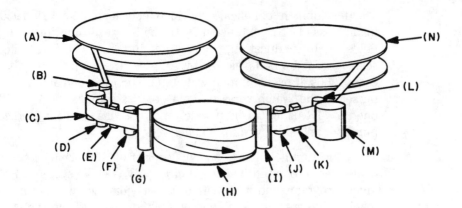

(A) SUPPLY REEL	(G) LEFT TAPERED GUIDE
(B) TAPE GUIDE (ON TENSION BRAKE)	(H) VIDEO HEAD DRUM
	(I) RIGHT TAPERED GUIDE
(C) TAPE GUIDE ROLLER	(J) TAPE GUIDE 2
(D) TAPE GUIDE (ON TENSION SERVO DECK)	(K) AUDIO HEAD
	(L) CAPSTAN
(E) ERASE HEAD	(M) PINCH ROLLER
(F) TAPE GUIDE 1	(N) TAKE-UP REEL

■ **1-11** *Early model Sony helical scan video recorder.*

recorded magnetic tracks on the tape during the playback mode of operation. Control track pulses are compared to the pulses developed by rotating heads.

Each slanted track on the recorded tape represents a TV picture field (two fields represent a complete frame). This early Sony machine included a skip field system, the result of recording by only one of the two heads at a time. During playback, both heads play the same track. This action provides the missing field (to complete the frame). However, the vertical resolution is only half the normal

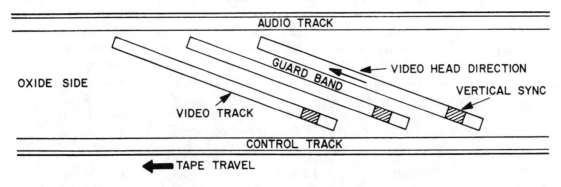

■ **1-12** *Slant track tape diagram.*

Video cassette recording

amount. It is necessary to switch between the heads so that the signal preamplifiers are always connected to the proper head—the one in contact with the tape. The switching is performed at a 60 Hz rate, which is a possible source of time-base errors in the playback signal. This reason is why it is desirable to have faster automatic phase control (APC) circuitry in the horizontal system of TV receivers that are used with VCR machines.

Shown in Fig. 1-13 are details of the signal frequency spectrum as processed by the early model Sony black-and-white recorder. For both recording and playback, the signal frequency band was moved to a higher range. The luminance or brightness signal frequency-modulated a carrier during the recording. Variations in brightness (white through shades of gray to black) became variations in frequency. This FM signal was present in the recorded slant tracks on the tape.

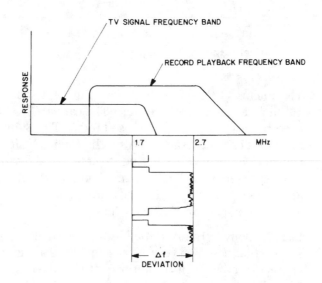

■ **1-13** *Early Sony black-and-white recording method.*

Guard bands are positioned between adjacent recorded slant tracks. These guard bands contain no recorded information and represent wasted space on the tape. They are necessary to prevent crosstalk between the tracks produced by the VCR.

Some measure of standardization came into the picture of videotape recording around the year 1968 as the previously mentioned system evolved into the EIAJ type 1 format for 1/2-inch reel-to-reel videotape recorders. EIAJ is an abbreviation for Electronic Industry Association-Japan.

Figure 1-14 shows the signal frequency spectrum for the color recorder system. The FM luminance signal was raised in frequency to the 3.2-MHz to 4.6-MHz region. The color subcarrier was shifted down during recording to 767 kHz, thus placing it below the luminance carrier. This conversion produces a "color under" system. The video tracks were reduced in width, and the guard bands narrowed. Both video heads recorded, achieving full-frame recording. These machines are used mostly in the industrial market.

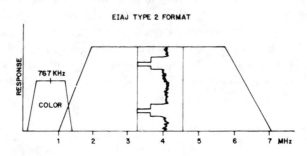

■ **1-14** *EIAJ color recorder.*

The Philips VCR system, introduced in 1961, was one of the first 1/2-inch tape cassette recorders. The cassette loading mechanism used by this model is widely used today. The Sony Betamax VCR machines use one example. Also, the Philips VCR used a rotating transformer instead of slip rings to couple the rotating head signal to the preamplifier. This design eliminates the need for slip rings and the need for wear and pressure adjustments. This feature is also used on current VCR machines.

In 1971, Sony introduced the U-MATIC format, a 3/4-inch cassette system that has been very successful in the industrial market. It is characterized by high-resolution pictures and easy operation. This format represents a refined application of earlier technology, including the tape loading and rotating head transformer features of the Philips system.

In 1972, Cartrivision introduced a 1/2-inch tape cassette system capable of 2 hours of playing time. This format used a skip field, three-head system, resulting in a reduction of tape consumption. Only every third TV video field was recorded. On playback, each of the three heads was played in order on the same track, producing a proper TV signal. Vertical resolution was one-half of a standard TV signal. Fast motion of the picture content could not be reproduced because of the loss of two out of three TV fields.

Cartrivision, in addition to recording live TV programs "off the air," promoted the playing of prerecorded cassettes. These cassettes could be rented or purchased. Cartrivision was marketed by Sears, Admiral, and Wards, but sales were slow. The time probably was not quite right.

One RCA format used four video heads, a 90-degree tape wrap of the head drum, and a 3/4-inch-wide tape cartridge. The head wheel extended into the inserted cartridge to push against the tape, permitting the head-to-tape contact. One significant result of RCAs investigation of videotape recording at the time was a customer survey that determined a 2-hour uninterrupted playing time was desired by the TV viewers.

The next round of formats represent today's home videotape recorders. All use 1/2-inch tape cassettes and have built-in tuners and program turn-on systems. At this time, the VHS and Betamax are the standard format VCR systems.

The Sanyo V Cord Two format was introduced in 1976. It uses a 1/2-inch cartridge and is capable of 1-hour, full-frame recording or 2-hour, skip-field recording. The latter method results in one-half vertical resolution pictures. Tape loading is of the Philips type. The track widths and guard bands are reduced over earlier systems, thus reducing tape consumption. The video track width is 60 microns with a guard band of 37 microns.

The Matsushita VX2000 is a one-head machine using an "alpha wrap" of the tape around the head drum. Alpha wrap means that the tape almost completely encircles the head drum, permitting the use of only one head. This cassette-type video recorder tapes up to a maximum of 2 hours per cassette. The first machine of this type in the United States had a Quasar brand name.

Through the years of development and improvement of videotape recording since the late 1940s, many variations have been tried in systems and component parts of the system. Several widths of tape have been used: 2 inch, 1 inch, 3/4 inch, 1/2 inch, and 1/4 inch. The 2-inch size has been a mainstay of the broadcast industry, although 1-inch tape width recorders have found a place in this market in recent years. Industrial and educational applications have employed the 3/4-inch tape size. The 1/4-inch-width tape is found in special machines such as portable color recording units. Of course, for the home videotape recorder market, the 1/2-inch-wide tape is now the standard.

The helical scan format is the most used system in video recording. Transverse recording is still used in the quad-head machines used by the TV broadcasters. But even this field includes recorders with the helical scan format.

The reduced tape width, packaging it in small cassettes and the helical scan format together make size reductions of the complete videotape recording machine possible, which in turn makes it very suitable for in-home use.

The success of the Sony Betamax-1 proves this point. An important element for consumer acceptance of the VCR is the reduction in the cost of tape usage. As Fig. 1-15 shows, the left curve, depicting the consumption of tape measured in square feet per hour, has been greatly reduced since 1968. The dots on the graph pinpoint this usage for the various videotape recorders indicated.

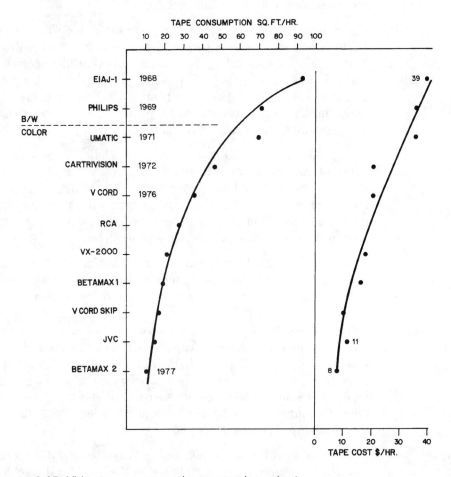

■ 1-15 *Video tape consumption comparison chart.*

Thus, a machine using EIAJ-1 format in 1968 consumed more than 90 square feet of tape per hour.

The later-introduced Sony Betamax-2 consumes only about 10 square feet per hour. This reduced tape consumption, plus improvements in the tape, increased tape production, and the 2-hour recording system have led to a sharp reduction in tape cost per hour. This curve appears on the right in Fig. 1-15. The graph shows a cost of $39 per hour in 1968 but only $8 per hour for the 2-hour Betamax in 1977. Thus, the purchase price for blank tape cassettes that can hold two hours of video recording costs about $16. The Beta-3 speed machines consume 50 percent less tape than the Beta-2 VCR machines.

Sony's Betamax-1 was introduced in the U.S. market in 1975. It was a 1/2-inch-wide tape cassette system, capable of 1 hour playing time. Two video heads are used, with full-frame recording. The tape threading (the method of pulling the tape out of the cassette and around the head drum) is of the Philips type. The Betamax-1 recorded tape pattern is shown in Fig. 1-16.

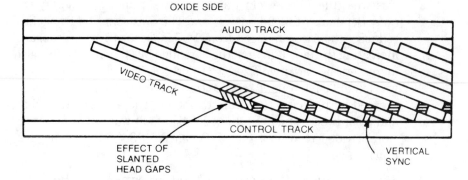

OXIDE SIDE

AUDIO TRACK

VIDEO TRACK

CONTROL TRACK

EFFECT OF
SLANTED
HEAD GAPS

VERTICAL
SYNC

■ **1-16** *Betamax-1 recorder tape video tracks.*

The recorded, slanted tracks of this helical scan machine are 60 microns wide with no spacing between tracks and therefore no guard bands. This is possible because of two well-devised arrangements that reduce crosstalk effects. Luminance signal crosstalk from track-to-track is avoided by slanting the two head gaps relative to the tracks. This is called the *azimuth technique*. Color crosstalk cancellation occurs because of a more involved signal phasing system involving a comb filter. The waveform drawing (Fig. 1-17) is the recorded signal spectrum for Betamax-1 tape speed.

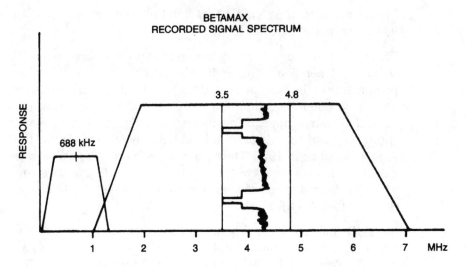

BETAMAX
RECORDED SIGNAL SPECTRUM

■ **1-17** *Betamax-1 recorder signal spectrum.*

Tape format considerations

Based on the fact that RCA had determined a 3-hour VCR system was what the TV viewers wanted, JVC set about to modify the Sony Betamax-1 to a 2-hour system. This was achieved by slightly reducing the capstan speed from 1.57 in./sec. to 1.34 in./sec. and increasing the tape length in the cartridge by 63 percent. This also increased the cartridge size by 31 percent. The video track width was left at 60 microns. To realize this, the track slant angle was increased by reducing the drum diameter. The azimuth recording and color processing are retained. The result was a two-hour version of video recording that JVC called VHS (Video Home System).

Sony in the meantime had produced its own 2-hour Betamax. Sony chose to keep the same cassette and reduce the capstan speed by a factor of two. The slant track pitch was now 30 microns instead of 60. The head track was reduced from 60 to 40 microns. This results in a negative guard band and produces overlapping tracks. The azimuth and comb filter techniques, however, still provide adequate crosstalk rejection. The result is lower tape consumption.

The Sony threading produces very gentle tape handling. Notice this tape threading path in Fig. 1-18. The tape guides are fixed and the guides are at 90 degrees to the tape motion and can be rotated to lower tape friction. The tape is simply wrapped around the drum. A longer piece of tape is removed from the cassette in the

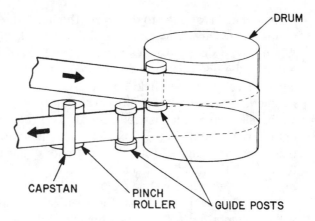

DRUM

CAPSTAN   PINCH ROLLER   GUIDE POSTS

■ **1-18** *Sony Betamax recorder tape path.*

Sony machine, isolating the tape from the cartridge feed irregularities and producing less time-base error. The Sony tape has a life of approximately 200 passes or plays. With the Sony VCR design, the tape does not leave the drum for fast forward or rewind, and precise location of a section of the tape is very easy to find.

VHS recording operations

Before delving into the VHS format, let's take a brief look at some video recording principles.

Video recording basics

Like audio tape recording, video information is stored on magnetic tape by means of a small electromagnet, or *head*. The two poles of the head are brought very close together but they do not touch. This creates magnetic flux to extend across the separation (gap), as illustrated in Fig. 1-19.

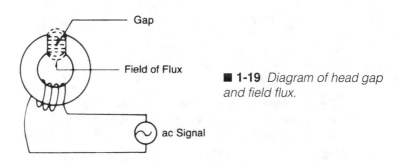

Gap

Field of Flux

ac Signal

■ **1-19** *Diagram of head gap and field flux.*

If an ac signal is applied to the coil of the head, the field of flux expands and collapses according to the rise and fall of the ac signal. When the ac signal reverses polarity, the field of flux will be oriented in the opposite direction and continues to expand and collapse. This changing field of flux is what accomplishes the magnetic recording. If this flux is brought near a magnetic material, it will become magnetized according to the intensity and orientation of the field of flux. The magnetic material used is oxide-coated (magnetic) tape. Using audio tape recording as an example, if the tape is not continually moved across the head, just one spot on the tape will be magnetized and constantly remagnetized. If the tape moves across the head, specific areas of the tape will be magnetized according to the field of flux at any specific moment. A length of recorded tape will therefore have on it areas of magnetization representing the direction and intensity of the field of flux.

As an example, the tape has differently magnetized regions that can be called north (N) and south (S), in proportion to the ac signal. When the polarity of the ac signal changes, so does the direction of the magnetization of the tape, as shown by one cycle on the ac signal in Fig. 1-20. If the recorded tape is then moved past a head whose coil is connected to an amplifier, the regions of magnetization on the tape will set up flux across the head gap that in turn induces a voltage in the coil to be amplified. The output of the amplifier, then, is the same as the original ac signal. This is essentially what is done in audio recording, with other methods for improvement like bias and equalization.

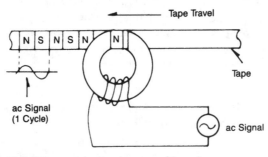

■ **1-20** *ac polarity change of head.*

Some inherent limitations in the tape recording process affect videotape recording. As shown in Fig. 1-20, the tape has north and south magnetic fields that change according to the polarity of the ac signal.

If the speed of the tape past the head (head-to-tape speed) is kept the same, the changing polarity of the high-frequency ac signal would not be faithfully recorded on the tape, as shown in Fig. 1-21.

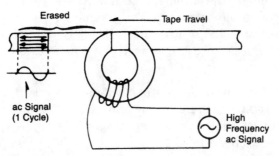

■ **1-21** *High-frequency ac signal considerations.*

As the high-frequency ac signal starts to go positive, the tape starts to be magnetized in one direction. But the ac signal quickly changes its polarity, and this will be recorded on much of the same portion of the tape, so north magnetic regions are covered by south magnetic regions and vice versa. This results in zero signal on the tape, or self-erasing. To keep the north and south regions separate, the head-to-tape speed must be increased.

When recording video, frequencies in excess of 4 MHz might be encountered. Through experience it is found that the head-to-tape speed must be about 10 meters per second to record video signals.

The figure of 10 meters per second was also influenced by the size of the head gap. Clearly, the lower head-to-tape speed, the easier it is to control that speed. If changes in head gap size were not made, the necessary head-to-tape speed would have been considerably higher. How the gap size influences this can be explained by looking at Fig. 1-22.

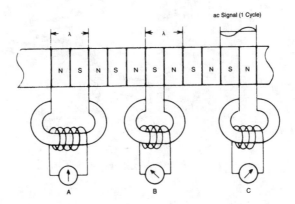

■ **1-22** *How gap size influences tape recording.*

Assume a signal is already recorded on the tape. The distance on the tape required to record one full ac signal cycle is called the *recorded wavelength*. Head A has a gap width equal to one wavelength. Here, there is both north- and south-oriented magnetization across the gap. This produces a net output of zero, because north and south cancel. Heads B and C have a maximum output because there is just one magnetic orientation across their gaps.

Therefore, maximum output occurs in heads B and C, because their gap width is 1/2 wavelength. Heads B and C would also work if their gap width is less than 1/2 wavelength. The same is also true for recording. A head-to-tape speed of 10 meters per second is a very high speed—too high in fact to be handled accurately by a reel-to-reel tape machine. Also, tape consumption on a high-speed machine is tremendous.

The method used in video recording is to move the video heads as well as the tape. If the heads are made to move fast across the tape, the linear tape speed can be kept very low. In two-head helical video recording, the video heads are mounted in a rotating drum or cylinder, and the tape is wrapped around the cylinder. This way the heads can scan the tape as it moves. When a head scans the tape, it is said to have made a track. This is shown in Fig. 1-23.

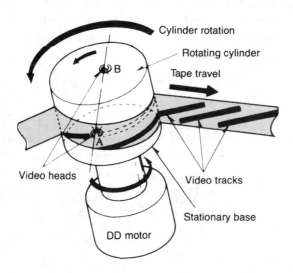

Cylinder rotation

Rotating cylinder

Tape travel

B

Video heads

Video tracks

Stationary base

DD motor

■ **1-23** *Head scan and tape tracks.*

In 2-head helical format, each head records one TV field, or 262.5 horizontal lines, as it scans across the tape. Therefore, each head must scan the tape 30 times per second to give a field rate of 60 fields per second.

The tape is shown as a screen wrapped around the head cylinder to make it easy to see the video head. There is a second video head 180 degrees from the head shown in front. Because the tape wraps around the cylinder in the shape of a helix (helical), the video tracks are made as a series of slanted lines. Of course, the tracks are invisible, but it is easier to visualize them as lines. The two heads A and B make alternate scans of the tape.

An enlarged view of the video tracks on the tape is shown in Fig. 1-24. The video tracks are the areas of the tape where video recording actually takes place.

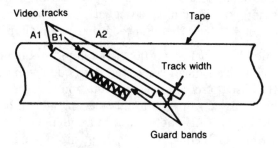

■ **1-24** *Video tracks and guard bands on the tape.*

23

There is one more point about video recording. Magnetic heads have characteristics of increased output level as the frequency increases, which is determined by the gap width. In practice, the lower frequency output of the heads is boosted in level to equal the level of the higher frequencies. This process is also used in audio recording and is called *equalization*.

Video frequencies span from 30 Hz to about 4 MHz. This represents a frequency range of about 18 octaves. Eighteen octaves is too great of a spread to be handled in one system or machine. For instance, heads designed for operation at a maximum frequency of 4 MHz will have very low output at low frequencies. Because there is 6 db/octave attenuation, $18 \times 6 = 108$ dB difference appears. In practice, this difference is too great to be adequately equalized. To get around this, the video signal is applied to an FM modulator during recording. The modulator changes its frequency according to the instantaneous level of the video signal.

The energy of the FM signal lies chiefly in the area from about 1 MHz to 8 MHz, just three octaves. Heads designed for use at 8 MHz can still be used at 1 MHz, because the output signal can be equalized. Actually, heads are designed for use up to about 5 MHz. Therefore, some FM energy is lacking. It does not affect the playback video signal, however, because it is resumed in the playback process. Upon playback, the recovered FM signal must be equalized and then demodulated to obtain the video signal.

Converted subcarrier direct recording method

To avoid visible beats in the picture caused by the interaction of the color (chrominance) and brightness (luminance) signals, the first step in the converted subcarrier method is to separate the chrominance and luminance portions of the video signal to be recorded. The luminance signal, containing frequencies from dc to about 4 MHz, is then FM recorded, as previously described.

The chrominance portion, containing frequencies in the area of 3.58 MHz, is down-converted in frequency in the area of 629 kHz. Because there is not a large shift from the center frequency of 629 kHz, this converted chrominance signal can be recorded directly on the tape. Also note that the frequencies in the area of 629 kHz are still high enough to allow equalized playback. In practice, the *converted chrominance* signal and the FM signals are mixed and then simultaneously applied to the tape. Upon playback, the FM and converted chrominance signal are separated. The FM is demodulated into a luminance signal again. The converted chrominance signal is reconverted back up in frequency to the area of 3.58 MHz. The chrominance and luminance signals are combined, which reproduces the original video signal.

Other VHS recorder functions

Search-forward (cue) and search-reverse (review)

In order to quickly find a particular segment on the tape during playback, the user can speed up the capstan and reel tables to nine times the normal speed, either forward or reverse, by pressing *cue* or *review* buttons. At this time, noise bars will appear in the picture because of head crossover. This is normal on some model VHS and Betamax machines. For example, some show four noise bars in cue and five noise bars in the review mode. The bars for the cue and review modes are shown in Figs. 1-25 and 1-26.

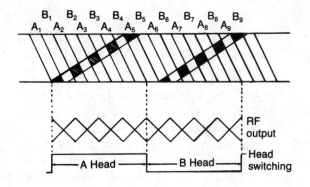

■ **1-25** *The cue mode.*

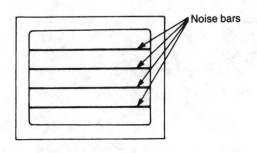

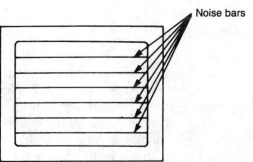

■ **1-26** *The review mode.*

Skew correction

The various VHS machines have speeds that are referred to as SP, LP, and SLP. In the SLP mode and with the proper VHS cassette the playing time will be 6 hours.

Horizontal sync alignment on the tape occurs in the SP and SLP modes, but not in the LP mode as shown in Figs. 1-27 and 1-28. Thus, when using cue or review on LP recordings, severe skew or picture bending will occur at the top portion of the screen. Also, the color AFC will malfunction for this same reason. To correct this, the playback video is delayed by 0.5 H to compensate for skew, and the AFC frequency is shifted to maintain color lock.

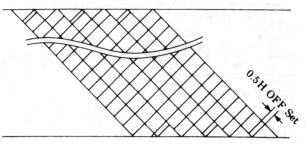

■ **1-27** *The SLP recording mode.*

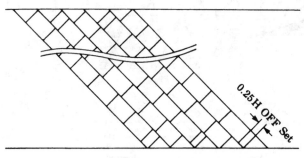

■ **1-28** *The LP recording mode.*

Add-on recording (transition editing)

Most VCRs allow pause during recording. But because of the arbitrary timing, there is most likely a disturbance of the picture during playback at the place where the pause was used. To eliminate this disturbance, *transition editing recording* is used, which backs up the tape for 2.2 seconds during pause recording. When the pause is released, the deck will play back for about 1.2 seconds while aligning the control pulses already on the tape to the incoming vertical sync. After 1.2 seconds, the deck switches to the record mode, with the overall effect of no synchronization loss during playback (Fig. 1-29). Therefore, there will be no sync disturbance during playback regardless of how many times pause was used during recording. Refer to Fig. 1-30.

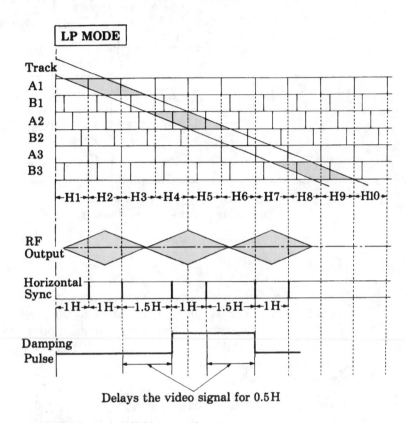

■ **1-29** *LP mode 1/2-H correction.*

About the video head

Reduced track width requires the use of smaller video heads. But just making them smaller does not make them better. With less actual head material to work with, the magnetic properties of the head suffer. To offset this, a change in head material is in order. Because the VHS recorder is designed to be small, a reduction in the size of the head cylinder is called for.

A reduction in the size (diameter) of the head cylinder changes the head-to-tape speed. Remember, the head-to-tape speed affects the high-frequency recording capability of the head. To offset this problem, the head gap size was reduced.

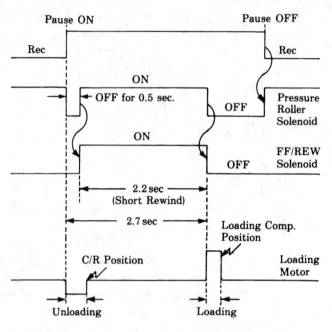

■ 1-30 *Timing diagram for VTR.*

The use of *hot pressed ferrite* as video head material in the VHS recorder helps improve the characteristics of the smaller heads. The hot pressed ferrite also has uniform domain orientation that further improves the head characteristics. It has been proven in many tests that the use of hot pressed ferrite material produces a superior video head.

From the preceding explanation, the need for smaller head gap size becomes apparent. In VHS recorders, the video head gap width is a mere 0.3 micrometer. This is quite a contrast from ordinary video heads used in other helical applications whose gap widths are typically in the area of micrometer.

Head azimuth

Azimuth is the term used to define the left-to-right tilt of the gap if the head could be viewed straight on. As azimuth recording is utilized in the VHS systems, the heart of the azimuth recording process is in the video heads themselves. This requires still another change in head design. In most VCR applications, the azimuth has always been set perpendicular to the direction the head travels across the tape, or more simply, the video track. Figure 1-31 helps to explain this. The gap is perpendicular (90 degrees) to the head's movement across the tape. Think of this standard as a perfect azimuth of 0 degrees.

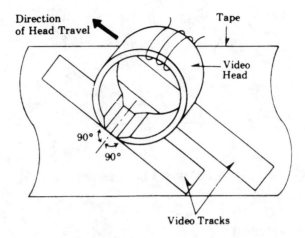

■ **1-31** *Video head alignment tape.*

In VHS, the video heads have a gap azimuth other than 0 degrees. Also, one head has a different azimuth from the other. The two values used in VHS machines are azimuth of +6 degrees and −6 degrees. Refer to Figs. 1-32 and 1-33. These heads make the VHS format different from most other VCR formats. Exactly how the azimuths of +/−6 degrees help to keep out adjacent track interference is explained next.

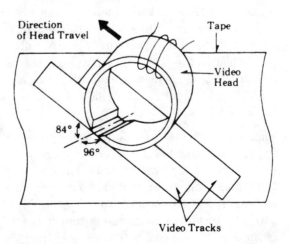

■ **1-32** *Head and tape alignment.*

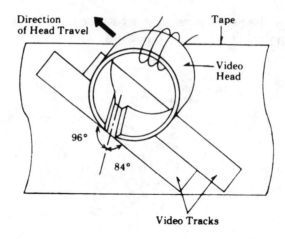

Direction of Head Travel

Tape

Video Head

96°

84°

Video Tracks

■ **1-33** *VHS video head format.*

Azimuth recording

Azimuth recording is used in VHS to eliminate the interference or *crosstalk* picked up by a video head. Again, because adjacent video tracks touch, or cross talk, a video head can pick up some information from the adjacent track when scanning. The azimuth of the head gaps assure that video head A only gives an output when scanning across a track made by head A. Head B, therefore, only gives an output when scanning across a track made by head B. Because of the azimuth effect, a particular video head will not pick up any crosstalk from an adjacent track. Let's examine this more closely.

Figure 1-34 shows a VHS video recorder system in the SLP mode with the video tracks in a not-to-scale north and south magnetized regions. It also can be seen that these N and S regions are not perpendicular to the track; they have −6 degrees azimuth in tracks A1 and A2, and +6 degrees azimuth in tracks B1 and B2. If we take track A1 and darken the N region, it becomes easier to see. Refer to Fig. 1-35.

Figure 1-36 shows the information on track A made by head A. Imagine now that head A is going to play back this track by superimposing the head over the track. Clearly, the gap fits exactly over the N and S regions, so that at any moment there is either an N region or an S region or an N-to-S (or S-to-N) transition across the gap. This produces maximum output in head A. Now, visually superimpose the B head over the track. Here there are N and S regions across the gap at the same time, at any given moment.

Azimuth in the LP Mode

A1 B1 A2 B2

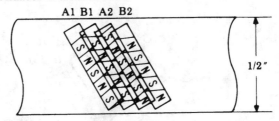

1/2″

Azimuth in the SLP Mode

A1 B1 A2 B2

1/2″

■ **1-34** *Azimuth in the SLP mode.*

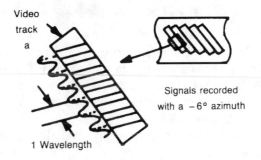

Video track a

1 Wavelength

Signals recorded with a −6° azimuth

■ **1-35** *Signals with a -6 degree azimuth.*

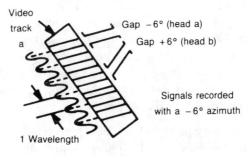

Video track a

Gap −6° (head a)
Gap +6° (head b)

Signals recorded with a −6° azimuth

1 Wavelength

■ **1-36** *Information made by A track.*

Remember that simultaneous N and S regions across the gap cause cancellation, and therefore, no output. Looking back at Fig. 1-33, see that the gap width is equal to one-half the recorded wavelength. Recall that this occurs at the highest frequency to be recorded. Therefore, the azimuth effect works at these high frequencies.

But what happens at lower frequencies? Figure 1-37 is a diagram similar to Fig. 1-36, except the recorded wavelength is longer, which represents a lower frequency.

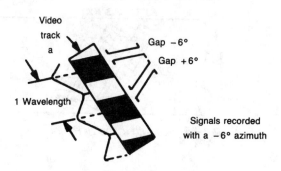

■ **1-37** *A longer recorded wavelength.*

Again, visually superimpose the heads over the track. Head A is the same as before. But look at head B. There is much less cancellation across the gap, and its output is close to that of head A. Therefore, you see where the azimuth effect is dependent on frequency. The higher the frequency, the better the azimuth effect. The lower the frequency, the lower the separation by azimuth effect.

VHS color recording system

Because there is insignificant azimuth effect at lower frequencies, a different type of color recording system was adopted. The fact that crosstalk occurs at lower frequencies cannot be changed, as it occurs right on the tape during playback. The method adopted processes the crosstalk component signals from the heads so that they are eliminated. It is important to realize that the crosstalk *does still occur*. It is the recording/playback circuitry that performs the crosstalk elimination.

In ordinary helical VCRs using converted subcarrier direct recording, the phase of the chrominance signal is untouched and is recorded directly onto the tape. The chrominance signal and its phase can be represented by vectors. Vectors graphically represent the amplitude and phase of one frequency. To keep it simple, assume the chrominance signal is only one frequency. For an example of vectors, see Fig. 1-38. The length of any vector represents its amplitude.

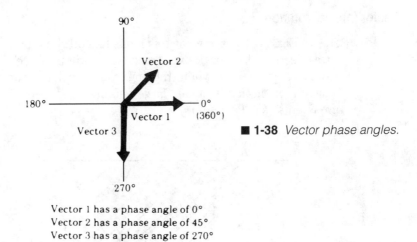

■ **1-38** *Vector phase angles.*

Vector 1 has a phase angle of 0°
Vector 2 has a phase angle of 45°
Vector 3 has a phase angle of 270°

The azimuth effect does not work at the lower frequencies. And because the color information in VHS is recorded at low-converted frequencies, another technique for color recording was adopted.

Vector rotation in recording is actually a phase shift process that occurs at a horizontal rate of 15,734 Hz, or 15.734 kHz. The chrominance signal can be represented by a vector, showing amplitude and phase. In ordinary helical scan VCRs, the vector is of the same phase for every horizontal line on every track, as shown in Fig. 1-39.

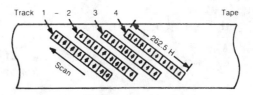

■ **1-39** *Normal helical scan for VCR.*

In VHS, the 3.58 MHz is still converted down to a lower frequency, namely 629 kHz, but the color technique used in VHS format is a process of vector rotation. During recording, the chrominance phase of each horizontal line is shifted by 90 degrees. For head A (channel 1), the chrominance phase is advanced by 90 degrees per horizontal line (H). For head B (channel 2), the chrominance phase is delayed by 90 degrees per H.

☐ Channel 1 +90°/H

☐ Channel 2 −90°/H

Vector (phase) rotation

Refer to Fig. 1-40 to see what this looks like on the tape. Assume that as head A plays back over track A1, it will produce a vector output as such: head A when tracking over A1 will have an output consisting of the main signal (large vectors) and some crosstalk components (small vector).

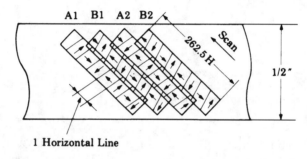

■ **1-40** *VHS helical scan.*

Figure 1-41 is a vector representation of the playback chrominance signal from the head. One of the most important things done in the playback process is the restoration of the vectors to their original phase. This is done by the balanced modulator in the playback process. Note the vector representation shown in Fig. 1-42. This restored signal is then split two ways. One path goes to

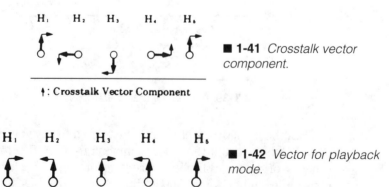

■ **1-41** *Crosstalk vector component.*

■ **1-42** *Vector for playback mode.*

one input of an adder. The other path goes to a delay line, which delays the signal by 1 H. The output of the delay line goes to the other input of the adder. This can be more easily seen in Fig. 1-43.

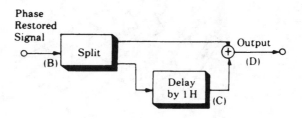

■ **1-43** *Split of the restored signal.*

As can be seen in Fig. 1-44, the crosstalk component has been eliminated after the first H line. The chrominance signal is now free of adjacent channel crosstalk.

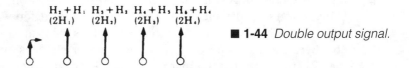

■ **1-44** *Double output signal.*

The double output is not a problem because it can always be reduced. The process of adding a delayed line to an undelayed line is permissible because any two adjacent lines in a field contain nearly the same chrominance information. So, if two adjacent lines are added, the net result will produce no distortion in the playback picture.

In conjunction with the crosstalk elimination is the reconversion of the chrominance 629 kHz to its original 3.58 MHz. Now the color signal is totally restored.

LP tape speed mode

The recorded signal in the LP mode is considerably different from that used in the other VHS system tape speed modes. Like the SP mode, the chrominance and luminance signals are separated as covered earlier. However, from here on things are treated differently. Let's examine again the video tracks on the tape of an LP recording.

Notice in Fig. 1-45 that the tracks do overlap, and that any picture area of any track does not line up perfectly with the picture area of the adjacent tracks. (No horizontal sync alignment).

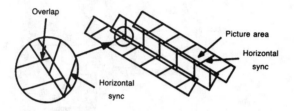

Overlap

Picture area

Horizontal sync

Horizontal sync

■ **1-45** *Overlapped tracks.*

Let's now pull several horizontal line segments off of the track for greater detail. As can be seen in Fig. 1-46, the horizontal sync Portion of track B lies somewhere in the picture area of track A, for any given horizontal line segment. Assume that track A was recorded first. Then, as track B is laid down, the 3.4 MHz horizontal sync section of A will produce a beat with the portion of track B that covers it. Although the entire overlapping region will produce beats, the beat caused by the horizontal sync is most noticeable because the sync tip FM frequency never changes, whereas the FM frequency for the picture portion is constantly changing. This beat is visible on the screen, so measures must be taken to eliminate it. The method employed is called *FM interleaving recording.*

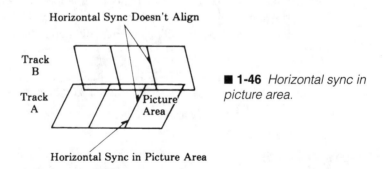

Horizontal Sync Doesn't Align

Track B

Track A

Picture Area

Horizontal Sync in Picture Area

■ **1-46** *Horizontal sync in picture area.*

Note that beats are not the same as adjacent track crosstalk. Azimuth recording prohibits crosstalk pickup. But, the beat produced is a new frequency. It was not present in the video signal; it is the result of laying one track over another. The beat signal has no true azimuth, therefore, it will be detected by both video heads.

The FM interleaving recording method does not actually eliminate the beat but rather places it at such a frequency so that no beat can be detected on the screen.

In the National Television Systems Committee (NTSC) color TV system, video frequencies that are an odd multiple of half the horizontal line frequency have the property called *interleaving*. Interleaving signals appear on the TV screen in a rather special way. Between any two adjacent lines, the signals are out of phase, as shown in Fig. 1-47 by the solid lines. Because the two lines are very close, the human eye tends to integrate them. The out-of-phase signals will virtually be canceled (invisible to the viewer).

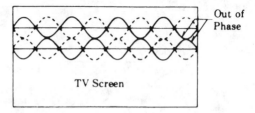

Out of Phase

TV Screen

■ **1-47** *Adjacent lines are out of phase.*

Now, when the frame is completed and the next frame begins, the signal on the top line will be out of phase with what was previously scanned. This is shown by the dotted line. This cancels any phosphor persistence from the previous scan. Thus, interleaved frequencies, for all purposes, do not create interference on the TV screen.

This interleaving is accomplished by raising the sync tip FM frequency in channel 1 by 15734/2 MHz, or 7,867 Hz. For channel 1, then, sync tip frequency is 3.407867 MHz and peak white becomes 4.407867 MHz. Channel 2 remains the same as before—sync tip is 3.4 MHz, and peak white is 4.4 MHz. This displacement by 7.867 kHz causes the beat produced by the overlapped horizontal sync to become an interleaving frequency, which solves the problem.

Recovery of this shifted FM signal, although somewhat different, is essentially the same as before. The chrominance and FM signals that are mixed and then applied to the tape occupy a spectrum that is shown in Fig. 1-48.

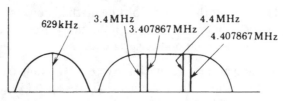

■ **1-48** *Chrominance and FM spectrum.*

SLP mode for VHS system

Like the other mode speeds, the video track on the tape of an SLP recording is shown in Fig. 1-49. Notice that the tracks do overlap, and any track picture area of any track will line up perfectly with the picture area of the adjacent tracks (horizontal sync alignment). Let's pull several horizontal line segments off of the track for greater detail.

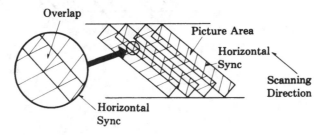

■ **1-49** *Video tracks in SLP mode.*

As can be seen in Fig. 1-50, the horizontal sync portion of track B is in alignment with one of track A. Assume that the SLP recorded tape is played back. When the A head scans the A track, the A head picks up the B track signals on the overlapping region. Although the entire overlapping region will produce beats as in other modes, the beat is eliminated by the FM interleaving recording.

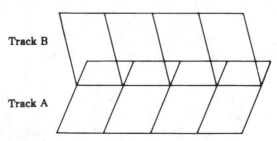

■ **1-50** *Horizontal sync is now in alignment.*

Tape M loading

In this type of VHS machine, the tape path is out of the cassette, across the stationary heads and around the video head cylinder forming a letter M, thus the name M loading. Refer to this M type loading diagram in Fig. 1-51. The M loading has several advantages over previous, more complex loading formats.

☐ Less tape is pulled out from the cassette. This reduces the chances of tape spillage and tangles.

☐ Because the tape path in the M load pattern is short, loading time is only 3 seconds, including video muting.

☐ Fast forward and rewind are performed inside the cassette, further reducing the chances of tape damage.

	8 mm	VHS-C(SP)
Dynamic Range	75dB	45dB
Frequency Response Line in:	20Hz—20kHz	50Hz—11kHz(SP)
Wow & Flutter	0.05%	0.5%

■ **1-51** *Comparison of 8 MM and VHS audio playback performance.*

8-mm format

The 8-mm video format has been designed to meet the requirements of smaller size and lighter weight. The 8-mm videotape is only 8-mm wide as compared to the VHS-C videotape, which is 12.6-mm wide. As a result the mechanism is much smaller. In the 8-mm video system, the tape runs at 14.3 mm per second and utilizes a 40 mm drum with a 221-degree wrap.

The 8-mm format features an FM hi-fi audio recording system. The audio signal is recorded as an FM carrier. Although most 8-mm systems can only record mono audio, provisions have been made in the format to record stereo *pulse code modulation* (PCM) audio. As Fig. 1-51 shows, this gives 8 mm a far better performance than VHS-C, with a 75 dB dynamic range and a low 0.05 percent wow and flutter rating.

Using an advanced metal tape formulation, the 8-mm metal particle (MP) boasts four times the magnetic energy of the conventional cobalt ferric oxide tape used by VHS-C. This tape enables 8 mm to record and play back video signals with much less dropout.

The 8-mm video format does not use a conventional control track (CCT) head, but the automatic track finder (AFT) circuit. With this circuit, the VCR monitors the video head position and compensates for error in the tape path. This new format and design results in a compact and low-weight high-quality camcorder.

Signal spectrum

The major portion of the 8-mm video format contains the IF-IN luminance signal and occupies the spectrum from 2.4 MHz to 5.04 MHz, and produces a deviation of 1.2 MHz. The major portion of the spectrum is for the FM luminance signal that occupies the spectrum from 4.2 MHz to 5.4 MHz, and produces a deviation of 1.2 MHz.

The chroma is down-converted to 743 kHz, which is slightly different than the 629 kHz of VHS-C. Although the frequency is different, the principle is the same. The down conversion is done to avoid beats with the FM luminance signal.

The audio is a 1.5-MHz frequency-modulated carrier located between the color and luminance signals. This carrier produces hi-fi quality audio with minimum wow and flutter.

A fourth signal is the tracking pilot signal. Tracking signals consist of four pilot signals recorded onto the tracks. The frequency of these signals is between 102 kHz and 165 kHz. The signals are sequentially changed and repeated after four consecutive fields.

8-mm cassette

The 8-mm cassette (see Fig. 1-52) is somewhat thicker, but a little narrower in width than a conventional audio cassette. The total volume is only slightly more than the audio cassette. The 8-mm video cassette is capable of 120 minutes of record time.

The cassette designation has also been standardized along with the cassette. An example is the P6-120MP cassette. In the first letter-number combination, 6 stands for NTSC system using 60 fields. The next three digits, 120, indicate the record time of 120 minutes. The last two letters stand for the type of tape: MP for metal powder, EP for evaporated powder.

By turning the cassette over, you can see that the tape is protected by a double door on both the outside and the inside of the tape.

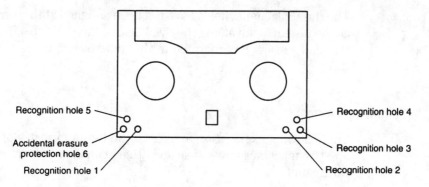

Recognition hole 5

Recognition hole 4

Accidental erasure
protection hole 6

Recognition hole 3

Recognition hole 1

Recognition hole 2

■ **1-52** *The 8-MM cassette layout format.*

The door can be released by pulling back the lid lock and folding the door up to expose the tape.

A reel lock is at the bottom center of the tape. When the tape is loaded into the mechanism, a pin fits above the reel lock tab. When the tape is loading, the reel lock is pulled back, releasing the locking mechanism. With the reel lock pulled back, the reels are free to turn.

The 8-mm tape uses two alignment holes toward the front of the cassette to rigidly hold the cassette in position when it is seated on the mechanism.

The large hole in the middle of the cassette is for the cassette light-emitting diode (LED). It works in the same manner as regular VHS or VHS-C. It detects the transparent leader at both ends of the tape to stop the mechanism when the supply or take-up reel is empty.

Six sensor holes are in the lower-left and lower-right corners of the cassette. At this time three of these holes have been assigned in the format.

The first one is the record-proof sensor. When the record-proof switch is in the record position, the record-proof sensor hole will appear red. If the record-proof switch is moved to the record-inhibit position, then the red will no longer be available.

The second hole is used to sense metal powder or metal evaporated tape. When the hole is closed, metal powder tape is loaded in the cassette; when open, metal evaporated tape is in the cassette. This is used to change to equalization and relevant recording circuits to produce optimum results with both types of tapes. The third hole is for the tape thickness.

The three remaining holes have not been assigned at this time, but as the format is advanced they will be assigned to a specific function and sensors incorporated in the units having these new features.

8-mm tape format

The following explanation applies to the 8-mm format. Some will apply to machines now in use and others can be incorporated into future units.

The 8-mm video format uses the helical scan system for video recording that is used in virtually all video recorders. This system produces the high video speed with a relatively low tape speed that is essential for recording the high-frequency (HF) video information. But the 8 mm-video format has been designed with many additional features, and therefore the tape is divided into four areas for recording information. Two of these areas are scanned by the rotating video heads producing the slanted tracks characteristic of all video recorders. The other two areas are conventional longitudinal tracks.

The recorder does not have fixed heads for recording or playback of longitudinal information. All information recorded onto the tape is accomplished with the rotating video head. An 8-mm tape format layout is shown in Fig. 1-53.

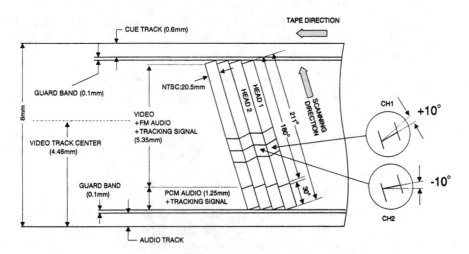

■ **1-53** *With some 8-MM VCRs there are no fixed heads for recording or playback of longitudinal information. All information recorded onto the tape is accomplished with the rotating video head.*

The first and largest area contains the video, both luminance and chroma, the FM mono audio signal, and the tracking signal. In the 8-mm format the major portion of the information is recorded here. The remaining three areas are optional and used for special features.

In this format two video heads are used, and they are distinguished from each other by a $+/-10$-degree azimuth difference to reduce crosstalk between the adjacent channels. The width of the tape track is 20.5 microns.

The second area of information extends for 1.25 mm below the video track. This area is for the pulse code modulation (PCM) audio and tracking signal, and is reserved for the recording of stereo PCM-encoded audio. It is not a requirement of the 8-mm format that all units play PCM audio, but this portion of the tape must be available on all 8-mm recordings. A tape recorded with PCM audio information will also have the FM audio mixed with the video signal. This tape can be played on any 8-mm machine without the PCM capability, which will produce sound from the FM audio. In addition to the PCM audio signal, the tracking signal is also recorded in the area of the tape.

Rotational head drum

The recording of the two areas of information by the rotating video heads is made possible by scanning the tape for approximately 221 degrees as shown in Fig. 1-54. In some machines the heads scan only 180 degrees of tape, because this is all that is necessary for one video head to be in contact with the tape at all times. For 8-mm machines with PCM audio, the tape is wrapped an additional 40 degrees. During this additional tape contact the entire audio signal is recorded as PCM information in addition to video information.

The third area of information is an optical cue track, 0.6-mm wide at the tape top. This cue track is separated from the video tracks by a 0.1-mm guard band. This cue track is optional, like the PCM audio area. It can be used on future machines to record editing information. Like the PCM signal, this cannot be recorded without also recording the full video, FM audio, and tracking signals in the scanning portion of the tape. So any tape made with or without a cue signal can be played on any 8-mm machine.

The fourth area of information on the tape is a longitudinal audio track at the bottom of the tape. This track is 0.6-mm wide and, like

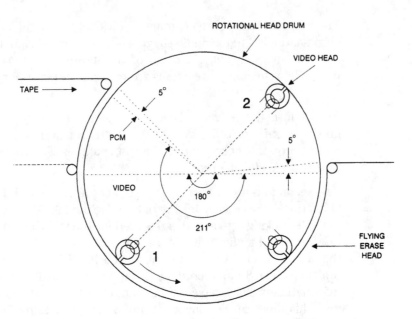

ROTATIONAL HEAD DRUM

VIDEO HEAD

TAPE

5°

PCM

2

5°

VIDEO

180°

211°

FLYING
ERASE
HEAD

1

■ **1-54** *The location of video heads and flying erase head on an 8-MM drum.*

the cue track, is separated by a 0.1-mm guard band. Like the cue track, this longitudinal audio track is optional. A camcorder that uses this track must still record using the FM audio mixed in with the video and tracking signal on the same tape. This is done so that all units will be able to play 8-mm tapes no matter which machine it was recorded on.

The RF switching pulse is used to switch between the video heads and produce a continuous RF envelope. In the 8-mm format, with the additional wrap for PCM audio, the RF switching pulse is still used to select video information from the heads. However, this switching pulse occurs between the PCM and video information, in the same relative position as it would occur in a conventional video recorder.

The PCM audio information is pulled off the tape by a different select pulse.

Tape erasing

Some 8-mm machines do not use a full erase head to remove previously recorded information from the tape. Some units use a flying erase head such as those found on professional editing machines. The *flying erase head* is positioned on the video drum and moves with the video heads. At the start of a recording, the erase head erases the two video tracks directly after the old recording, thus leaving no gap. Therefore, no color rainbows are produced. The blank tracks behind the video flying erase head are properly filled in by the rotating video heads. When the recording is ended, it is timed so that the flying erase head erases only two tracks of information. This eliminates the formation of blank areas on the tape, and a perfect edit will exist between the old recording and the new recording that follows.

The flying erase head is located between the CH1 head and CH2 head on the video drum assembly. The drum rotates counter-clockwise so that the erase head precedes the track 1 head by 1/4 field. The erase head is twice the width of the video heads, which results in the erasing of two tracks simultaneously. These tracks are then properly filled in by the head 1 track and the head 2 track before the flying erase head contacts the tape again and begins to erase another two tracks. The timing of the flying erase head and video head switching is controlled by the system control micro-processor. In order to maintain the balance of the drum assembly, a counterweight is positioned on the video drum assembly oppo-site the flying erase head.

In these machines, because there is no longitudinal CTL or audio track required, the use of the flying erase head eliminates the need for all fixed heads. Units with flying erase heads contain all the heads necessary to record all information on the video drum as-sembly.

Automatic tracking finding circuit

Some 8-mm systems do not incorporate a CTL head. A system des-ignated automatic tracking finding (ATF) is used to correctly track the heads on the videotape. This system provides continu-ous correction of the video tracking, and eliminates the need for a custom tracking adjustment.

In this system, four pilot signals are recorded on the tape along with the video information. As shown in Fig. 1-55, the pilot signals are designated F1 to F4, and are switched for each track with the sequence repeating every four tracks. Refer to Fig. 1-56 for the frequency of the signals.

Pilots F1 and F3 are set for recording on CH1 (+ azimuth) and F2 and F4 for recording on CH2 (− azimuth).

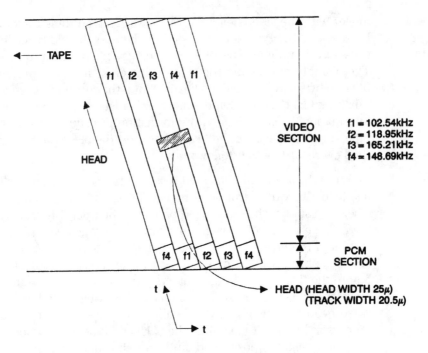

TAPE

f1 f2 f3 f4 f1

HEAD

VIDEO
SECTION

f1 = 102.54kHz
f2 = 118.95kHz
f3 = 165.21kHz
f4 = 148.69kHz

f4 f1 f2 f3 f4

PCM
SECTION

HEAD (HEAD WIDTH 25μ)
(TRACK WIDTH 20.5μ)

t

t

■ **1-55** *Frequencies for video and PCM are offset by one track.*

Pilot (fi)	Frequency (kHz)	Recording head	Frequency difference (kHz) (Approx.)	* N	SEL	
					1	2
f1	102.54	CH1		58	H	H
			16			
f2	118.95	CH2		50	L	H
			46			
f3	165.21	CH1		36	H	L
			16			
f4	148.69	CH2		40	L	L
			46			
f1	102.54	CH1		58	H	H

■ **1-56** *The 4 pilot signal frequencies.*

Detection of the trace position of the video head is performed during playback by a special circuit. A pilot signal with the same or different frequency as was recorded is used as the reference pilot (REF). Assume that REF is F3 and that the head is in the A position (recording F1) as shown in Fig. 1-57. At this time the play-

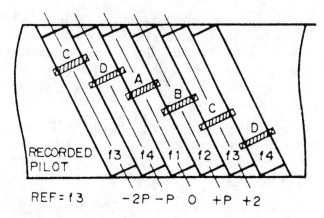

■ 1-57 *Recorded pilot tracks.*

back pilot (PB) will be the mixed signal of the F2 and F4 neighboring tracks in addition to F1.

The neighboring pilot frequencies are selected for a mutual spectrum interval of 16 kHz to 46 kHz, as shown in Fig. 1-58. In this case, the following three components are obtained:

☐ F1 to F3 REF = 62 kHz

☐ F2 to REF = 46 kHz

☐ F4 to REF = 16 kHz

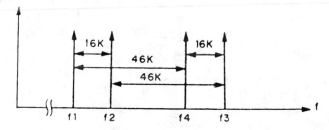

■ 1-58 *Selection of mutual spectrum intervals.*

The 46-kHz to 16-kHz level corresponds here to the F2/F4 level. The 46-kHz component is extracted by the next BPF and the level is detected. The 16-kHz level increases when the video head is shifted in the F4 direction (left) and the 46-kHz increases when shifted in the F2 direction (right).

A shift to the left or right can be detected by determining the 46-kHz and 16-kHz level difference. The signal detected in this manner is called the ATF error signal.

Automatic tracking finding circuit

The error detection characteristics with the A position as reference (origin) are obtained with a special circuit. The output polarity is plus (+) when shifted to the right and minus (−) when shifted to the left. If REF is changed, the reference point will change. When REF is F2, the B position is the origin for obtaining the detection characters, C for F1, and D for F4.

Thus REF cycles between F3, F2, F1, and F4 for each track (field), and the detection reference point (origin) moves to A, B, C, and D at the switching point. When the tape speed decreases and there is a shift to the left in the vicinity of the origin, the 16-kHz level increases. When the tape speed increases and there is a shift to the right, the 46-kHz level increases. This is shown in Fig. 1-59.

The 16 kHz/46 kHz frequencies are compared at −/+ polarity and power supplied to the motor is altered with that polarity, and tracking centered on the origin is possible. This is the principle of tracking by ATF.

ATF circuit operation

ATF processing is carried out by IC371. The circuit consists of two sections, one that generates the recording pilot and playback REF pilot, and another for error detection.

Recording

The 5.9-MHz pulse generated by the crystal oscillator (X3701) is divided into four kinds of pilot signals. Division rate is determined by a 2-bit selecting signal (SEL 1/2) at pins 23 and 21. SEL 1/2 is output from pins 49 and 50 of IC351.

The recording pilot is recorded along with the YC (video) and A (audio) signals.

Playback

After the luminance/chroma and audio components are removed from the playback amplifier signal by the LPF, it is input to pin 40 of the IC. It then passes through a mixer, a BPF, a comparator, and other elements, finally becoming output from pin 12 as the ATF error signal. This signal passes through the RC secondary filter to remove its 15-Hz components, then it is input to pin 3 of IC351, converted from analog to digital inside the IC, and added to the motor speed data. ATF data are thus used instead of the mode phase data during recording. After this, tracking is adjusted by servo control as in the recording mode.

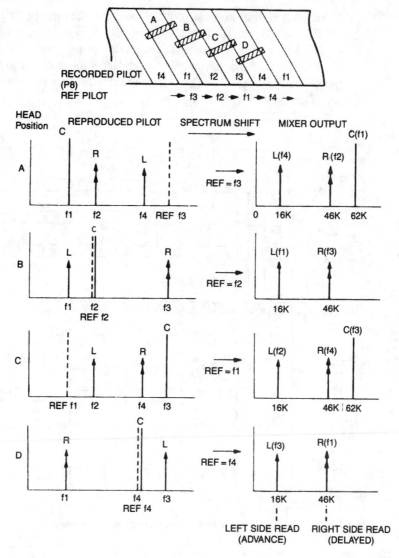

RECORDED PILOT (P8)
f4 f1 f2 f3 f4 f1

REF PILOT
f3 → f2 → f1 → f4 →

1-59 *When the tape speed increases, and there is a shift to the right, the 46-KHz level increases.*

Detection by time sharing, ATF-lock signal

In Fig. 1-60, the head is tracked over F1, F2, F3, and F4 on the recorded pattern by switching of REF (1) to F1, F2, F3, and F4. At the same time, there is REF (2), which is a separate F2, F3, F4, and F1 one field delayed, and if error detection is performed between the PB pilots, the 16-kHz and 46-kHz components will be alternatively detected in the output. Accordingly, the polarity of the 46-kHz side is inverted by pin 37 of IC371 so as to always output the 16-kHz component. As the inversion is performed inside the IC, when tracking is properly locked, this should be fixed at a low level. The lock state can be determined from this signal. This signal is called ATF-LOCK signal.

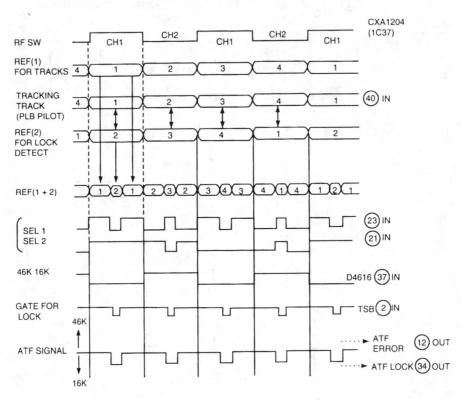

■ 1-60 *The REF and signal frequency pattern comparison.*

In reality, two types of detection are performed by switching REF (1) and REF (2) and are output within a single field by time sharing. The sample pulse TSB (output from pin 52) output from IC351 is input to pin 2 of IC371 LOCK when TSB = LOW (pin 34) is separated and output. The LOCK signal is set to the logic level by the *Schmidt circuit*, and is output.

The signal level obtained by TSB is proportional to the playback level of the head that is tracing and being used. The output voltage from pin 35 is fed back to the gain control terminal (pin 42 of the previous stage amplifier) to configure an automatic gain control (AGC) level.

VCR test equipment requirements

This section covers the types of test equipment needed for VCR servicing, test set-ups, and VCR test tapes.

The new home VCR market developed a new service market. Videotape recorder service is no longer confined to a few service centers that specialize in only industrial recorders, as was the case a few years ago. The VCR owner now expects to receive local service from the service shop that provides TV and stereo sales and service. Many dealers that sell VCRs believe that they must be able to service what they sell to have an advantage over their competition. This service is usually part of the same service network that services the TV receivers sold by the same dealer.

The VA48 Sencore analyzer

Before Sencore introduced the VA48 video analyzer, TV service technicians usually avoided VCR servicing. Or, most could only afford to equip the shop for VCR service and work with other older test equipment for TV work. The Sencore video analyzer has eliminated this dilemma because it updates the shop for both TV and VCR servicing with one piece of test equipment.

Special considerations made at the time the VA48 was designed ensured that the signals would be compatible with the new VCR systems. Extensive testing, both in Sencore's own lab and in working with the service managers of leading VCR manufacturers, has shown that these signals produce service results that are equal to, and in some cases superior to, those produced by an NTSC signal generator. This is especially true if the service shop is not equipped with an NTSC vectorscope for evaluating the output signals produced by this very high-cost NTSC generator.

Using the VA48 analyzer patterns

The VA48 Sencore analyzer is shown in Fig. 1-61. The two patterns used for VCR troubleshooting on the VA48 are the *bar sweep* and *chroma bar sweep*. There are a few special features built into each of these patterns that make them even more versatile in VCR service than you might find in standard TV service, because the TV receiver does not have all the circuits that are part of a VCR machine.

The bar sweep pattern should be used for all luminance (black-and-white) circuit testing. The first reason that this pattern is important is that it does not have a color burst. This gives a positive

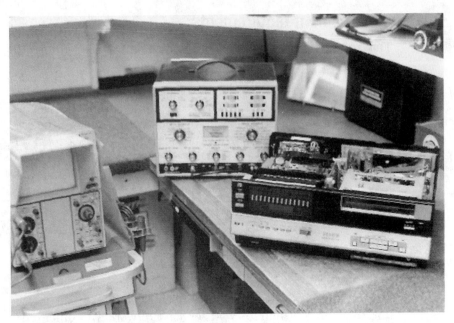

■ **1-61** *The Sencore VA48 video analyzer.*

test of all automatic color-killer circuits found in both the record and playback VCR circuits. These circuits should automatically switch to the black-and-white mode when the color burst is not present.

The bar sweep pattern produces various amounts of black-and-white information. The gray scale at the left of the switchable frequency bars, for example, checks the black, gray, and white signal levels for proper amplitudes, linearity, and clipping. Each frequency bar after the gray scale alternates between pure black and pure white levels to test the circuits for proper frequency response. The bar sweep pattern looks much like the familiar "multi-burst" pattern used in many types of video testing. One difference between the bar sweep and the multi-burst is that the bar sweep is made up of square waves while the multi-burst is made up of sine waves. The key difference is that the bar sweep can show circuits that ring on fast signal transitions, where the multi-burst might be passed by these stages without any noticeable ringing.

The various frequencies of the bar sweep pattern then check the video frequency response of the entire record/playback system. Certain applications of the bar sweep involve recording this pattern on a tape using a known good recorder. Many times you can use this tape in place of the expensive alignment tape and save wear and tear on your special alignment tape.

The second pattern from the VA48 is the chroma sweep bar. This pattern contains a standard level color burst for operation of the color-killer circuits and the automatic color control (ACC) circuits. The color information is phase-locked back to the horizontal sync pulses so that this pattern produces a line-by-line phase inversion just like that of an off-the-air signal. This is very important because the comb filters used in almost all playback circuits require this phase-inversion to detect the proper color signal as opposed to the unwanted crosstalk information from an adjacent video stripe on the magnetic tape. The fact that the chroma bar sweep covers the entire frequency range of the color information (0.5 MHz either side of the color subcarrier) makes this pattern ideal for testing the operation of all the chroma conversion circuits to make sure that they are not restricting some of the color detail information.

Recording your own personal test tape

If you have a VA48, you will want to record your own personal VCR test tape using the patterns from the analyzer. This test tape will not replace the need for the VCR test and alignment tape but rather supplements the alignment tape. The advantage of having a tape that you recorded yourself is that it will save much wear and prevent the possibility of damaging the expensive alignment tape. Should a defective machine damage your own test tape, you can replace it for the cost of a blank tape rather than having to purchase a new, costly alignment tape. However, the alignment tape allows you to confirm the various operations of the color circuits as well as the performance of the luminance circuits. A defective video playback head or preamplifier, for example, can be quickly determined by using your own test tape.

Any adjustments that affect compatibility with other machines should be made with the manufacturer's own test tape. This tape has been carefully prepared to provide a standard from one machine to another so that a tape recorded on one machine will play properly on another machine. This is especially important if your customer has more than one VCR and plans to play tapes on a different machine than was used for recording. The same goes for commercially prepared prerecorded programs on tapes that can be purchased and played back.

When you record your own VCR test tape, be sure to use a machine that you know is operating properly. New VCR machines are generally set up and aligned properly. Be sure to feed the signals directly into the CAMERA input jack of the VCR rather than feeding them through the antenna input terminals. This will assure you that an improperly adjusted IF stage or fine-tuning adjustment will not cause the quality of your recording to be less than ideal for test purposes. Take advantage of the tape counter to allow you to locate specific portions of the tape so you can easily find the pattern you need to use. It is also desirable to include an audio signal for at least a portion of the tape to test the audio playback circuits. This audio signal is available from the "Audio 1000 Hz" position of the drive signal output from the Sencore VA48 video analyzer.

Perform the following outline of steps to record your VCR test tape.

1. Connect the VCR standard output from the analyzer to the CAMERA input of a known good VCR. Switch the input switch to the CAMERA position.

2. Set the tape counter to 000.

3. Select the faster tape recording speed on the machines that have more than one speed (i.e., Beta III, X1, or SP modes).

4. Select the BAR SWEEP position of the video pattern switch on the analyzer.

5. Place the VCR in the RECORD mode and record the pattern until the tape counter has progressed 50 counts.

6. Switch the video pattern switch on the analyzer to the CHROMA BAR SWEEP position and record this pattern for the next 50 counts.

7. Switch the video pattern switch on the analyzer to the COLOR BAR position and record this pattern for the next 50 digits on the counter.

8. Switch the video pattern switch on the analyzer to the CROSS HATCH position and record this pattern for the next 50 counts.

9. Switch the VCR to the slower tape recording speed (the Beta I, II, LP, or X2 mode).

10. Repeat steps 5 through 8 and record the same video patterns at each tape speed.

Your test tape is now recorded. If you want the audio tone recorded during one or more of these test sections, simply connect the drive signal output to the auxiliary audio input jack of the

VCR while recording the pattern or use the AUDIO DUB function to add the audio after the video is recorded.

The test tape now has the following patterns:

Tape counter	Pattern
000 - 050	Bar sweep
050 - 100	Chroma bar sweep
100 - 150	Color bars
150 - 200	Cross hatch
200 - 250	Bar sweep (slow speed)
250 - 300	Chroma bar sweep (slow speed)
350 - 400	Color bars (slow speed)
450 - 500	Cross hatch (slow speed)

VCR servo alignment

The Sencore video analyzer signals are ideal for performing servo alignment adjustments on a VCR. The phase-locked, broadcast quality sync makes it perfect for this alignment application. Inject the signal from the VCR standard jack on the analyzer directly into the video or camera input jack on the VCR to eliminate the fine tuning errors that can occur when using an RF air signal.

There are several adjustments that call for counting the horizontal sync pulses just before the vertical sync pulse. When using an off-the-air signal, the equalizing pulses are present and can cause an error in the counting, resulting in setting a critical control setting wrong. The VA48 analyzer phase-locked sync signals do not have the equalizing pulses, and the counting of the horizontal sync can be done correctly without the equalizing pulses causing confusion, so the proper setting can be obtained easier. Equalizing pulses are not used in the VCR and are not important to servo alignment procedures.

VCR audio checks

The 1-kHz audio signal from the drive signal section of the analyzer can be used when making checks of the record and playback functions of the VCR machine. The manufacturer's alignment tape should be used for making the playback equalization adjustments. The following procedure allows injection of the audio test signal.

1. Adjust the drive level control with the drive signals switch in the audio 1,000 Hz position to about 2 volts peak-to-peak reading on the VA48 meter.

2. Inject the audio from the drive output jack on the analyzer into the audio input jack on the VCR. For VCRs without this jack, the signal can be injected into the audio circuits at a test point that is usually noted in the service data.

3. The audio test signals can then be traced with an oscilloscope.

The newer model Sencore VA62A video analyzer is shown in Fig. 1-62.

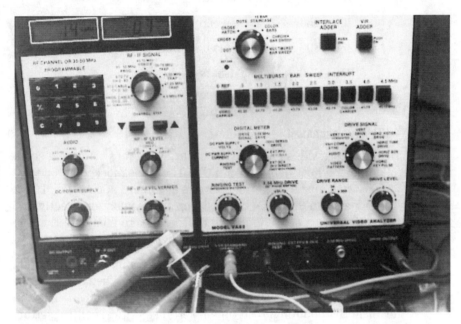

56

■ **1-62** *The newer model Sencore VA62 video analyzer.*

Other equipment requirements

Besides a video analyzer or NTSC color bar generator, you will need a dual-trace oscilloscope, frequency counter, digital voltmeter, transistor checker, and perhaps a capacitance-inductance checker. Here is an overview of a few of these test instruments that are now available for VCR troubleshooting.

Servo alignment procedures

In some VCR servo alignment procedures, a dual-trace scope and ADD channel capabilities might be called for. The ADD mode means that the two channels are added together in the scope. This feature is not found on some scopes, but it might not actually be needed in the alignment procedures. A Leader model LBO-515 dual-trace scope, shown in Fig. 1-63, is being used for servo alignment checks.

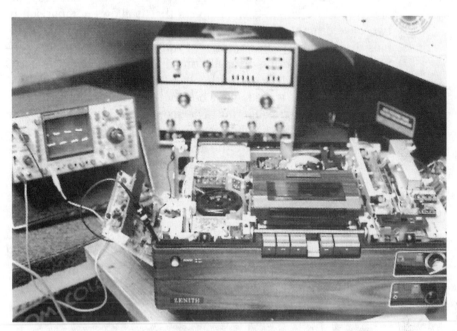

■ **1-63** *A dual-trace scope set-up for servo alignment.*

On scopes with no ADD mode, you can utilize the dual-chopped mode to provide proper waveform information for aligning the VCR servo systems as follows:

1. Set both channels to ground and adjust the trace position control so that the two traces align in the center of the screen.
2. Connect the A and B scope channels for the servo adjustment procedure and set the scope to the dual-chopped mode.
3. Now make the adjustments for the servo alignment as stated in the service data.

Note: It is recommended that external sync be used in these procedures so that the oscilloscope will be locked to the channel that is not being adjusted. All you need do is determine which channel is stationary and which one is adjusted, and connect the external sync jack to the stationary channel test point.

Dual-trace scopes

Many of the modern electronics service shops are now using wide-band dual-trace triggered sweep scopes for their everyday TV electronics troubleshooting. If you don't have a dual-trace scope, perhaps now would be a good time to review the need for one for VCR servicing. The greatest advantage of a dual-trace scope as compared to a single-trace scope is that it allows comparison of two waveforms at the same time. This is very useful for tracing signals through inputs and outputs of various stages, but it is essential in troubleshooting critically timed circuits such as the servo stages. A dual-trace scope with a bandwidth of 20 to 50 MHz that has a good bright trace is your best bet for VCR and overall electronic circuit troubleshooting. The scope also should have a very stable trace with rock-solid lock-in of the waveforms. If you purchase a new scope, be sure the instrument has true TV sync separators built into the trigger circuits for both vertical and horizontal sweep rates. Some scopes that have a "TV" position do not use integrator filters to allow triggering on vertical sync pulses. One scope that has the features for VCR servicing and a 60 MHz bandwidth is the Sencore SC61 shown in Fig. 1-64. Another scope that can cover your electronic service needs is the Tektronix model

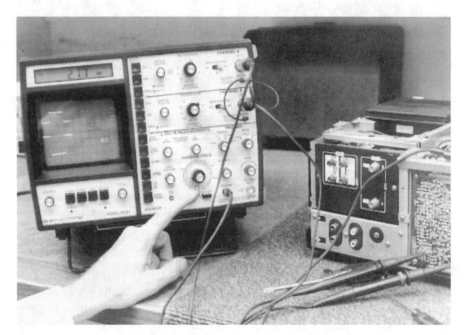

■ **1-64** *Using the Sencore scope for VCR troubleshooting.*

5403 (shown in Fig. 1-65), which has a bandwidth of 70 MHz. The one pictured is set up for dual-trace operation with one vertical plug-in unit (there is space on the front panel for another plug-in vertical amplifier).

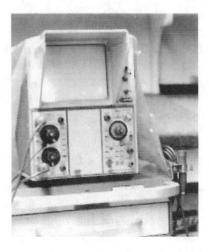

■ **1-65** *A Tektronix dual-trace oscilloscope.*

Frequency counters

A stable frequency counter is required for setting the reference oscillators found in the color circuits of the VCR and for adjusting the servo circuits. The Sencore FC45 counter shown in Fig. 1-66 offers a high 25-mV sensitivity for measuring low-level signals with a full 8-digit readout for direct readings of the oscillator frequencies down to 1 Hz resolution.

■ **1-66** *Sencore frequency counter.*

The versatility of the FC45 can be expanded even more with the use of the PR50 audio prescaler, which lets you adjust the 30-Hz control track signal to an accuracy of 0.01 Hz. Some frequency counters use a "period measurement" mode for measuring these low frequencies. This can be very time consuming, because the frequency is shown as a time interval instead of a frequency. You must then use a calculator to figure the actual frequency by dividing the time measurement into 1 (frequency = 1/time). The PR50 provides a direct readout of frequency, which is updated every second, or 10 times a second with 0.1-Hz resolution, to eliminate the time-consuming calculations necessary with a period counter. The PR50 also has filtering built in to prevent false double-counting due to signal noise that is often present in these low-frequency signals.

Transistor tester

Many of the circuits found in VCRs are controlled by discrete transistors. Some VHS video recorders, for example, have over 150 transistors that are used for various control functions. The use of an in-circuit transistor checker, such as the Sencore TF46 Super Cricket (shown in Fig. 1-67) greatly simplifies the troubleshooting of a transistor circuit, because the suspected transistors can be checked in-circuit to confirm whether they are defective or not. The TF46 has a high in-circuit testing accuracy for both transis-

■ **1-67** *Sencore transistor checker.*

tors and FETs that helps to speed up circuit diagnosis. Also, the TF46 has an automatic gain test to allow grading or matching of transistors used in critical circuits.

Other VCR test equipment

A capacitance and inductance checker can help you to locate leaky or off-value capacitors and check out any coils or transformers. The Sencore model CA55 shown in Fig. 1-68 will do a good job.

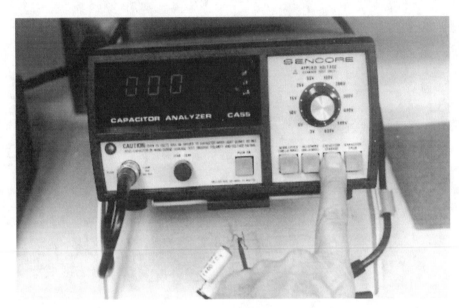

■ **1-68** *Sencore capacitor analyzer.*

Troubleshooting tips and techniques

THIS CHAPTER CONTAINS GENERAL VCR TROUBLESHOOTING tips to help locate problems in various machines. It is not practical to provide troubleshooting information for all VCR makes and models, so this is an overview of troubleshooting techniques and how to use various test instruments. The same service techniques can usually be applied to both the VHS and Beta format VCRs.

This chapter features Sencore test instruments, such as the SC61 waveform analyzer, the VA62 video analyzer, the VC63 VCR test accessory, and the NT64 NTSC color pattern generator. I explain how to use the test equipment for quick VCR alignment, trouble diagnosis, and VHS servo system analysis. A photo of the Sencore SC61 waveform analyzer is shown in Fig. 2-1.

Getting started

One of the important first steps is to have proper service information, such as schematics and block diagrams. These should include scope waveforms, alignment data, and the parts list. You also should look at the owners operating manual if you are not familiar with that particular brand of VCR.

The next step is to connect the VCR to a TV monitor and verify the complaint. Test the normal operation of the machine, using all of the controls. Make any required adjustments. Play back a good prerecorded or test tape, and then make a recording on a blank tape.

Once the problem is identified, use logic to try and determine which section or block of the machine is at fault. The block diagrams for the unit, such as those in Fig. 2-2, will be helpful in pinpointing the trouble.

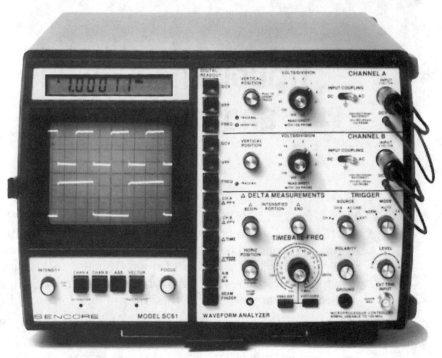

The Sencore SC61 waveform analyzer.

When you have determined which block is at fault, you can pinpoint the faulty circuit and component. Remove the cover. A good way to begin is to use the old reliable look, listen, touch (carefully), and smell technique. Look for any obvious parts defect, listen for any unusual mechanical noise (or lack of such), carefully touch components that might be too hot or cold, and use your nose to ferret out any burned or overheated components. Always be on the alert for smoke.

After any obvious faults are cleared, use your test equipment to zero in on the defective component. Test equipment should include a video analyzer to inject signals, a triggered-sweep oscilloscope to look at waveforms and peak-to-peak voltages, a voltmeter, and a frequency counter.

Observing symptoms

Use the following tips to determine the operating condition of the VCR. If one check does not produce the desired results, mentally (or physically) note some symptom of fault. Performing additional checks might add to a list of symptoms. This compiled information then can direct you to the area of the machine that is defective. What operating mode is not functioning properly? Is the playback mode or the record mode at fault?

Sometimes both record and playback processes are affected. The playback of a known, correctly recorded tape cassette in the suspected machine can aid in this determination. If the playback is normal, the record section should be investigated. However, if the attempted playback is not normal, check the playback system. A tape that has gone through a record process on the faulty machine will be able to be played back on a known good VCR. This test tells you whether the machine in question is recording properly. Some of the symptoms will be obvious; others will not be.

Plug-in modules and chips

With some VCR brands, you can obtain new or rebuilt plug-in modules. With these types, you might want to isolate the problem by changing various modules. You then have the option of repairing or replacing the defective module. This technique allows you to quickly isolate the problem. Do not overlook loose or corroded module plug-in contacts.

Some VCRs have plug-in ICs that contain the complete playback video, record amplifier, and the chroma processing system. These are usually large-scale integration (LSI) chips. Thus, should you suspect trouble in these chips, a new IC can be easily plugged in for a quick check. However, in most VCRs, these ICs are soldered into the board. In this case, use a scope to check input and output signals or use a signal injection technique. Also, check the dc voltages at the chip pinouts. Only when you are certain a chip is defective do you want to desolder it and install a new one.

The divide and conquer technique

The "divide and conquer" technique can be used to quickly pinpoint defects in VCR circuits. This technique, sometimes referred to as *functional analyzing*, can save many steps in circuit diagnosis and bench time and will increase your troubleshooting efficiency. The D and C method also can be applied to all types of electronic troubleshooting.

To begin this technique, concentrate on the circuit operation to narrow the problem down to one stage or block, rather than individual circuit components. This method differs from other types of troubleshooting in that the first step is not finding which stage is defective, but finding which ones are operating properly. Do this in a logical sequence until the defect is pinpointed to one block of the VCR block diagram.

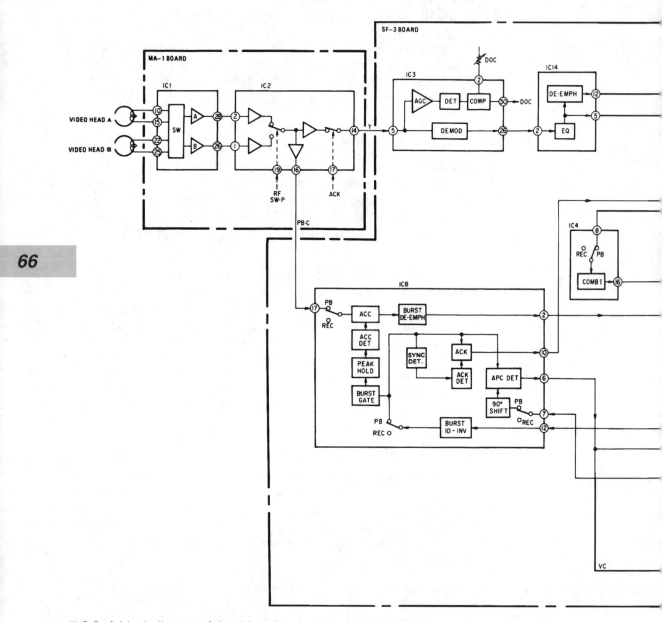

■ 2-2 *A block diagram of the video signal system in playback.*

Once the problem has been isolated with signal tracing or signal injection, use scopes, meters, transistor testers, etc. to isolate the faulty part within the block. Thus, signal tracing and signal injection can be time-savers for troubleshooting all types of VCR cir-

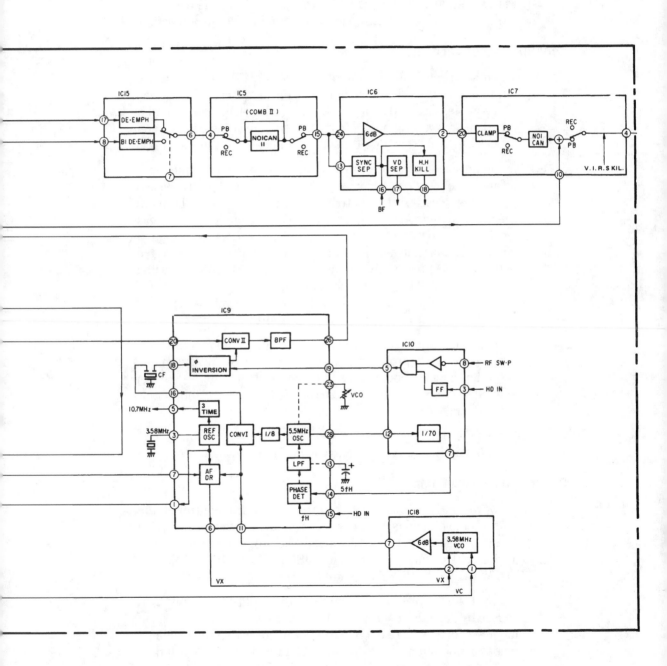

cuits. Also, keep in mind that signal tracing and injection techniques can be used simultaneously, for example, in the RF, IF, and video amplifier stages.

The block diagram for a selected VCR amplifier stage is shown in Fig. 2-3. It illustrates the D and C technique. Note the test points 1 through 4 that can be used for signal tracing or signal injection.

Servicing cautions and tips

When servicing VCRs or any other electronic device, be careful not to add to the original complaint. This can occur if a clip lead, for example, slips and shorts a signal or voltage level to ground or B+, damaging a component. Careful positioning of test instrument probes can avoid this time-consuming and costly service situation.

In measuring a signal or voltage at an integrated circuit pin, using an alternate test point is advised, because IC pins are very close to each other. Observe the foil trace that leads away from the pin to see if it is connected to another nearby component lead that is easier and safer to put a test probe on. Be sure there are no components in the circuit between the chip pin and the alternate test point. The best test probes to use in these cases are those with a spring-clip end. The plastic shroud insulates the small probe lead from adjacent pins or terminals. Finally, remove your wrist watch or other jewelry to help prevent circuit shorts.

Do not change an IC or other component without good reason. Determine the most probable fault in the circuit by measurements and other checks. The chip itself might not be at fault. Also keep in mind that a chip failure could have occurred due to another circuit defect.

Remember to make sure the TV set you are using as a monitor is operating normally before troubleshooting the VCR.

Control function operation

All of the early model VCRs used the mechanical control system that resembles a piano keyboard. The keyboard control moves leaf spring contacts together that activate the proper control function. These control functions include EJECT, PLAY, RECORD, FAST FORWARD (FF), REWIND, and PAUSE.

Many of the control function faults in these older machines were caused by defective contacts and bent springs. However, most late model VCRs now use a microprocessor to control all the machine's modes of operation. Just tap a button for the desired mode and the

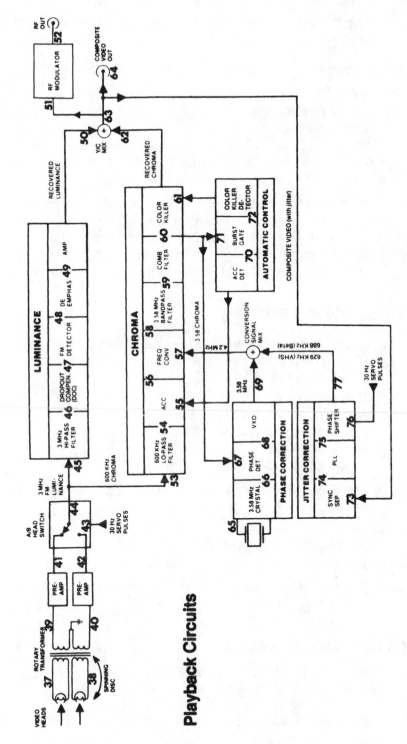

Playback Circuits

2-3 Block diagram of VCR playback circuit. Inject the signals to "Divide and Conquer."

The divide and conquer technique

microprocessor transmits digital pulses to activate transistors to turn-on motors, relays, etc.

Automatic shut-down

All VCRs have sensors that automatically shut down the machine in case of a malfunction to prevent any further damage. In newer machines, these automatic shut-down modes are controlled by the microprocessor. Automatic shut-down occurs when activated by sensors that detect slack or broken tape, end-of-tape, take-up reel problems, and dew (moisture).

These stop sensors are usually microswitches, photo transistors, or sensor lamps. In many of the older model VHS machines, all you might have to do is replace a burned-out sensor lamp. Check a suspected automatic shut-down circuit by bypassing the microswitches and/or covering the sensor lamps with black tape.

Service procedures summary

☐ Collect as many symptoms as possible. Go through a checklist of the functions for the VCR to determine which operations are normal and which ones are not.

☐ If possible, have a normally operating machine available for comparison purposes and for the interchange of tape cassettes with the faulty machine.

☐ Have a known good recorded (in all modes) tape available as a reference.

☐ Use block diagrams to relate to the observed symptoms and to localize the area to be investigated.

☐ Study the circuit diagrams to determine which components or test points should be checked.

☐ Refer to the layout diagram to locate circuit test points.

☐ Determine if a mechanical rather than an electrical failure is involved.

Your factory expert: Sencore

Say you are an employed technician in a TV/VCR repair shop. On your last VCR repair, you replaced the cylinder head, drive belts, idler, and capstan pinch roller. The mechanical tape tension, tracking, and head switching all checked OK. The work went off quickly with no hitches. Two days later the customer comes back with the unit and the complaint, "It just doesn't look as good as it used to."

Is it the VCR or the customer? If you challenge the VCR owner by saying, ". . . that's the best it's gonna do . . ." or ". . . it looks good to me," you stand a good chance of losing a customer and any of his or her referrals. Is there a way to positively eliminate doubt and prove to your customer that the VCR does perform the same or better? Perhaps the only acceptable opinion would be that of a factory expert.

In the case of Sencore, your VCR factory expert is the Sencore Tech-Pak: the SC61 video analyzer (Fig. 2-1), the VA62 Universal Video Analyzer (Fig. 2-4), the VC63 VCR Test Accessory (Fig. 2-5), and the NT64 NTSC Pattern Generator (Fig. 2-6). All of these instruments have patents and exclusive tests that allow you to completely test VCRs to see if they meet manufacturers' specifications. In fact, the VA62 is accepted by all major manufacturers for factory-authorized VCR warranty work.

How can the Sencore Tech-Pak prove to the customer that the VCR really is as good as it used to be? There are many ways.

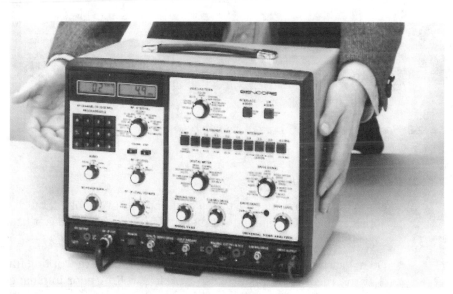

■ **2-4** *Sencore VA62 universal video analyzer.*

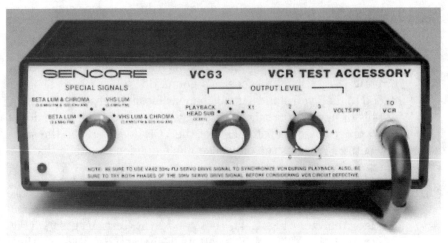

■ **2-5** *Sencore VC63 VCR test accessory.*

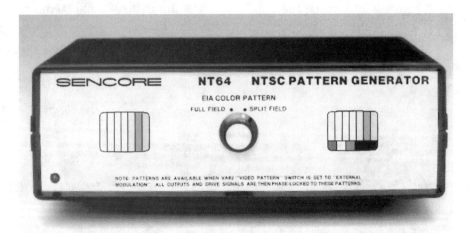

■ **2-6** *Sencore NT64 NTSC pattern generator.*

Record current bias adjustment

One VCR performance test area that is often overlooked, but has considerable influence on whether a VCR looks as good as it used to, is the *record current bias adjustment.*

Do not assume that because the playback circuits are working fine that the record circuits are also working well. A poor luminance signal-to-noise ratio, degraded chroma, video overload, and "diamond beats" on the TV monitor screen can all be caused by inaccurately set record current bias adjustments.

The introduction of the new high-fidelity VCRs makes record current adjustments more important to accurately check and set than before. These new hi-fi VCRs deep bias the stereo audio (150 to 200 mV) right in the normal videotape path, but just before the video luminance information (100 mV average) is laid down (see Fig. 2-7). The azimuth recording principle (audio/video recording and playback heads placed at opposite angles) enables the deep bias system to work.

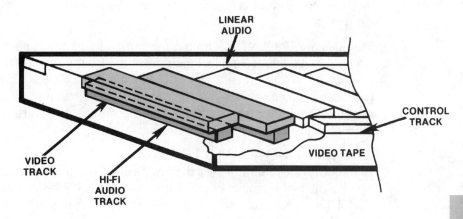

LINEAR AUDIO

CONTROL TRACK

VIDEO TAPE

VIDEO TRACK

HI-FI AUDIO TRACK

■ **2-7** *Video hi-fi audio track locations shown from edge of tape.*

A small change in the record bias level can alter the video, chroma, and hi-fi quality. The exclusive features of the VA62 video analyzer and the SC61 waveform analyzer enable you to quickly and accurately check VCR record bias.

Most manufacturers recommend that you set your record bias in the sequence of FM audio, chroma, and then luminance record level for the best performance. Because color is a common complaint, let's take a closer look at the chroma record level adjustment.

Chroma level adjustment

Chroma level bias adjustment accuracy is critical because too high of a level can cause heterodyning or diamond beats in the picture, and too low of a level causes poor color. You are working with a small visual area for the adjustment—less than 1 centimeter of scope signal (30 to 40 mV). But what makes chroma record level checking and adjustment even more difficult is the NTSC signal required.

The NTSC split field produces a maximum 100 percent white level positive pulse, which along with the standard negative sync pulse drives most scopes nuts trying to decide which pulse to sync onto. At 35 mV, it's even tougher because the pulses are removed. Now your scope has to decide which low-level color bar or color burst signal to sync onto. Your adjustment accuracy is questionable at best. Plus, you have to adjust chroma bias current for the color bar having the highest modulation level, which is cyan at 75 percent modulation. Check your chromaticity chart first, because if you pick the wrong chroma bar, your chroma record bias level will be set wrong.

The Sencore VA62 NTSC chroma bar sweep saves the time you would spend guessing and looking. The chroma bar sweep pattern generates the 75 percent level cyan bar plus two 500-kHz sideband bars. You can switch out the two sideband bars and just use the one cyan bar for fast, easy lock-on for accurate chroma record percent adjustment.

First connect the VA62s VCR 1-volt peak-to-peak standard output to the VCR video input. Select the VA62's chroma bar pattern, and remove the two sideband bars. Then connect the SC61 scope with the DP226 direct probe to the chroma record test point.

The chroma record adjustment procedure is a simple three-step process:

1. Load in a blank tape. Put VCR in SLP record mode.
2. Turn the luminance record level adjustment down to view chroma record level signal.
3. Push the SC61 peak-to-peak button and reset chroma record current level (30 to 40 mV average level).

That's all there is to it, and it's quick and accurate. Two hookups and three steps are all you need to do to know whether the VCR's critical chroma record current level is correct.

The patented chroma bar sweep pattern (shown in Fig. 2-8) generates a 100 percent white level plus 75 percent cyan and two 500-kHz sideband bars.

Sencore VC63 test accessory

The VA62 brings the speed and effectiveness of signal substitution to VCR servicing. VCRs have two special signals that are quite different from signals in TV receivers, however. For this reason, the

■ **2-8** *The chroma bar sweep pattern generates a 100 percent white level, plus 75 percent cyan and two 500-KHz sideband bars.*

VC63 VCR accessory to the VA62 is the main signal source for VCR servicing. Let's now look at why VCRs have special signals, what these signals are, where to use the special VCR signals provided by the VC63, and how to use the VC63 signals effectively to isolate VCR problems. Figure 2-9 shows the engineering models of the VA62 and VC63 that the author of the troubleshooting manual used to evaluate these units in writing the manual.

Special VCR signals

The special VCR signals were developed to overcome the limitations of the magnetic tape recording process. A basic problem with using magnetic tape to store information lies in transferring information onto and off the tape. When signals are read back off the magnetic tape, the amplitude at the playback head increases as the frequency of the recorded signal increases, as Fig. 2-10 illustrates. This occurs because the amount of flux produced across the playback head gap varies with the wavelength of the recorded signal. For each octave or doubling in frequency of the recorded signal, the output voltage increases by 6 dB (quadruples). At some point, determined by the size of the playback head gap, the output sharply drops to zero.

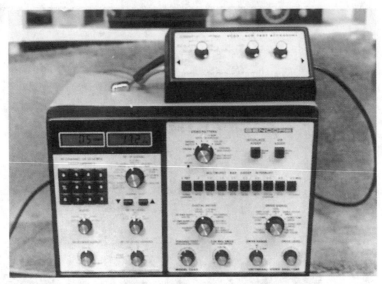

■ **2-9** *Photo of the original engineer's model of the Sencore VA62 video analyzer and VD63 VCR test unit.*

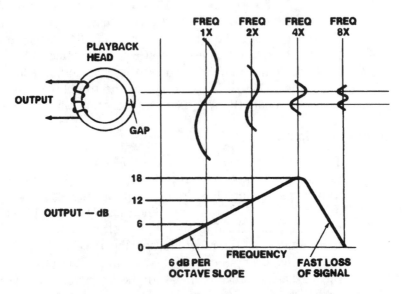

■ **2-10** *The voltage induced in the playback head increases at a 6 dB/octave rate as the frequency of the recorded signal increases. When the wavelength equals the head gap, the induced voltage is zero.*

If only a relatively narrow range of frequencies is to be recorded, this limitation can be tolerated. Audio engineers have worked around this problem for many years. Recorded audio frequencies range from about 40 Hz to 20 kHz, encompassing 10 octaves. The 6 dB/octave playback response produces a 60 dB change in output. Although recorders use elaborate equalization circuits to minimize the effects of tape playback response, 10 octaves is the limit for tape recording.

FM luminance signal

The answer to this dilemma is to reduce the octave range of the recorded signal. This is accomplished, as Fig. 2-11 illustrates, by using the composite video signal to frequency modulate a carrier. Recording the luminance video information using FM is the first special VCR signal. In this FM conversion, the sync tips are the "at rest frequency" of the carrier, and the peak white video signals are maximum deviation. How fast the carrier deviates corresponds to the frequency of the video signal. The range of video frequencies modulating the carrier is limited to about 2.5 MHz. This FM signal is recorded directly onto the magnetic tape.

77

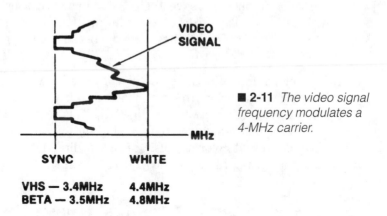

■ 2-11 *The video signal frequency modulates a 4-MHz carrier.*

Both VHS and Beta formats use FM to reduce the number of recorded octaves. With the FM modulation sidebands extending from about 1 to 6 MHz, less than 3 octaves are recorded. Though slightly different FM frequencies are used in VHS and Beta, frequencies around 4 MHz were selected as the best compromise between head gap size, writing speed, and bandwidth.

The FM signal offers a further advantage. An FM system is not affected by moderate signal amplitude changes. Thus, small amplitude variations in the FM signal picked up off the tape, caused by erratic tape-to-head contact, do not affect picture quality. During playback, the FM signal is converted back to a standard video signal.

Down-converted color

Using FM to record the luminance solves bandwidth problems, but it creates problems for recording color. The reason is conventional NTSC video signals use phase-referenced color information. The phases must be accurately maintained, or portions of the video picture will be the wrong color or continually change in color. All phase information, however, is totally lost in an FM process. The color information in a VCR, therefore, cannot be recorded as FM along with the luminance signal. Recording the color information directly as it appears in the 3.5-MHz NTSC signal would cause very noticeable and undesirable interference beats between the 4-MHz luminance and 3.58-MHz color signals and sidebands.

Because neither FM nor direct recording of color will work, the alternative is to convert the frequency of the recorded color, placing it either higher or lower than the FM luminance frequency. Placing the color frequency above the FM luminance causes problems with the head gap, so the color information is lowered in frequency. This becomes the second special VCR signal-down-converted color.

The color information in a VCR is converted down to about 600 kHz, placing it below the luminance frequency (thus the term *color under*) as illustrated in Fig. 2-12. This signal is still amplitude- and phase-modulated for color saturation and tint. The down-converted color signal is recorded directly onto the magnetic tape using the same heads that record the FM luminance

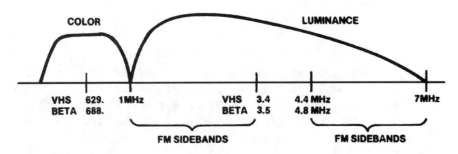

■ **2-12** *Color information is converted down to 600 Khz, placing it below the FM luminance.*

signal. As with the FM luminance, the color signal is converted back to the standard video signal during playback.

Down-converted color answers a second problem of recording color information on magnetic tape. As the tape moves through a VCR, variations in tape speed occur that cause severe problems for the critically phased color signal. During the playback process of converting the 600-kHz signal back to 3.58 MHz, these errors are easily corrected.

Color recording requires one other change to the conventional color signal besides frequency conversion. To reduce color crosstalk from adjacent tape tracks, the phase of the recorded color information is shifted. During playback, color signals in the adjacent tracks have the "wrong" phase and are removed by a comb filter. Thus, only the correct color for each picture element is recovered off the tape.

Special signals cause problems

The FM luminance signal presents a major problem for VCR technicians who rely on signal tracing. To begin with, the FM signal picked off the tape by the spinning video heads is a minute 500 microvolts—well below the measuring capabilities of an oscilloscope. Yet, unless this signal is good, the VCR will not play back a good picture. Second, you cannot determine much by looking at it as you can an FM signal with a scope.

Most of the symptoms caused by a problem in the circuits before the FM detector are the same—a raster filled with snow or a raster that is completely blank. How do you troubleshoot the luminance stages between the video heads and the FM detector? After measuring a few voltages and using a little head cleaner, you are faced with the possibility that the heads could be defective. So, to put an end to your doubt, you swap heads. If the VCR is now fixed, the heads were defective. But what if it's not okay? At least you know that the heads probably were not the problem (you hope). Head swapping is a 50/50 proposition at best.

Using the recorded signal of the tape as the reference when troubleshooting VCRs leaves too many variables. You do not know at what point the signal goes bad, or if the signal is even being properly picked up off the tape. Is the problem electrical, or is the tape path alignment severely off, preventing the heads from picking up any signal? Perhaps the tracking is completely out of whack or the heads are not switching properly. The answer to these problems is

79

to inject a known good reference signal into the circuits rather than trying to follow impossible-to-trace signals. This is the same efficient method of signal substitution that the VA62 makes possible for television servicing.

With signal substitution, you inject a good signal into the circuits. If you inject the good signal *after* the defect, the VCR operates properly, telling you that the circuits from that point forward are working. Then simply trace backward, step by step, into preceding circuits until the VCR stops functioning properly. At that point, you are injecting into the defective stage.

Where to use the VC63

As you have learned, VCRs have special FM luminance and color-under signals. The VC63 VCR Accessory works with the VA62 Video Analyzer to provide these special signals for VCR troubleshooting. Not all the circuits in a VCR require the use of the VC63. The signals after the FM detector are baseband video and are substituted for by using the VA62. Most of the chroma circuits involve important timing relationships and require the use of the SC61 Waveform Analyzer.

For review summation, the FM luminance and down-converted color signals begin at the spinning video heads during playback. The signal picked up by the heads is coupled through a rotary transformer to a head preamplifier. The output from the head amps is switched by the A/B head switcher so only the signal from the head that is in contact with the tape continues. This keeps the noise picked up by the other head out of the picture.

After the switcher, the 4-MHz FM luminance signal is separated from the 600-kHz color-under signal by a combination of highpass and lowpass filters. Both special VCR signals continue down separate paths until they are converted back to standard NTSC format signals and combined by the Y/C mixer. All of these circuits, from the video heads through to the FM detector and color-under frequency converter, require you to use the special signals provided by the VC63.

You troubleshoot a VCR by first connecting it to a television monitor and playing back a test tape. (Chapter 1 explains how to make a test tape.) Playing the tape can reveal one or more important symptoms on the monitor, depending on the defective stage in the VCR. Carefully analyzing the symptom tells you if the problem is in the servo, luminance, or chroma circuits.

Servo symptoms look like picture tearing or rolling. Sound that is "too fast" or "too slow" also indicates a servo problem. Symptoms other than these are called *video playback* and are caused by any of the nonservo circuits. Video playback symptoms include no video, poor video, no color, and poor color. The "trouble tree" in Fig. 2-13 summarizes these symptoms and possible causes. Use it as a guide for isolating playback problems.

To better understand how to use the VC63, let's use a Panasonic PV1225 VCR and the common VCR playback symptom of a blank raster as an example. A portion of the actual luminance block diagram supplied by the manufacturer for this deck is shown in Fig. 2-14. Always troubleshoot from a block diagram and trouble tree to help keep you on the right track.

The first step is to bypass the video mute circuit. This circuit (not used on all VCRs) mutes the video and audio if the servos are not locked in. Bypassing the muting allows you to determine if the

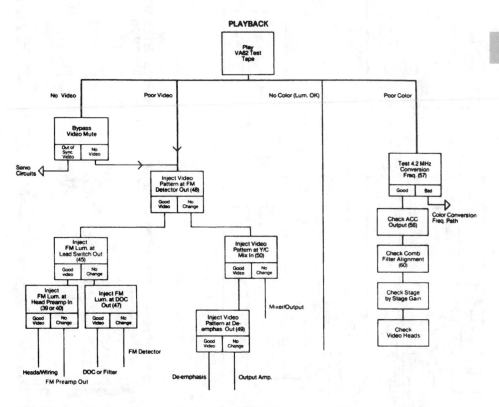

■ **2-13** *Troubleshooting trouble tree for VCR playback problems.*

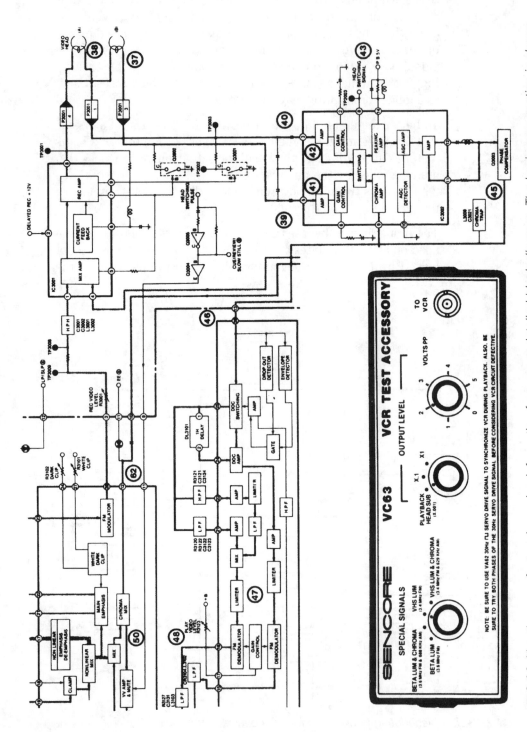

2-14 Use the manufacturer's service information for a more detailed block diagram. The luminance section is shown above.

problem is servo related or video circuit related. In this example, the monitor remains blank. Does this mean the problem is in the direction of the heads, or the video stages?

To determine which way to go, inject a video signal at the output of the FM detector, point (48). The signal here is a baseband video, rather than one of the special VCR signals, so use the VA62 for the signal source. Select the cross hatch video pattern on the VA62, because this pattern is different from any recorded on the test tape. Next, connect the VA62 DRIVE OUTPUT to the FM detector output. Increasing the drive level just above 0.3 volts brings a picture onto the monitor. This is a very important clue. It means the remaining video circuits function correctly and the problem lies in the direction of the heads.

The circuits in the path toward the heads require you to use the special VCR signals from the VC63. Because this is a VHS VCR, set the SPECIAL SIGNALS switch on the VC63 to VHS LUM, as shown in Fig. 2-14. This setting provides an FM luminance signal without the down-converted color signal (the presence of the color-under signal could cause misleading luminance symptoms). The LUM & CHROMA position provides the special color-under signal for troubleshooting color circuits. The setting of the VA62 VIDEO PATTERN switch determines the modulation of the VC63 signal that is set to provide a contrasting signal with those recorded on the work tape. Why? So you can quickly tell if the picture on the monitor is the tape signal or the VC63 test signal.

Now you want to see which of the circuits between the detector and heads are working, so inject an FM luminance signal into the output of the A/B Head Switcher, point (45). Injecting here cuts the remaining stages in half, allowing you to quickly narrow down the problem. Point (45) is the combined output of both video heads after the preamplifiers. The substitute signal is supplied by the VC63 with an OUTPUT LEVEL control setting of X.1. Increasing the output level vernier between 1 and 2 in this example brings a pattern onto the monitor. Now you know that all the stages between test points (45) and (48) are also working.

As the trouble tree shows, the next logical step is to inject a signal into points (39) and (40). These points are the outputs of the video heads. The signal level here is unamplified and very small. The PLAYBACK HEAD SUB (X.001) setting of the VC63 OUTPUT LEVEL switch provides this low-level signal. With the X.001 setting, some setting of the OUTPUT LEVEL vernier should produce a picture. But in this example, increasing the vernier all the way to

5 mV does not produce a picture on the TV monitor. Can you see that this means the problem lies between points (39), (40), and (45)?

You could further isolate the problem to either the preamps or the A/B head switch by injecting a signal into test points (41) and (42), but as the block diagram for this particular VCR shows, the preamps and the head switch in this VCR are part of IC3002. Before you replace the chip, double-check that it is receiving B+ and the 30 Hz head-switching signal.

What if injecting at points (39) and (41) had returned the picture? This leaves the heads, rotary transformer, or associated connections in question. You can use the VC63 to isolate problems here as well. By using the VC63 and signal substitution to prove what stages work, VCR troubleshooting is reduced to a few, quick signal injections.

Sencore VA48 video analyzer

Let's now perform some actual VCR electronic checks with the Sencore VA48 video analyzer. VCR electronic circuits require correct reference signals to perform the proper troubleshooting techniques. Various output signals and video patterns from the VA48 can simplify VCR servicing techniques.

The VCR record circuit

The first VCR recording stage is the AGC amplifier. This stage must be first tested with a reference input level of 1 volt p-p to make sure that the output is of the proper amplitude. To do this, inject the output of the VCR standard signal directly to the camera input of the VCR and adjust the AGC control for the proper output level using the bar sweep video pattern. The circuit is further tested by using the adjustable output supplied from the drive signals output jack instead of the VCR standard 1-volt jack. This adjustable output should be varied from 0.5 to 2 volts (negative polarity) while the output of the AGC is viewed with an oscilloscope. The top waveform in Fig. 2-15 should result if the AGC circuit is operating properly. The bottom trace is of a waveform produced by a faulty AGC circuit.

If the AGC stage is not working properly, the first step is to determine if the AGC stage, the dc amplifier, or the AGC detector is the cause of the defect. To do this, connect the bias and B+ sub supply voltage from the analyzer to the AGC line and vary the voltage.

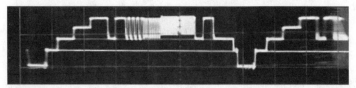

■ **2-15** *The top trace shows the normal AGC output while the bottom trace is that of an improper output.*

At the same time, feed the standard VCR signal into the camera input jack. The test setup is shown in Fig. 2-16. The signal level at the output of the AGC stage should change as you vary this voltage. If it does not, you know that the trouble is in the gain-controlled stage. If the voltage does produce a change, you know to go back to the AGC detector and inject the dc voltage at its output.

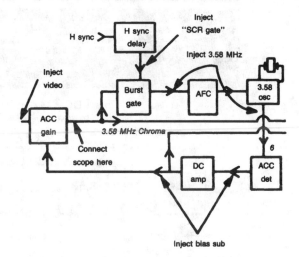

■ **2-16** *The driver signal's output from the analyzer allows each stage of this AGC circuit to be checked.*

Finally, you can substitute for both of the signals feeding the AGC detector itself by injecting the composite video signal at the VCR CAMERA input (using the VCR standard jack) and then feeding the composite sync pulses supplied by the drive signal output in place of the VCR's own sync separator output. This example lets you pinpoint an AGC defect to a single stage, which is especially important when you are troubleshooting defects caused by defective ICs or poor solder connections in the signal path.

By using the bias and B+ voltage subbing supply, you can tell if the color-killer signal is properly switching the color/B&W filters. Set the bias supply to 4 volts and inject it into the IC that controls these filters. The injection points and scope waveforms in Fig. 2-17 show the difference in the bar sweep pattern with the color killer activated. Normal operation of this circuit should allow more high-frequency response during a black-and-white program as compared to one in full color. Thus, you now have proof that the circuits are working properly.

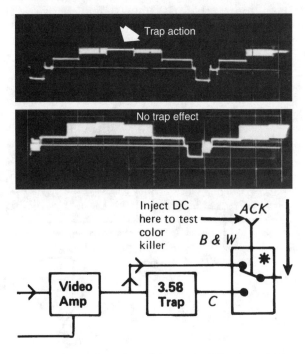

■ **2-17** *The operation of the color killer is quickly confirmed by using the bar sweep pattern and the bias and B+ subber voltage.*

Preemphasis circuits

The bar sweep allows the recording preemphasis circuits to be checked for proper operation. The newer two-speed VCRs require different amounts of preemphasis for each speed. The scope waveforms shown in Fig. 2-18 show what the bar sweep looks like in properly adjusted preemphasis circuits.

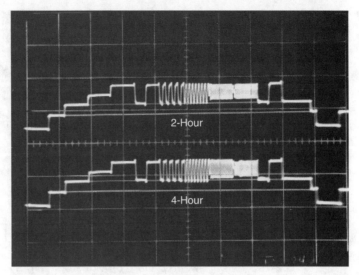

■ **2-18** *The recording preemphasis circuits boost the high-frequency content to reduce video noise.*

White and dark clipping

The clipping circuits located between the preemphasis network and the FM modulator must be properly set to prevent overmodulation, which causes the picture to tear out during playback. The adjustment of these circuits requires both a reference white level and a reference black level to make sure the limiters are not favoring one portion of the signal over another. Use the bar sweep pattern for these adjustments. This pattern has a 3-step gray scale to check for proper video linearity and different frequency bars for a dynamic check of the clipping circuits at different video frequencies. This is needed when the signal is preemphasized, because the higher frequency content is boosted in amplitude. It is possible for the clipping circuits to be operating at the low-frequency range of the signal (like that produced by a 10-step gray scale) but they provide too much limiting to the compensated high-frequency information. Thus, the bar sweep pattern allows all frequencies of the video signal to be checked at the same time (Fig. 2-19).

The VCR circuits from the FM modulator to the video heads are best analyzed by tracing the signals with the VA48 signal tracing meter, or with a scope. For general signal tracing, the high-frequency response of the meter is usually faster, because the shape of the waveform is not as important as the peak-to-peak amplitude.

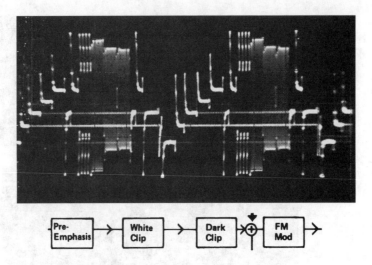

■ **2-19** *The bar sweep pattern provides both a gray scale and a multi-burst signal to allow checking the clipping circuits at all operating frequencies.*

Color circuit analyzing

Phase-locked signals produced by the VA48 let you quickly troubleshoot the chroma processing circuits. Direct signal substitution is used for checks of any signal up to the stage that converts the frequency of the chroma signal down from 3.58 MHz. Let's look at a few examples of how to find a defect in the VCR color processing stages.

The *automatic chroma control* (ACC) requires two input signals for proper operation. The first is the composite chroma signal. The important part of this signal that is required for ACC operation is the color burst. The amplitude of this burst signal is used to control the gain of the chroma circuits to maintain a constant color level with changing input signals. The second signal required is the *burst flag*. This flag signal is the horizontal sync pulse that is delayed a small amount to place its timing exactly in line with the burst signal riding on the back porch of the horizontal blanking interval.

The timing of this signal is very important because it determines what portion of the color signal is used to control the gain of the color circuits. If the burst flag arrives too late, for example, the burst gate will separate the first part of the picture (just after the blanking interval) instead of the color burst. The result is that the color levels will be constantly changing because the amount of chroma information will be different in each color scene.

Now use the drive signals from the analyzer to check out some faulty VCR color circuits. Start with a symptom of changing color levels when a tape is being played. First find out if the color levels are changing during the recording or playback of the color program. This can be easily confirmed by playing back a tape that has been recorded on the suspected faulty machine on a machine that is operating properly. If the color levels remain the same, the defect is in the playback circuits. In this case, however, the levels are changing when the signal has been recorded on the suspected machine and then played back on the good VCR. Thus, the recording circuits of the machine in question are faulty.

Changing color levels could be caused by a defect in any of the seven circuits shown in Fig. 2-20. These seven circuits make up the ACC circuit. A defect (such as a faulty IC or poor solder connection) anywhere in the stages would produce almost the same symptom—changing color levels with different input signals. A scope can be used to trace down the missing signals, but a substitute signal from the analyzer will give you a more positive check as it will duplicate the signals that should be produced at the output of each stage. The first step is to provide a reference signal at the input to the ACC stages. For this check, just use the VCR standard output of the analyzer and feed it into the VCR camera input jack. Now select the chroma bar sweep pattern to provide a reference color pattern these chromas check.

Then check the circuits that produce the burst flag signal. The input to this stage is the composite sync pulses that have been separated from the luminance signal. The V and H composite sync test signal from the analyzer is ideal for this injection signal check. Use the solid-state mode of the drive signal switch to prevent the possibility of feeding in too much signal that could damage a solid-state device. The impedance of this mode is also matched to drive the low-impedance solid-state circuits found in these stages. Use the drive signal meter to monitor the amount of the injected test signal.

These checks are started by injecting the composite sync signal into the horizontal sync delay circuit to see if proper VCR operation returns. The best place to monitor the operation of this circuit is at the output of the ACC controlled stage.

Should the injected test signal *not* return the proper amplitude at the output of the ACC circuit, you can move one stage and substitute for the burst flag. For this test, change the drive signal switch from the composite sync position to the SCR gate drive signal.

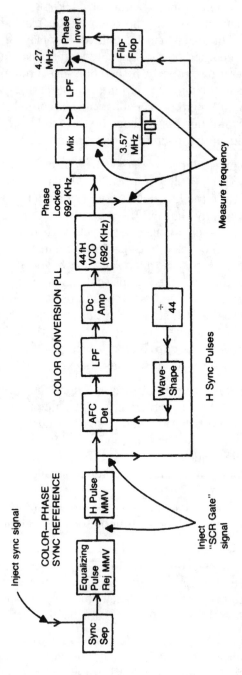

Inject sync signal

COLOR—PHASE
SYNC REFERENCE

COLOR CONVERSION PLL

Sync Sep

Equalizing Pulse Rej MMV

H Pulse MMV

AFC Det

LPF

Dc Amp

44 fH VCO (692 KHz)

Phase Locked 692 KHz

Mix

LPF

Phase invert

4.27 MHz

3.57 MHz

Flip-Flop

Wave-Shape

÷ 44

H Sync Pulses

Measure frequency

Inject "SCR Gate" signal

■ 2-20 *Each stage of the AGC circuit can be substituted with the drive signals from the video analyzer.*

This signal provides a proper substitute for the burst flag because the pulse produced by the SCR gate signal is "stretched" the same amount as the burst flag. The pulse is present during the color burst and will therefore operate the burst flag just the same as the signal produced by the circuits inside the VCR's chroma processing stages. If the operation of the ACC circuits returns to normal, you know that the trouble is in the horizontal sync delay stage.

If the signal does not return proper operation, just continue the stage-by-stage injection at the output of the burst gate. This time, use the 3.58-MHz (phase-locked) signal. You can also check the dynamic operation of the ACC circuit with this substitute signal by just varying the amplitude of the injected 3.58-MHz test signal. As an example, if you increase the amplitude slightly, the ACC output signal should reduce in amplitude. If you see this dynamic change, you know that the ACC circuit is working and the trouble is in the burst gate stage.

The same 3.58-MHz signal can then be used at the output of the 3.58-MHz oscillator. Varying the amplitude of the substitute signal should again produce a change in the ACC output level. If not, go forward to the next stage.

The output of the ACC detector is a dc voltage whose amplitude is related to the amplitude of the burst signal. For this check, use a variable B+ or bias supply as a test of both the dc amplifier and the ACC controlled stage. Begin this check by setting the output of the bias and B+ sub to the level indicated on the schematic. Then raise and lower this voltage about 10 percent. The results should be a change in the ACC output signal level. Should you still not obtain proper operation, inject the dc voltage at the output of the dc amplifier (input to the ACC controlled stage) and again vary the voltage. If the output level still does not change, you know that the trouble is in the ACC-controlled stage itself. With these checks, you can tie down all seven of the circuits in this rather complicated feedback system to locate a defect in one stage.

This might appear to be a long way to go, but in actual troubleshooting you would not have to substitute for every signal. As an example, you could start at the output of the ACC detector and feed in the dc test voltage. This divides the circuit in half. If proper operation is obtained, you know that the defect is somewhere in front of this stage. If correct operation is not obtained, you know that the trouble is in either the dc amplifier or the ACC-controlled stage. Thus, you now know which direction to go to further analyze the stages. The key point to keep in mind is you have a signal

91

to sub-in for every input and output stage so you won't have to guess the cause of the circuit fault.

Signal substitution is handy when combined with oscilloscope signal tracing. You inject the substitute signal at the input to a stage and monitor the resulting signal at the output of the same stage or one that is supposed to be controlled by the substitute signal. Then, use the scope for circuits that require both the amplitude and the waveshape of the signals to be correct. The combination of scope signal tracing and signal substitution is the best team to use for VCR circuit analyzing.

Before leaving the VCR chroma processing stages, let's look at one more example where signal substitution can be used to locate a defective stage. In this case, use a frequency counter (in addition to scope and analyzer meter) to confirm that the stages are working properly. Recall that the frequency conversion stages mix the incoming 3.58-MHz chroma signal with a second signal that is referenced back to the horizontal sync pulses via a phase-locked loop arrangement. These steps follow the signals through a VCR Beta format machine, although the operations of the VHS conversion is similar. You might want to follow along with the block diagrams shown in Fig. 2-20.

The first step of our frequency conversion is to separate the horizontal sync pulses from the incoming composite video signal. These pulses are then formed into a series of pulses with a fixed amplitude (and pulse width) in two multivibrator stages. These clean pulses are then fed to the phase-locked loop (PLL) to maintain the proper conversion frequency at the output.

The composite sync signals provided by the V and H comp sync output can be fed directly to the input or output of the sync separator stage. These pulses (being phase-locked to the composite video) will then replace the signals that should be at these two points. You could take the composite signals past the equalization pulse rejection multivibrator, but doing so would result in the wrong frequency at the output of the PLL. The reason is that the PLL would try to lock up to the vertical sync pulse (as well as the horizontal sync pulses) and change frequency of the equalizing pulse rejection multivibrator, which is to provide a constant pulse rate during the vertical blanking and vertical sync pulse intervals.

To eliminate this error, use the horizontal output (SCR gate) signal for injection after the multivibrator stages. This signal works well because it does not contain the vertical sync pulses. It is just a series of pulses that are phase-locked to the horizontal sync pulses. Therefore, it is an exact duplicate of the output of the "horizontal pulse MMV" and can be injected directly into the AFC detector, which is used to keep the PLL output frequency an exact multiple of the horizontal frequency. Remember when using the drive signals from the analyzer to feed in signals of the same polarity and amplitude as the signals normally found in the circuit.

The total operation of the PLL is determined by checking the output frequency with a frequency counter. The PLL output frequency should be exactly 44 times the horizontal sync pulse frequency of 15,734 Hz or 692,307 Hz. If this frequency is not correct, the VCR will record and reproduce color, but a color tape that has been recorded on another machine will not play back in color.

If you do not find the correct frequency at the output of the PLL, check the frequency at output of the divide-by-44 stage. At this point, you should find the horizontal sync frequency of 15,734 Hz. If this stage is dividing properly, the trouble could be in the low-pass filter (LPF) or the dc amplifier. The dc subber supply can be connected to the output of these stages to see if the adjustment of the dc voltages changes the frequency of the PLL output. If there is no change in output frequency when the bias voltage is changed, you know the defect is in the *voltage controlled oscillator* (VCO) and that the IC is defective. As soon as you get the proper output frequency (with an injected signal) you know that the injection point is after the defective stage. Now, check inputs and outputs until you find the stage that provides no improvement, and that is the one that is defective.

The operation of the remainder of the frequency conversion stages is analyzed with the frequency counter or scope. The second conversion frequency oscillator (in the Beta format, that is the 3.57-MHz crystal controlled oscillator) is adjusted until you have the proper conversion frequency. The output of the mixer stage is measured with the frequency counter, and should provide 4.267918 MHz in the Beta format. If you have a scope with vector measuring capabilities, you can check for proper phase shifts in the frequency conversion stages. To make this check, connect the A channel of the scope to the 4.27-MHz signal before it is phase inverted and the B channel to the output of the phase inversion stage. The waveform shown in Fig. 2-21 shows what the patterns should look like if the stages are processing the phase properly. If they are not, the VCR will have "noisy" color, or no color at all on playback.

93

Proper operation phase   Improper phase shift

■ **2-21** *Vector mode can be used to test for correct phase shifting of the chroma conversion frequency.*

Converted color

The fact that each of the two VCR formats (Beta and VHS) uses a different converted chroma frequency means that direct substitution is not practical. The use of the chroma bar sweep pattern, however, allows you to check the resulting response at the output of the color conversion stages to be sure that you are not going to lose color detail in the converting processes.

An important point is that the pattern produced by an NTSC generator does not provide a check of the total color bandwidth of the color subcarrier information. The actual frequencies occupied by the color subcarrier sidebands are determined by the size of the color information being represented. Several small colored objects in the picture, for example, will represent a higher color sideband frequency than a large object. Remember that the chroma bandpass amplifier of a TV receiver is designed to accept all color information from 3.08 to 4.08 MHz or 500 kHz on either side of the color subcarrier. This frequency range determines the amount of color detail that can be reproduced properly on the color picture. The output of the videotape recorder should be able to record and play back the same amount of color detail for good color reproduction.

The signals from the analyzer's chroma bar sweep produce a dynamic check of the entire color frequency response necessary for a good color picture with color detail in even the smallest objects on the TV screen. The three bars of the chroma bar sweep represent the color subcarrier and the points 500 kHz above and below the subcarrier frequency. Each of the bars is generated at the same amplitude, so they can be used as a reference of the total system's frequency response. Use a scope to trace the converted chroma bar sweep through the amplifier stages to make sure that the machine is not losing some of its color detail during the recording process. The scope waveform shown in Fig. 2-22 illustrates how these patterns are processed in a properly operating VCR. Note that there is a loss of the high-frequency color signal detail.

The following sections look at some more uses of the chroma bar sweep in troubleshooting the playback portions of the VCR. The

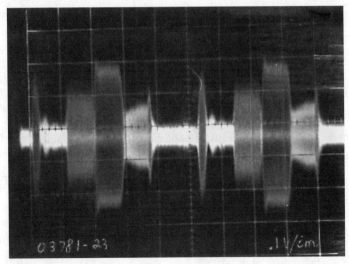

■ **2-22** *The chroma bar sweep tests full chroma bandwidth of the color circuits.*

key requirement is to have these signals phase-locked to the horizontal sync pulses for troubleshooting or alignment of the playback comb filters used to eliminate color crosstalk.

The VCR playback circuits

The test signals produced by the VA48 analyzer provide important checks of the playback circuits of the VCR. Some common VCR defect areas and some methods for troubleshooting them follow. A reference tape from the VCR manufacturer should be used for most of these playback checks. You'll see how to use this reference test tape with a tape recorded with the VA48 test signals.

The big advantage of using your own test tape is the lower cost compared to the prerecorded alignment tape. Thus, you can easily record another tape if it is accidentally damaged during repair of the VCR. The chroma bar sweep pattern provides an additional check of the color processing circuits that is not found on the prerecorded tapes. Finally, as you use the same patterns for testing the playback circuits as those used for the record circuits, you learn to interpret the different patterns. Let's look at some tests you can make with your own reference tape.

Video frequency response

The most important test of the playback system is to make sure that the entire system is providing the best possible frequency response. The use of the bar sweep pattern will produce a dynamic test of the entire system's video response. All you need do is connect a scope to the output of the VCR. Remember to terminate this output with a 75-ohm resistor to make sure that the signal levels at the output are at the proper amplitude.

The top scope tracer in Fig. 2-23 shows the output of a properly operating VCR. Notice that the output is flat to the 3-MHz bar and then drops off at frequencies above this level. If the bars dropped off more quickly as shown in the bottom trace, this would indicate a loss of frequency response somewhere along the line. The first place to suspect is the adjustment of the head equalization circuits that are used to compensate for the nonlinear output of the video heads. The best way to check for proper equalization is to use the manufacturer's alignment tape and follow the recommended procedures for that VCR.

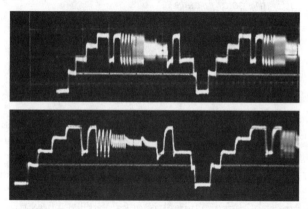

■ **2-23** *Top trace of bar pattern is of a properly operating VCR. Bottom trace shows a defective head alignment that causes loss of high-frequency detail.*

The chroma bar sweep provides a check of the chroma frequency response. This test is one of the advantages of using the analyzer patterns to record a test tape because no other test is as complete for testing all of the circuits that are used to process the color signals. The chroma bar sweep checks the color circuits at both the upper and lower frequency limits that are necessary for good color details. The center (3.56-MHz) bar provides a reference level for a comparison of the frequency detail 500 kHz above and below the subcarrier. A key point about all three of the bars produced by the chroma bar sweep is that they are phase-locked back to the horizontal sync pulse and have a 180-degree phase shift every horizontal sweep line. This means that they will be properly separated by the comb filters used to cancel color crosstalk during the playback mode of operation. This test is not found in the NTSC color generator, but it is very important for good color detail.

Setting the comb filter used in the playback portion of the color circuits is easy to do if you use a scope with good vertical amplifier sensitivity. Start by using a direct scope probe to connect the scope to the output of the comb filter bridge that *is not* connected to the chroma amplifiers. This output point will show the crosstalk rather than the chroma output. The signal at this point does not have sync pulses, so the external trigger input should be connected to the video output jack for triggering. Set the scope to trigger at the horizontal rate. Scopes with built-in sync separators will give you a stable trace with the composite video signal used as a reference.

Now, play back the portion of the alignment test tape that has the chroma bar sweep pattern. Adjust the comb filter's mixer control until the amplitude of the signal has the least amount of the second (3.56-MHz) bar as shown in Fig. 2-24. Be sure to use a direct scope probe and have the scope set for maximum sensitivity, as the signal level is very low at this point. Do not attempt to align the two phasing coils used in the comb filter of the VHS tape systems, because a broadcast vectorscope is required for alignment.

Locating video head problems

Most VCR service technicians say that one of the most difficult stages to analyze is the low-level input circuits associated with the video playback heads. The reason for this difficulty is that the signal levels produced by the spinning playback heads are so low that an oscilloscope is not effective in tracing a signal. The symptom for a defective head is the same as a dirty switch contact, a bad rotary transformer, or a defective head preamp. Thus, a technique is needed to determine which of these components in the low-level head signal circuit is actually causing the problem.

The symptom for a defective head circuit is easy to recognize—the picture (on playback) has a severe flicker and is very noisy. The cause of this symptom is that only every other video field is being viewed on the screen. Recall that the two-head system uses one of the heads to pick up every odd field and the other head to pick up every even field. When one of the signal paths is defective, you have one complete field followed by a period of noise information. The result, in an interlaced picture, is the symptoms just described.

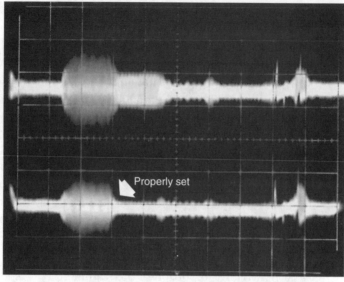

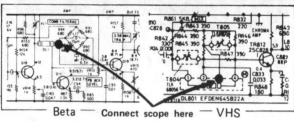

Beta — Connect scope here — VHS

■ **2-24** *The chroma bar sweep lets you quickly check the combfilter in VCR playback circuits.*

Each of the two video heads has its own rotary transformer that transfers the signal from the head output to the preamplifier. This rotary transformer is actually made up of two coils—one that is part of the moving head disk and the other that is part of the stationary portion of the video head assembly. As the video head picks up the signal from the tape, it is inductively coupled to the stationary coil by the moving coil. This eliminates the need for slip rings or brushes that could cause intermittent operation as they wear down.

The signal picked up by the stationary portion of the rotary transformer is passed on to the head preamplifiers. There are two of these preamplifiers, one for each head. Each preamplifier has a set of adjustments that allows any differences in the frequency response of the two heads to be compensated. The signal level at the input to these preamplifiers is approximately 1 mV. This low signal that was diagnosed as a defective head, but it turned out to be faulty switching contacts.

The same output test signal used to troubleshoot the head preamps is also used (without the attenuator) to troubleshoot the dropout compensator circuit. The injector test points are shown in Fig. 2-25. All you need is to feed a signal to the input of the DOC detector and look at the output of the DOC circuit with a scope. This circuit is designed to switch to the delay line output any time the signal level coming from the video heads drops to a certain level.

Then you inject the 4.5-MHz signal, the DOC detector should switch the signal around the delay line. When the signal level drops below the detector trigger level, the circuit should switch back to the delay line. Because you are no longer feeding a signal into the delay line, the output quickly drops to zero.

The DOC trigger circuit is tested by increasing the RF IF control to full output and then reducing the signal level. When the level is about 0.1 V, the output signal should suddenly disappear. Increasing the signal level should then return the output. If the DOC circuit is not operating properly, use a dc voltmeter to check the output of the DOC detector. The detector should provide a dc voltage to control the switching circuits inside the IC. If this voltage changes as you change the signal level at the input, you know that the detector is working properly and the defect is in the switching circuits. If the voltage does not change, you know that the defect is in the detector itself. The delay line is also checked by feeding in the 4.5-MHz signal to its input and checking for an output.

One other playback circuit that is tested with the adjustable 4.5-MHz signal is the limited circuit. The function of the limiter is to compensate for changing levels in the playback signal so the FM demodulators always have enough signal to operate properly. To

101

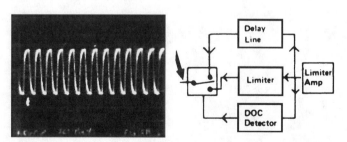

■ **2-26** *The limiter output should remain about the same with different input signal levels.*

Sencore VA48 video analzer

test the limiters action, just feed the 4.5-MHz signal into the limiter input and look at the output level. The output level should remain almost the same over the full range of the input signal (Fig. 2-26). If the limiter is defective, you will have a playback signal that varies in detail and noise content.

The color-processing circuits in the playback stages are treated the same as those found in the recording circuits. In fact, the same circuits are often used for both record and playback. The use of the bias and B+ subber output is the best way to check any of the automatic circuits as you substitute for the feedback voltage and see if you notice some change in the condition you are trying to correct.

3

All-format VCR analyzer

IN THIS CHAPTER, YOU'LL LEARN HOW TO USE THE SENCORE[1] VC93 VCR analyzer for its fullest capabilities. The chapter is divided into four sections that cover the major functional portions of a VCR or camcorder:

☐ Luminance

☐ Chroma

☐ Audio

☐ Servos

The information in this chapter can be used in two ways. First, you can use it as a reference for how to do a specific test. For example, details on how to test rotary transformers are in the section entitled **Testing rotary transformers**.

You can also use this information as a step-by-step troubleshooting guide. You can consult the *Trouble Tree* (Fig. 3-1) to help you decide what tests and signals you need for a given symptom. It outlines a logical troubleshooting sequence that will lead you to the defective stage in the least amount of time. This chapter provides you with specific details and procedures for troubleshooting defective machines. After you have become familiar with the test procedures, you troubleshoot following only the steps outlined by the *Trouble Tree*. Begin your troubleshooting by selecting the path that best fits the symptom you observe.

Some starting tips

The VC93 improves troubleshooting effectiveness through a technique called *functional analyzing*. With this technique you inject known good signals, supplied by the VC93, into the functional blocks. If the output returns to normal, you are injecting after the defective stage; if the output remains the same, your injection is before the defective stage.

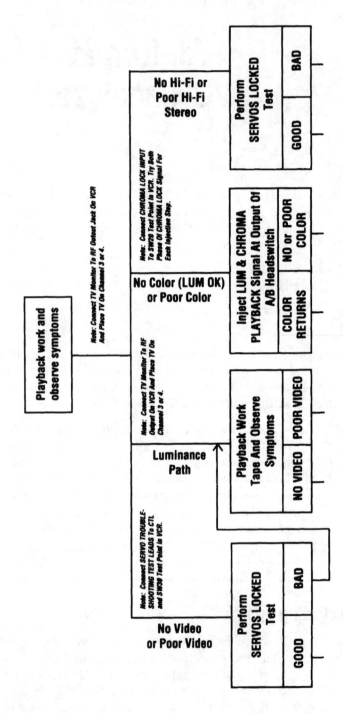

■ 3-1 Select the path on the trouble tree that most closely matches the VCR's trouble symptoms.

Begin your troubleshooting by checking the machine for obvious operator problems. Check the position of the controls for a clue to the problem. A *tracking* control that is all the way to one end of its range, for example, is a clue of a possible tracking problem.

After checking for obvious operator problems, place a work tape into the machine and observe the playback audio and video on a monitor connected to the machine's output.

Always troubleshoot a VCR or camcorder in the playback mode first. Many circuits are common to both playback and record. Ensuring that the playback circuits work eliminates most of the potential circuits that could be at fault. Additionally, to check the record circuits you must record a signal and play it back.

Caution: Do not use the servo performance test tape or factory alignment tape for the initial checkout. Doing so could ruin a valuable tape. Always check the machine first using a work tape to ensure that it will load and play the tape without disastrous results.

VCR overview

A VCR contains five major sections: luminance, chroma, audio, servo, and system control (Fig. 3-2). The following is a brief description of each.

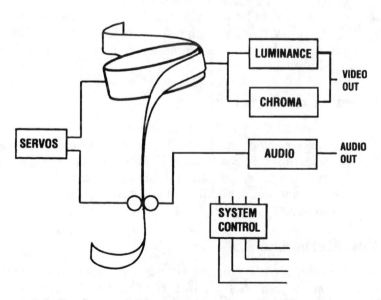

■ **3-2** *The five overall functional sections of a VCR.*

The *luminance section* receives the recorded information from the tape and processes it into a form that can be used by a TV or monitor. Inject the VC93 playback signals in the circuits between the video heads and the FM detector. Use the drive signals between the FM detector and the video output.

The *chroma section* processes the down-converted chroma signal so it can be combined with the luminance signal to create a color playback signal. The VC93 playback signal contains a down-converted chroma signal and other key signals to troubleshoot the chroma conversion circuits.

The *audio section* converts the linearly recorded audio and FM-modulated stereo hi-fi signals to baseband audio. Use the VC93 playback stereo signals to troubleshoot the hi-fi stereo section. Use the audio drive signals to troubleshoot the linear audio section.

The *servo section* controls the tape movement and adjusts the speed and phase relationship of the heads with respect to information recorded on the tape. The VC93 servo analyzer tests evaluate the servo operation, and the servo sub bias supply is used to further troubleshoot the servos.

The last section is the *system control*. This section has one or more microprocessors that monitor the entire operation of the machine and tell the various circuits what to do at a certain time. The microprocessor program is unique to a particular model. This, combined with the digital signals involved, means signal substitution is not the best method for troubleshooting the system control section. Instead, troubleshoot this section with an oscilloscope and DVM.

Luminance troubleshooting

Use the VC93 playback and drive signals to troubleshoot the luminance section. Use the playback signals to substitute in any circuit from the video heads to the FM detector. Use the composite video and lum drive signals in the luminance circuits between the FM detector output and the video output jack/input or the RF modulator. Note setups in Fig. 3-3.

Video head troubles

There are two symptoms of a bad video head, depending whether one or both heads are bad. The first symptom, caused by two bad heads, is complete loss of video. In older VCRs the entire monitor

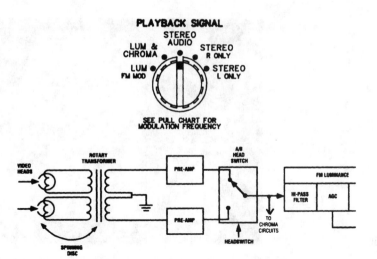

PLAYBACK SIGNAL

STEREO AUDIO

LUM & CHROMA

STEREO R ONLY

LUM FM MOD

STEREO L ONLY

SEE PULL CHART FOR MODULATION FREQUENCY

VIDEO HEADS

ROTARY TRANSFORMER

PRE-AMP

PRE-AMP

A/B HEAD SWITCH

TO CHROMA CIRCUITS

HEADSWITCH

FM LUMINANCE

HI-PASS FILTER

AGC

SPINNING DISC

DRIVE SIGNAL

AUDIO   3.58MHz

CHROMA

HEAD SWITCH

LUM

SW 30

COMPOSITE VIDEO

CHROMA KEY PULSE

LUMINANCE

BASEBAND LUMINANCE

Y/C MIX

RF OUT

FM DETECTOR

LO-PASS FILTER

DE-EMPHASIS

EQUALIZER

DROPOUT COMPEN. (DOC)

AMP

RF MODULATOR

DROPOUT DETECTOR

FROM CHROMA CIRCUITS

■ **3-3** *Use the VC93 PLAYBACK signals to inject before the FM detectors and the DRIVE signals to inject after the detector.*

107

screen is snow, as shown in Fig. 3-4. Newer machines often mute the video to produce a blank or solid-color raster as shown in Fig. 3-5.

The second symptom appears when one head fails and the machine produces a noisy playback picture. An important part of interpreting this symptom is that the noise *must* cover every part of the picture. If any section of the picture is clear (even if it is only a few inches somewhere on the screen) you do not have a defective head. The problem is likely in the servo or tape path.

■ **3-4** *Snow or streaks on monitor screen usually indicate a defective video head.*

■ **3-5** *On newer VCRs the screen is blanked out upon loss of video, possibly indicating a bad video head.*

Note: TV sets that use vertical countdown circuits will show several inches of clear picture at the bottom of the screen even if one video head is bad. These sets blank the noise from one field of video. For best results use a monitor that does not use a vertical countdown circuit.

Symptoms alone do not prove when video heads are bad. Defects in other luminance circuits can produce identical symptoms. If the symptoms suggest bad video heads, use the VC93 playback signal to substitute for the head signal. Because you cannot inject the signal directly into the heads, inject at the output of the rotary transformer. This will determine if the problem is in either the video heads or rotary transformer, or if the problem is after the injection point. To determine if the circuits after the rotary transformer are working:

1. Insert a blank tape into the machine and press PLAY.
2. Set the VC93 as follows: (Note Fig. 3-6.)
   - *VCR format* to match format being serviced.
   - *Modulation* to color bars or external.
   - *Playback range* to playback head sub.
   - *Playback signal* to lum.
3. Connect the *head substitution test lead* to the playback output jack.
4. Connect the head substitution test lead to the ch A and ch B head pre-amp inputs of the VCR.
5. Observe the playback monitor for an improved picture.
6. Adjust the *playback level* control for best picture.

A good picture shows that all the circuits after the injection point are good. This leaves either the rotary transformer or the video heads to be defective. Test the rotary transformer following the procedures in the next section. If the rotary transformer tests good, go to the servo analyzer tests. If they do not indicate a problem, then replace the video heads.

If you do not obtain a good picture, a stage after the injection point is bad. Refer to the **Testing Video Stages** section for information on troubleshooting those stages.

Testing rotary transformers

Rotary transformers fail when either the rotating or stationary windings develop an open or short. Test the transformer by injecting the lum playback signal. If it passes the signal, the rotary transformer is good. To test the rotary transformer:

1. Turn the VCR OFF.
2. Set the VC93 as follows:
   - *VCR format* to match format being serviced.
   - *Modulation* to color bars or external.
   - *Playback range* to 5 VPP.
   - *Playback* signal to lum.

Note set-up connections in Fig. 3-7.

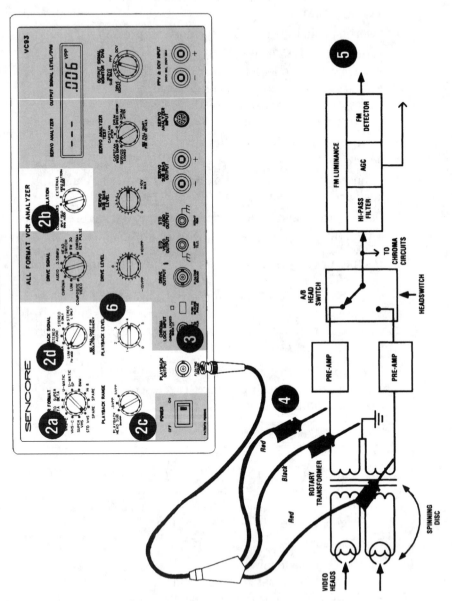

■ **3-6** *Set-up for injecting the PLAYBACK signal into the luminance sections.*

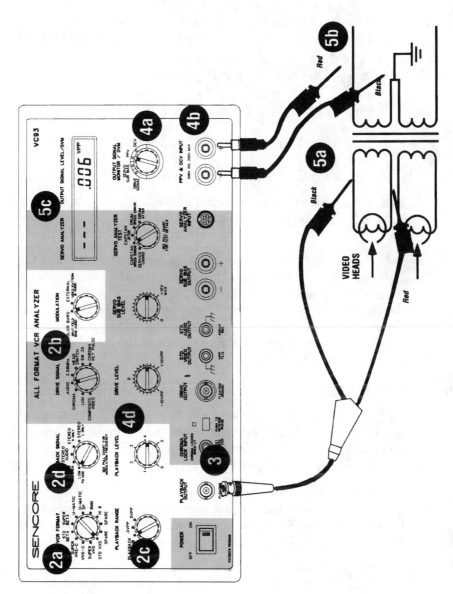

■ **3-7** Set-up for checking out the rotary transformer.

111

3. Connect the *direct test lead* to the *playback signal output* jack.

4. Set the playback output level:
   - Set the output signal monitor/DVM to PPV.
   - Connect the *DVM test leads* to the PPV & DCV input.
   - Connect the *DVM test leads* to the *direct test leads*.
   - Adjust the *playback level* control for an output of 2 VPP.

5. Perform the rotary transformer test:
   - Connect the *direct test lead* to the rotary transformer primary.
   - Connect the *DVM test leads* to the rotary transformer secondary.
   - Read the results on the *output signal level/DVM* LCD readout.
   - Repeat the last three steps for the remaining windings.

If the readings are similar for each winding, the rotary transformer is good. If any winding reads lower than the rest, or gives no reading at all, replace the rotary transformer.

Note: Some rotary transformer shorts occur only at certain spots in the rotation. This typically produces a streak in the playback picture. If you suspect this problem, connect an oscilloscope in place of the VC93 PPV meter. Then slowly turn the transformer while pressing down on the upper drum assembly. If the output signal disappears at any point in the rotation, the transformer has an intermittent short and must be replaced.

Testing four-head VCRs

Some VCRs use one set of video heads for the fastest tape speed (SP for VHS and Beta I for Beta VCRs) and another set for the other speeds. Testing these four-head machines requires a work tape recorded at the different tape speeds.

Play back the tape and observe the picture at each speed segment. If you see symptoms that suggest a bad head, note what speed the tape is playing at. The noise can be caused by a bad head, but it also can be caused by the video head selection circuitry.

To determine if the problem is in the head selector or in the heads, locate the rotary transformer windings that correspond to the video heads used for the speed in question. Refer to Fig. 3-8. Inject the VC93 lum playback signal before the video head selector for that winding. The picture will either stay bad or clear up. Note test setup in Fig. 3-9.

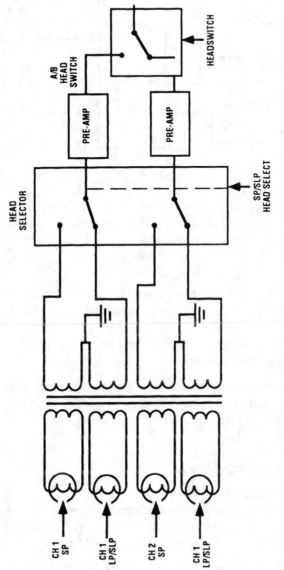

HEAD SELECTOR

A/B HEAD SWITCH

PRE-AMP

PRE-AMP

HEADSWITCH

SP/SLP HEAD SELECT

CH 1 SP

CH 1 LP/SLP

CH 2 SP

CH 1 LP/SLP

■ **3-8** *A four-head VCR has additional rotary transformer windings and switching circuits.*

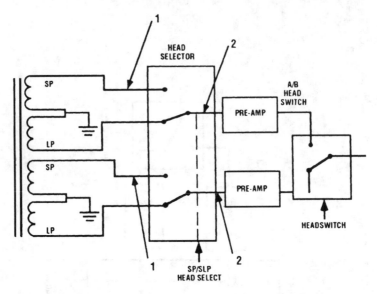

■ **3-9** *By injecting the VC93 PLAYBACK signal both before (1) and after (2) the video head selector switch should produce a good picture. This example shows injection into the SP windings.*

If the playback picture stays bad, either the head selector or the circuits after it are defective. To determine which, inject the playback signal after the head selector. If the picture clears up, the video head selector is bad. If the picture stays bad, the problem is after the head selector.

If injecting into the head selector produces a good picture, either the rotary transformer or the video heads are at fault. Follow the testing rotary transformers procedure to determine which component is defective.

Testing VCRs with "trick" heads

Some VCRs use a "special effects" or "trick" head to minimize playback noise when the machine is placed in the *pause, fast forward,* or *reverse* modes. A head select circuit similar to that used in four-head machines selects the "trick" head.

A problem in the "trick" head or in its selector circuitry causes a noisy picture when the machine is placed in *pause.* To determine if the problem is in the selector or in the "trick" head, determine which rotary transformer winding corresponds to the "trick" head. Also, identify the winding that is used for both the "trick" and the normal mode. Connect one of the red *head substitution test* leads to the rotary transformer "trick" head output and the other red lead to the common output. Place the VCR in *pause* and observe the playback monitor. Note connections shown in Fig. 3-10.

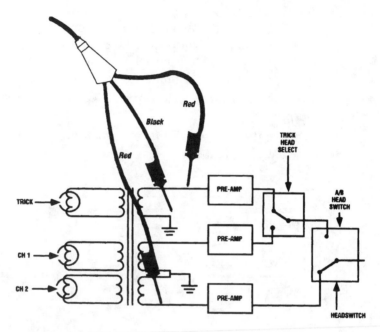

■ **3-10** *Place the VCR on pause and inject the PLAYBACK signal into the trick head selector input to determine if the switching circuits are working.*

A good picture indicates that all the circuitry after the injection point is good and the problem is either the rotary transformer or the "trick" head. Do the *rotary transformer test* to check the rotary transformer.

If the picture is bad, suspect the "trick" head switcher. Test the switcher by injecting the VC93 signal at the output of the "trick" head switcher. If the picture returns, the "trick" head switcher is defective.

Checking video head pre-amps

The video head preamplifiers amplify the weak signals from the rotary transformer. One pre-amp is used for each head. A defective pre-amp produces a snowy picture that can be easily misdiagnosed as a bad video head. Use the VC93 lum *playback signal* to positively identify and locate head pre-amp problems. To check video head pre-amps:

1. Insert a blank tape in the VCR and press *play*. (Note test setup in Fig. 3-11.)

2. Set the VC93 as follows:
   - *VCR format* to match format being serviced.
   - *Modulation* to color bars or external.
   - *Playback range* to playback head sub.
   - *Playback signal* to lum.

3. Connect the *head substitution test lead* to the *playback output* jack.

4. Connect the *head substitution test lead* to the input of both video head preamps.

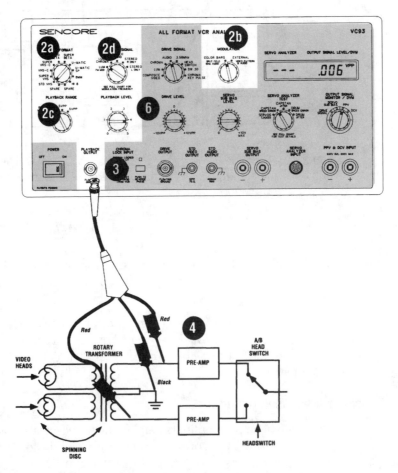

■ **3-11** *Set-up to inject into the head pre-amp input.*

5. Observe the picture on the playback monitor.

6. Adjust the *playback level* control for best picture.

If the injection produces a good picture, the circuits after the rotary transformer (including the video head pre-amps) are good. If the picture is snowy, either a video pre-amp, or the A/B headswitcher is defective.

Note: Use the playback head sub position for this test. If you must switch the playback range control to a higher range to get a good picture, one or both pre-amps are bad.

Most machines incorporate the video head pre-amps and the A/B headswitcher in the same IC (as shown in Fig. 3-12). If so, you do not need to isolate the problem further because both will be replaced with the IC. Before replacing the IC check that the A/B headswitcher is receiving a head-switching pulse.

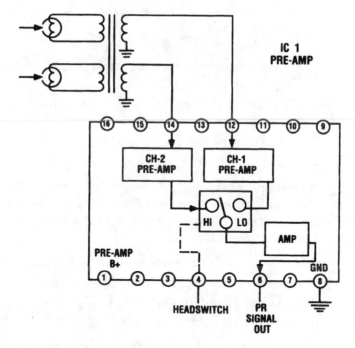

■ **3-12** *In many VCRs the video head pre-amps and the A/B headswitched are in the same IC chip.*

If the A/B headswitcher and the video head pre-amps are not in the same IC, inject the lum playback signal at the output of the video head pre-amps. You will need to increase the playback range control to the 0.5 VPP or 5 VPP range to obtain a signal level at this point.

Checking A/B headswitchers

The A/B headswitcher selects the output from the video head that is contacting the tape. A defective A/B headswitcher most likely provides a signal from only one head. The playback picture appears snowy just as it would if one video head were defective or if the rotary transformer or video head pre-amp were defective.

Headswitcher checks

To determine if a problem is before or after the headswitcher (note test setup in Fig. 3-13):

1. Insert a work tape in the VCR and press *play*.
2. Set the VC93 as follows:
    - *VCR format* to match format being serviced.
    - *Modulation* to color bars or external.
    - *Playback signal* to lum.
    - *Playback range* to 5 VPP.
3. Connect the *head substitution test lead* to the *playback output* jack.
4. Connect the *head substitution test lead* to the A/B headswitcher.
5. Observe the picture on the playback monitor.
6. Adjust the *playback level* control for the best picture.

A snowy picture means the problem is after the A/B headswitcher. Refer to Injection before the FM Detector. If the test produces a good picture, the problem is before the output of the A/B headswitcher.

To further identify the problem, do the following:

1. Move the *head substitution test lead* from the output of the A/B headswitcher to the input of A/B headswitcher.
2. Observe the picture on the playback monitor.

If the picture is still good, refer to **Checking video head pre-amps**. If the test produces a bad picture, the problem is associated with the A/B headswitcher. It could be caused by either a defective A/B headswitcher or a missing headswitch pulse. To prove the operation of the A/B headswitcher:

1. Set the VC93 as follows:
    - *Drive signal* to headswitch.

■ **3-13** Test set-up for checking the A/B headswitcher.

- *Output signal monitor/DVM* to drive signal.
- *Drive level* to 0.

2. Connect the *head substitution test lead* to the headswitch input.

3. Increase the *drive level* control while monitoring the picture.

Note: In many modern VCRs the A/B headswitcher, video head pre-amps and other circuits are contained in the same IC. If so, set the *playback range* switch to playback head sub and inject into the video head pre-amps.

If you do not obtain a good picture, try the opposite polarity of the *drive level* control. If you still do not obtain a good picture, replace the A/B headswitcher. If you do obtain a good picture, the headswitch pulse is missing.

Injecting before the FM detector

The VC93 *playback output* jack supplies the FM luminance signals to troubleshoot any problem up to the FM detector. To inject a single into the FM signal path refer to the set-up diagram in Fig. 3-14.

1. Insert a work tape in the VCR and press play.

2. Set the VC93 as follows:
   - *VCR format* to match format being serviced.
   - *Modulation* to color bars or external.
   - *Playback signal* to lum.
   - *Playback level* to 0.5 VPP.

3. Connect the *head substitution test lead* to the *playback output* jack.

4. Connect the *head substitution test lead* to the injection point.

5. Observe the picture on the playback monitor.

6. Adjust the *playback level* control for the best picture.

Note: If the picture remains snowy, switch the *playback range* to 5 VPP and adjust the *playback level* for the best picture. You should obtain a good picture if all the circuits are working properly. If you are using a work tape with a color pattern, it is normal for some color from the test tape to show through.

If the signal injection results in a snowy picture, a circuit after the signal injection point is defective. Repeat the above injection at a later test point that is still before the FM detector. If there are no other injection points available before the FM detector, go to the testing video stages.

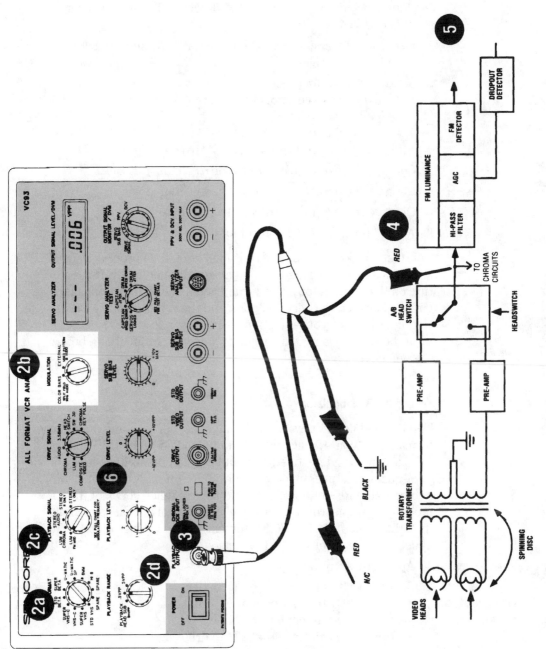

■ 3-14 Set-up for injecting into the luminance circuits.

Testing video stages

The VC93 *drive signal* produces signals to identify problems in the luminance circuits after the FM detector. To inject a signal in the video stages:

1. Insert a work tape in the VCR and press play. Refer to set-up connections in Fig. 3-15.

2. Set the VC93 as follows:
   - *Modulation* to color bars or external.
   - *Drive Signal* to lum.
   - *Drive level* to 0.
   - Set *output signal monitor/DVM* to *drive signal*.

3. Connect the *direct test lead* to the *drive output* jack.

4. Connect the *direct test lead* to the injection point.

5. Observe the picture on the playback monitor.

6. Adjust the *drive level* control for the best picture.

This test checks the ability of the video circuits to process a signal and pass it on to the video output jack or RF modulator. If the picture on the TV monitor is good when you make an injection, all the circuits between the injection point and the video output are good. Continue moving your injection point closer to the FM detector until the picture is no longer good. Now you are injecting before the defective stage.

Testing the RF modulator

The RF modulator converts the luminance, chroma, and audio information into an RF signal that can be picked up by the tuner of a TV receiver. A defective RF modulator can result in a snowy or poor-quality picture, or poor or no audio on a TV receiver connected to the RF output of the VCR. Make a quick check of the RF modulator by injecting a composite video signal or baseband audio into the inputs of the modulator as the symptoms dictate. To check the RF modulator (video portion):

1. Apply power to the VCR.

2. Set the VC93 as follows (see Fig. 3-16):
   - *Modulation* to color bars or external.
   - *Drive signal* to composite video.
   - *Output signal monitor/DVM* to drive signal.
   - Adjust *drive level* for 1 VPP on *output signal level* LCD readout.

3. Connect the *direct test lead* to the *drive output* jack.

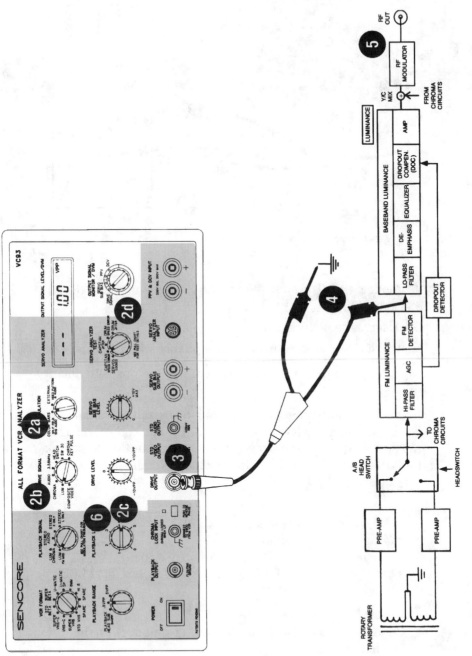

■ 3-15 *Set-up for injecting into the luminance path after the FM detector.*

123

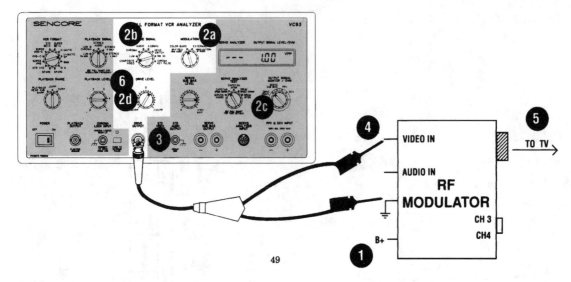

■ **3-16** *Set-up for checking out the RF modulator (video).*

49

4. Connect the *direct test lead* to the video input of the RF modulator.

5. Observe the picture on playback monitor.

6. Adjust *drive level* control for best picture.

Note: If the picture appears distorted, use the opposite polarity drive signal.

If you obtain a picture when injecting a signal, the modulator is good. If no picture is obtained, check for B+ and control voltage to the modulator. If these voltages are both present the RF modulator is defective. To check the RF modulator (audio portion):

1. Apply power to the VCR.

2. Set the VC93 as follows:
   - *Modulation* to color bars or external.
   - *Drive signal* to audio.
   - *Output signal monitor/DVM* to drive signal.
   - Adjust *drive level* for 0.400 on *output signal level* LCD readout.

3. Connect the *direct test lead* to the *drive output* jack.

4. Connect the *direct test lead* to the audio input of the RF modulator.

5. Listen to playback monitor for audio.

6. Adjust *drive level* control for best audio.

If you obtain audio when injecting a signal, the modulator is good. If no sound is obtained, check for B+ and control voltage to the modulator. If these voltages are both present, the RF modulator is defective.

Super-VHS, Super-Beta, and HI-8 testing

Super-VHS, Super-Beta, and HI-8 formats extend the FM deviation and frequency range of the FM luminance signal to enhance the picture quality. These VCRs will play videotapes recorded with the standard format signal and the enhanced signals. The following conditions must be present for these machines to go into their special enhanced mode:

Super-VHS VCRs

A special Super-VHS tape is required to record or play back in the Super-VHS mode. This tape has a small hole that is detected by Super-VHS machines. Super-VHS machines also contain a circuit that monitors the FM signal and detects the higher frequency Super-VHS signal. If both conditions are present, the machine will switch to the Super-VHS mode. Note block diagram in Fig. 3-17. To check Super-VHS circuits:

1. Insert a Super-VHS tape in the machine or defeat the Super-VHS tape detection circuit.

2. Press the *play* button.

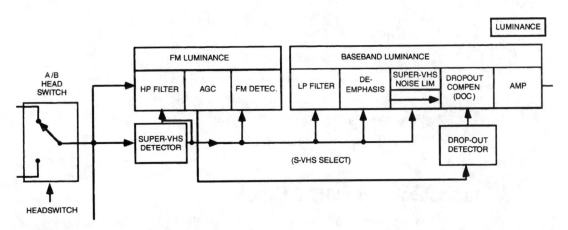

■ **3-17** *A Super-VHS VCR checks for presence of the Super-VHS signal before switching to the Super-VHS mode.*

3. Set the VC93 *VCR format* switch to Super-VHS.

4. Set the VC93 *playback signal* switch to lum.

5. Inject the signal before the Super-VHS detector takeoff point.

6. Check that the machine's Super-VHS indication comes on.

Note: The Super-VHS takeoff point is usually located right after the A/B headswitcher.

Super-Beta VCRs

Super-Beta machines do not use a special tape. The FM detector automatically detects the presence of the Super-Beta signal. To check Super-Beta circuits:

1. Insert tape into the machine.

2. Press the *play* button.

3. Set the VC93 *VCR format* switch to Super Beta.

4. Set the VC93 *playback signal* switch to lum.

5. Inject the signal before the FM detector.

6. Check that the machine's Super-Beta indication comes on.

If you do not see a Super-Beta indication, move your injection point closer to the video heads.

Hi-8 VCRs

Hi-8 VCRs use a special tape. The Hi-8 cassette contains a detect hole that tells the machine if the proper tape is being used. In the playback mode, a Hi-8 signal detector verifies that a Hi-8 signal was recorded on the tape. To check Hi-8 circuits:

1. Insert a Hi-8 tape in the machine or defeat the Hi-8 tape cassette detection circuit.

2. Press the *play* button.

3. Set the VC93 *VCR format* switch to Hi 8.

4. Set the VC93 *playback signal* switch to lum.

5. Inject the signal before the FM detector.

6. Check that the machine's Hi-8 indication comes on.

Troubleshooting the VHS-C format

The main difference between VHS and VHS-C machines is the configuration of the video heads. The VHS-C format was developed to make the mechanical section physically smaller for use in camcorders. To get this reduction in size, the size of the video head cylinder was reduced. To make the information recorded on the tape compatible with standard VHS machines, VHS-C uses four video heads instead of two.

Use the same procedures to troubleshoot and signal inject into VHS-C machines as used for standard VHS machines. The only procedure that is different is injecting a signal between the rotary transformer and the headswitcher. Because VHS-C machines have four outputs instead of two, you will need to jumper the corresponding video heads into pairs so you can inject the signal into all four heads simultaneously. Figure 3-18 shows the connections for injecting at the output of rotary transformers in VHS-C machines.

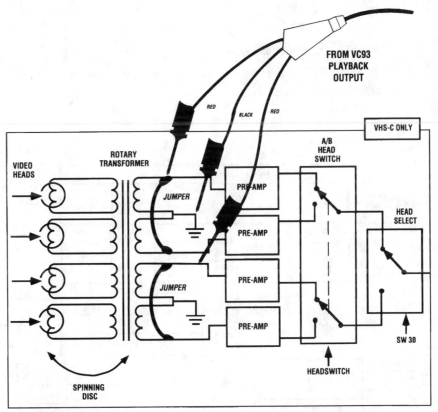

■ **3-18** *VHS-C VCRs use four heads and a combination of switches to select the correct video heads. Jumper the corresponding video heads into pairs to inject the signal.*

Chroma troubleshooting

Let's now look at the signals required for proper color operation. The chroma portion of the NTSC composite video signal is frequency converted and processed before being recorded onto the tape. This minimizes crosstalk between tracks and compensates for errors introduced in playback. The playback circuits convert the signal back to NTSC standards. The chroma playback circuits need four key signals to return the color to its original form. Each of these signals is available from the VC93.

Play back a test tape that has a color video signal recorded on it. The most likely color symptoms are no color or poor color. Either problem occurs when any of the four key signals is missing or incorrect. Refer to these key blocks in Fig. 3-19.

Down-converted chroma

Down-converted chroma is the converted chroma signal picked up from the tape. It is separated from the luminance information after the A/B headswitcher using either a lowpass or comb filter.

The 3.58-MHz reference signal

The 3.58-MHz reference is used to reestablish the correct NTSC standard video signal.

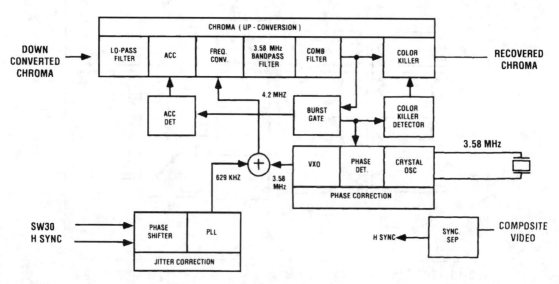

■ **3-19** *The chroma conversion section requires four key signals for proper operations.*

The 30 Hz

Several formats change the phase of the recorded color information to minimize crosstalk and interference between tracks. The vertical sync pulse provides the recorded phase reference. During playback, a 30-Hz signal is the reference to cancel the phase shifting.

Composite video (with jitter)

Formats that use phase shifting to minimize crosstalk also change the color phase for each horizontal line. Horizontal sync pulses, taken from the luminance circuits, reference the timing of the phase shifting during playback. The horizontal sync pulses are either separated before the chroma circuits or they are separated from composite video by the chroma circuits.

Checking overall chroma operation

The first step in troubleshooting a chroma problem is to be sure that the luminance circuits are working properly. If they are, supply a known good reference signal from the VC93 playback output to the input to the chroma circuitry. Because the color circuits also require composite video from the luminance circuits, you also must inject the playback signal before the point where the luminance and chroma signals separate. To make an overall chroma operation check:

1. Insert a blank tape in the VCR and press play. Refer to test setup in Fig. 3-20.
2. Set the VC93 as follows:
   - *Format* to match format being serviced.
   - *Modulation* to color bars or external.
   - *Playback range* to 5 VPP.
   - *Playback level* to midrange.
   - *Playback signal* to lum and chroma.
3. Connect the *head substitution test lead* to the *playback output* jack.
4. Connect the *head substitution test lead* to the output of the A/B headswitcher.
5. Connect the *chroma lock test lead* to the *chroma lock input* jack.
6. Connect the *chroma lock test lead* to the SW30 test point in the VCR. Note: The *chroma lock indicator* will light if you are connected to the correct test point.
7. Observe the monitor for a color picture.

129

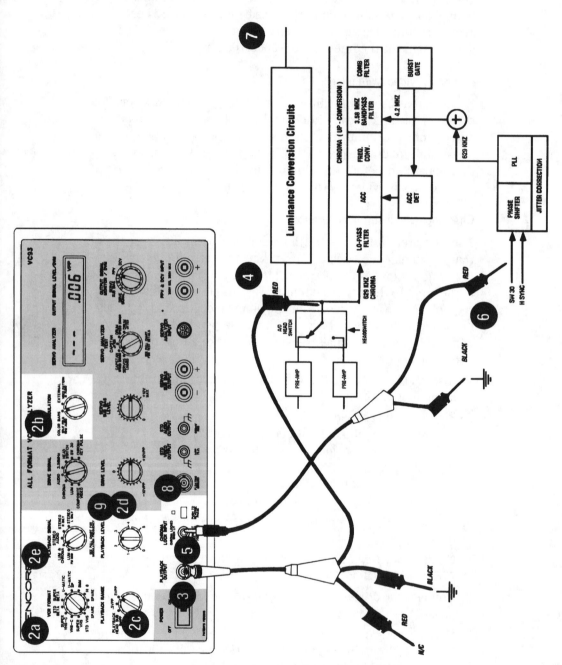

■ **3-20** *Set-up for checking overall chroma circuit operation.*

8. If you do not see color, press the *chroma lock phase* button to select the other SW30 phase.

9. Adjust the *playback level* control for best color.

If you see color, all the playback color circuits are working. If you cannot obtain color when playing a test tape that contains a color pattern, the chroma signal is getting lost before the VC93 injection point.

If the heads check out, then either the color circuits are defective or one of the chroma signals is missing or incorrect. For information on troubleshooting no-color symptoms, refer to the section on troubleshooting chroma conversion problems.

Note: Some VCRs separate the chroma and luminance signals inside the enclosure containing the head pre-amps. These VCRs have separate luminance and chroma signal output pins. Connect one of the red *head substitution test leads* to the luminance pin and the other red lead to the chroma pin. Refer to block diagram in Fig. 3-21.

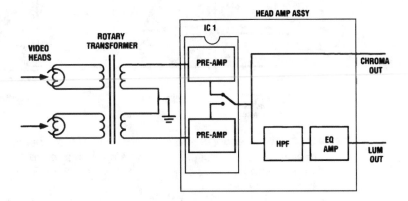

■ **3-21** *Some VCRs separate the luminance and chroma signals in the head pre-amp circuits. Connect one of the head substitution leads to the chroma output and the other to the luminance output.*

Identifying head-related chroma problems

A wrong or missing color problem can be caused by a frequency response problem in a stage before the chroma and luminance processing circuits. Before the chroma conversion circuits:

1. Set the VC93 as in the checking overall chroma operation test.

2. Set the *playback range* control to playback head sub.

3. Move the *head substitution test lead* to the rotary transformer output. Note test hookup in Fig. 3-22.

4. Observe the playback monitor for color.

5. Adjust the *playback level* control for best color.

If you see color, either the video heads or the rotary transformer are defective. A likely cause is a worn video head. If you do not obtain color, the problem is between your injection point for the overall chroma operation test. Use the VC93 to step through the remaining stages to isolate the defective stage.

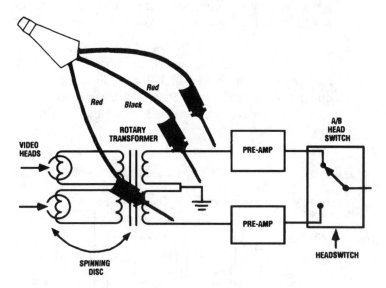

■ **3-22** *If you cannot obtain color from a VCR tape, but the VC93 proves that the color circuits are working, move the test leads to head pre-amp input to determine if the pre-amps are working.*

Note: Remember to increase the level of the Playback signal after the video pre-amps. Turn the *playback range* switch to 0.5 VPP or 5 VPP and adjust the playback level for the best picture on the monitor.

Troubleshooting chroma conversion problems

If the overall chroma operation test does not produce color, you need to troubleshoot the chroma conversion circuits. The following items should now be checked:

☐ Power supply voltage to the color conversion IC.

☐ Grounds to the color conversion IC.

☐ The record/playback selector signal.

If these check out, you need to troubleshoot the color conversion section. First, supply a down-converted chroma signal from the VC93 to the VCR chroma conversion circuits. Next, do the following tests until you locate the problem: Substitute the 3.58-MHz color oscillator, substitute the chroma key pulse and substitute the SW30-Hz pulse.

Substituting the 3.58-MHz color oscillator A missing or off-frequency 3.58-MHz reference oscillator will result in no or incorrect playback color. Use the VC93 3.58-MHz Drive Signal to substitute for the reference signal.

Note: Some VCRs use the color 3.58-MHz signal to control the servos. A missing or wrong 3.58-MHz reference will not allow the VCR to function. Injecting the 3.58-MHz signal into the chroma section will return the machine to operation.

To substitute for the 3.58-MHz reference:

1. Connect the VC93 to the chroma circuits as described in the section entitled checking overall chroma operations. (See Fig. 3-23.)
2. Set the *drive signal* switch to 3.58 MHz.
3. Connect the *direct test lead* to the *drive output* jack.
4. Connect the *direct test lead* to the 3.58-MHz oscillator.
5. Adjust the *drive level* control while watching the monitor for color.
6. If you do not see color, press the *chroma lock phase button* to select the other SW30 phase.

If color returns, the 3.58-MHz signal is either missing or defective. If color does not return, remove the 3.58-MHz drive signal and go to the substituting the chroma key pulse signal section.

Substituting the chroma key pulse signal VHS, Beta, and 8-mm VCRs use a horizontal sync pulse to operate the chroma phase shift circuits. Note block diagram shown in Fig. 3-24. No or poor playback color occurs if the phase shift circuits are not functioning. Some VCRs feed the composite video from the luminance circuits to the color IC and extract the horizontal sync pulses inside the IC. Other VCRs have the sync separator outside the chroma IC.

To substitute horizontal sync (Sync separator external to color circuits):

1. Connect the VC93 to the chroma circuits as found in the section on checking overall chroma operation. Note test setup shown in Fig. 3-25.

133

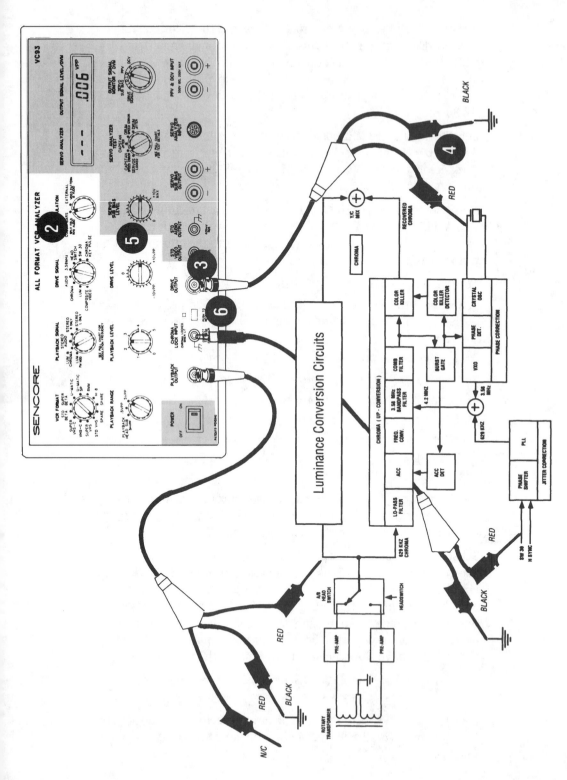

■ **3-23** Set-up for injecting the 3.58-MHz signal.

All-format VCR analyzer

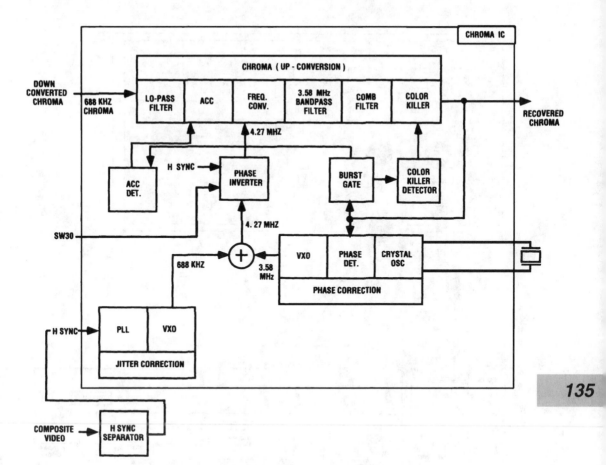

■ **3-24** *Some VCRs use a sync separator to obtain the horizontal sync pulse for the chroma circuits.*

2. Set the *drive signal* switch to chroma key pulse.

3. Connect the *direct test lead* to the horizontal sync input on the chroma IC.

4. Adjust the *drive level* control "clockwise" (+polarity) while watching the playback monitor for color.

5. If color does not return, rotate the *drive level* control "counterclockwise" (−polarity).

6. If you do not see color, press the *chroma lock phase button* to select the other SW30 phase and repeat steps 4 and 5.

If color returns, troubleshoot the cause of the bad or missing sync pulse. If color does not return, remove your injection signal and go to the section on substituting the 30-Hz phase switching signal.

■ 3-25 *Set-up for injecting horizontal sync pulses into the chroma circuits.*

To substitute composite video (Sync separator internal to color IC):

1. Connect the VC93 to the chroma circuits as described in the section on checking overall chroma operation. Refer to test setup in Fig. 3-26.

2. Set the *drive signal* switch to composite video.

3. Connect the *direct test lead* to the composite video input on the chroma IC.

4. Adjust the *drive level* control "clockwise" (+polarity) while watching the playback monitor for color.

5. If color does not return, rotate the *drive level* control "counterclockwise" (− polarity).

6. If you do not see color, press the *chroma lock phase button* to select the other SW30 phase and repeat steps 4 and 5.

If color returns, troubleshoot the cause of the bad or missing composite video signal. If color does not return, remove your injection signal and go to the section on substituting the 30-Hz phase switching signal.

Substituting the 30-Hz phase switching signal VHS, Beta, and 8-mm formats also use a 30-Hz signal to reference the phase rotation circuits. If this signal is missing you will get no or a wrong playback color. To substitute the 30-Hz phase switching signal:

1. Connect the VC93 to the chroma circuits as shown in Fig. 3-27.

2. Set the *drive signal* switch to SW30.

3. Connect the *direct test lead* to the SW30 input on the chroma IC.

4. Adjust the *drive level* control "clockwise" (+polarity) while watching the playback monitor for color.

5. If no color is seen, rotate the *drive level* control "counterclockwise" (− polarity).

Note: The SW30 signal is used throughout the VCR. Substituting for the SW30 signal might cause the VCR to shut down. This is normal. Simply remove your injection signal and press the play button on the VCR. If color returns, find the reason for the missing SW30 signal.

Audio troubleshooting

VCR and camcorder audio is recorded in several ways, so you must select the correct audio test signal. Audio was originally recorded only in a linear track. Linear audio is limited in frequency response and has wow and flutter. Stereo hi-fi recording uses spinning heads and FM carriers, much like the video signal, to improve audio fidelity.

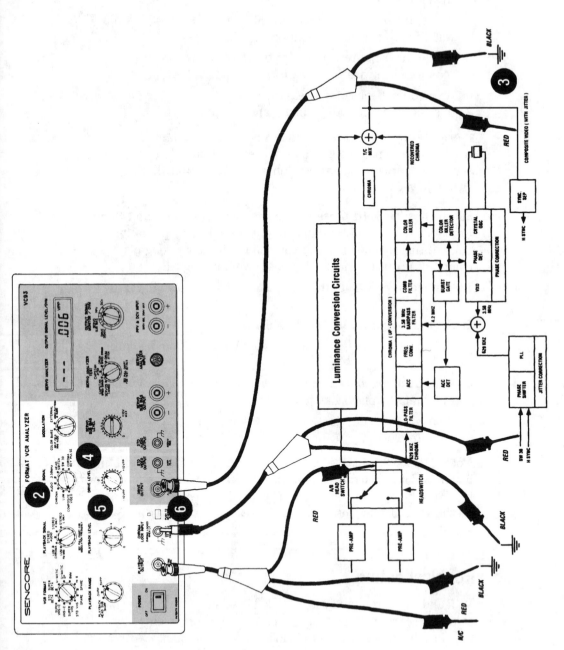

■ **3-26** Set-up for injecting composite video into the chroma circuits.

There are three different FM hi-fi stereo VCR systems: VHS hi-fi stereo, Beta hi-fi stereo, and 8-mm hi-fi stereo. Also, 8-mm mono is recorded using a similar FM system. The frequency waveforms are shown in Figs. 3-28 A, B, and C for different audio systems. Figures 3-29 A, B, and C give the block diagrams of the different audio recording schemes.

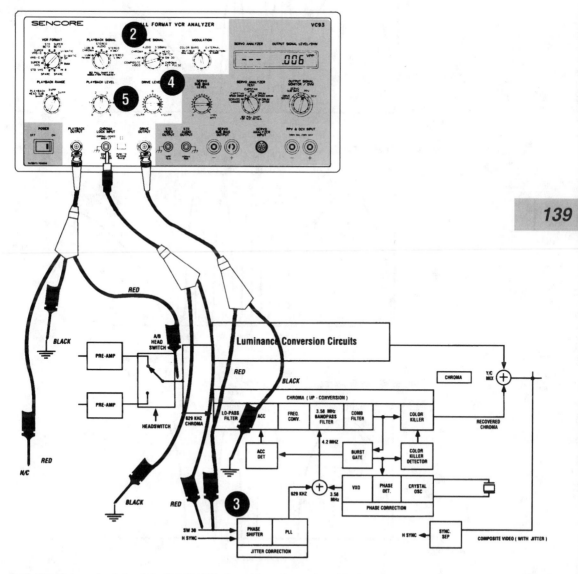

■ **3-27** *Set-up for injecting the SW30 signal into the color circuits.*

The VC93 playback signals duplicate the FM audio signals coming from the spinning heads. Use them to substitute in any hi-fi stereo circuit from the rotary heads to the FM detector. Use the VC93

a. **VHS Hi-Fi Stereo**

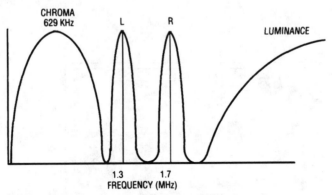

b. **Beta Hi-Fi Stereo**

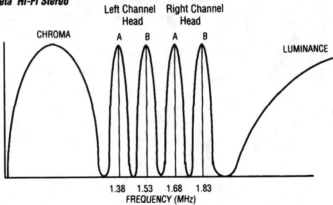

c. **8MM Hi-Fi Stereo**

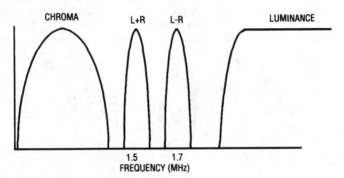

■ **3-28** *Various VCR audio signal formats.*

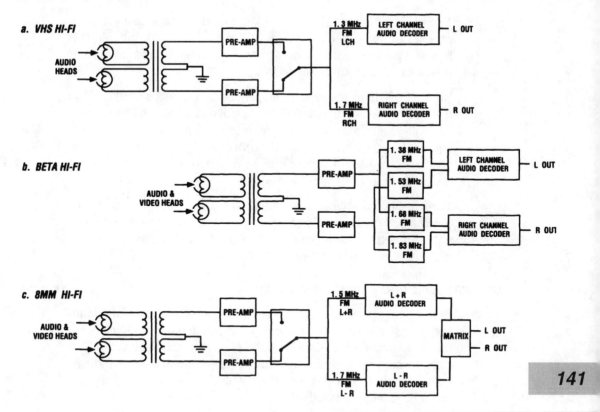

a. VHS HI-FI

AUDIO HEADS

PRE-AMP
PRE-AMP

1.3 MHz FM LCH → LEFT CHANNEL AUDIO DECODER → L OUT

1.7 MHz FM RCH → RIGHT CHANNEL AUDIO DECODER → R OUT

b. BETA HI-FI

AUDIO & VIDEO HEADS

PRE-AMP
PRE-AMP

1.38 MHz FM
1.53 MHz FM → LEFT CHANNEL AUDIO DECODER → L OUT

1.68 MHz FM
1.83 MHz FM → RIGHT CHANNEL AUDIO DECODER → R OUT

c. 8MM HI-FI

AUDIO & VIDEO HEADS

PRE-AMP
PRE-AMP

1.5 MHz FM L+R → L+R AUDIO DECODER

1.7 MHz FM L-R → L-R AUDIO DECODER

MATRIX → L OUT / R OUT

■ **3-29** *Block diagrams for VHS, Beta, and 8-MM stereo audio systems.*

drive signals to inject into any audio stage after the stereo hi-fi FM detectors and anywhere baseband (15 Hz to 20 kHz) audio is present.

Note: Some 8-mm VCRs use pulse code modulation (PCM) hi-fi stereo. This system uses digitally coded information. Troubleshoot these systems by using an oscilloscope to check for proper data flow.

Identifying defective hi-fi stereo audio heads

All three hi-fi stereo systems (VHS, Beta, and 8-mm) use rotating heads to pick up the FM-modulated hi-fi stereo signal from the tape. In Beta and 8-mm, the same head that picks up the video signal also picks up the audio signal. Thus, any head-related problem that affects the audio also affects the video. Before suspecting a head problem in these types of VCRs, check to be sure the video is correct.

The VHS hi-fi stereo signal is picked up off the tape using a separate set of audio heads and rotary transformer windings. In this format a problem with the audio heads or audio rotary transformer windings will not affect the video.

Note: The VHS hi-fi stereo audio headswitch signal is derived from the video headswitch signal. Thus, a video headswitch signal problem will affect the audio headswitch signal. The audio headswitching point is more critical than the video headswitching. Therefore, headswitching problems in VHS hi-fi stereo machines will affect the audio before the video signal.

Troubleshoot problems associated with the rotating audio heads the same way you would troubleshoot video head problems. The only difference is that you use the stereo audio playback signal and listen for improvements in the audio.

Checking audio rotary transformer windings

In the Beta and 8-mm formats the same rotary transformer windings transfer both the audio and the video signals. A defective rotary transformer will affect both the playback audio and video. The VHS format uses separate audio rotary transformer windings, meaning a bad rotary transformer will affect either the playback audio or video, but not likely both.

Check the audio rotary transformer using the same procedures used to check out video rotary transformers. The only difference is that you use the stereo audio playback signal.

Checking audio head pre-amps

Beta and 8-mm machines use the same head pre-amps to amplify the audio and the video. A defective head pre-amp will affect both the video and the audio playback signal.

Troubleshoot audio head pre-amp problems using the same procedures used to check out the video head pre-amps. The only difference is that you use the stereo audio playback signal and listen for improvements in the audio. The various audio pick-up heads and pre-amp circuits are shown in Figs. 3-30 A, B, and C).

Injection after the A/B headswitcher

Beta hi-fi stereo VCRs separate the audio signal from the luminance signal before the A/B headswitcher. The audio signal undergoes several processing steps before reaching the headswitcher.

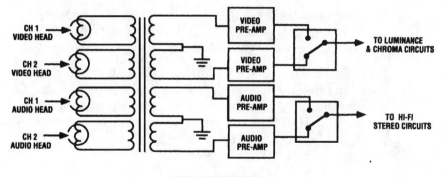

a. VHS HI-FI STEREO

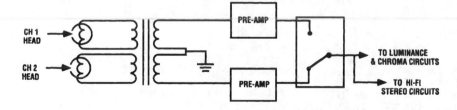

b. BETA HI-FI STEREO

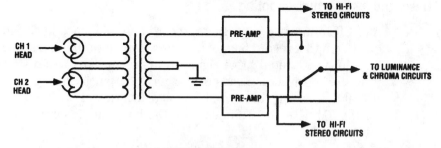

c. 8MM HI-FI STEREO

■ **3-30** *VHS hi-fi systems use separate rotary heads for the hi-fi audio signals. Hi-fi Beta and 8-MM systems use the same rotary heads for video and audio.*

The 8-mm and VHS-format VCRs, however, combine the FM audio signals using an A/B headswitcher before processing the signal. In either case, you can inject the VC93 stereo audio playback signal anywhere in the audio signal path before the audio FM detector.

Misadjustment of the audio headswitching can cause the symptom of non hi-fi stereo in VHS format machines. The VC93 stereo audio playback signal is a continuous carrier signal and is not affected by incorrect headswitching. If you obtain good audio when injecting the playback signal, check the headswitch adjustment before troubleshooting the rest of the hi-fi audio signals.

Note: Troubleshoot the FM portion of the hi-fi stereo circuits in 8-mm and Hi-8 machines by selecting the Hi-8 VCR format. This provides both FM carriers needed for hi-fi troubleshooting. If the 8-mm or Hi-8 VCR you are servicing has mono FM only, select the 8-mm VCR format to troubleshoot the mono FM circuits.

Baseband audio troubleshooting

All VCR formats, except 8-mm, record and play back linear audio. Hi-fi VCRs record linear audio as a backup to the hi-fi signal. Use the VC93 audio drive signal to inject into any audio stage from the linear audio heads to the audio output jacks. Also, use the audio drive signal to troubleshoot the hi-fi stereo circuits after the FM detector. To inject into the linear audio circuits (all formats):

1. Place a work tape in the VCR and press *play*. Refer to Fig. 3-31 for this test setup.
2. Set the VC93 as follows:
   - *Modulation* to color bars or external.
   - *Drive signal* to audio.
   - *Drive signal* to 0.
3. Connect the *direct test lead* to the *drive output* jack.
4. Connect the *direct test lead* to the injection point.
5. Monitor the playback audio.
6. Adjust the *drive level* for best sound.

If you obtain audio, the circuits between the injection point and the output are good. If you do not obtain audio, the problem is between the injection point and the output. Locate the defective stage by moving the injection point closer to the audio head until you no longer hear an output. You are now injecting before the defective stage.

Note: The audio will distort if you inject directly into the linear audio head due to the high gain of the audio amplifiers. Inject into the output of the first amplifier or build a voltage divider to reduce the amplitude of the signal if you need to inject directly into the audio heads.

Servo troubleshooting

Servo problems can be more difficult to identify and troubleshoot than other problems. Many servo problems produce symptoms of poor audio or poor video that can mislead you into troubleshooting circuits that have no defects.

The VC93 provides five servo analyzer tests. Do the tests in sequence, starting with the *servos locked* test. Use the results of

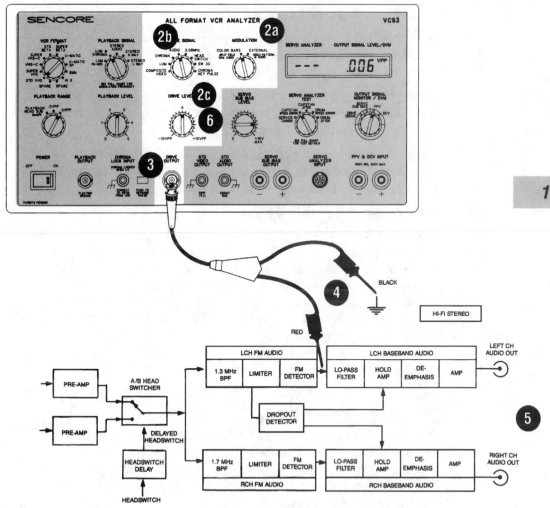

■ **3-31** *Set-up for injecting into the linear audio circuits, and the hi-fi stereo circuits after the FM detectors.*

each test to zero in on the defect. Here is a review of each servo test:

Servos locked test Determines if the servo capstan phase loop and drum phase loop are locked to the reference signal.

Capstan speed test Determines if the capstan servo is operating at the correct speed. It identifies speed select circuit problems and other speed related problems.

Capstan jitter test Measures how constant the capstan movement is. It helps identify capstan phase problems and mechanical problems including motor bearings, bad idlers, etc.

Drum speed test Determines if the drum is operating at the correct speed. It helps identify a drum that is operating too fast or too slow.

Drum jitter test Measures how constant the drum rotation is. It helps identify drum-related problems including bad motor bearings, bad drum phase loop, etc.

The next part of this chapter explains how to use the results of the servo tests for troubleshooting. A summary of the servo tests and possible causes of bad test results is provided in Fig. 3-32.

Understanding the servos locked test

The servos locked test compares the change in phase of the CTL pulse to the SW30 pulse when using the troubleshooting test lead or the phase of the audio and vertical sync pulse when using the performance test lead and performance test tape. VCRs lock the capstan and drum phase circuits to a common REF 30 source. Usually the REF 30 source is internal to the servo IC and cannot be viewed or measured. Because the capstan and drum phase loops are locked to the same REF 30, they also must be locked to each other. The VC93 uses this fact to check servo lock.

Note: The SW30 pulse is used for this test since it is derived from the PG pulse and is more universal to all VCRs.

Results of the servos locked test are displayed as a percentage reading that indicates how close the two servos are locked. This reading varies between brands and models. After testing several machines, however, you will develop a feel for what reading is typical for the machine you are servicing. Older VCRs, which use analog servo circuits, generally show a higher percentage reading indicating a looser lock. Newer, digital machines typically have a much smaller percentage reading indicating a more precise lock.

SERVOS LOCKED	CAPSTAN SPEED ERROR	CAPSTAN JITTER	DRUM SPEED ERROR	DRUM JITTER	MOST LIKELY DEFECT
GOOD	GOOD	GOOD	GOOD	GOOD	NO SERVO DEFECTS*
GOOD	GOOD	GOOD	GOOD	BAD	DRUM MECHANICAL
GOOD	GOOD	GOOD	BAD	N/A	REFERENCE FREQUENCY
GOOD	GOOD	BAD	GOOD	GOOD	CAPSTAN MECHANICAL
GOOD	BAD	N/A	GOOD	GOOD	REFERENCE FREQUENCY
GOOD	BAD	N/A	BAD	N/A	REFERENCE FREQUENCY
GOOD	BAD	N/A	GOOD	BAD	REFERENCE FREQUENCY
BAD	GOOD	GOOD	GOOD	GOOD	CAPSTAN PHASE LOOP or DRUM PHASE LOOP
BAD	BAD	N/A	GOOD	GOOD	CAPSTAN SPEED LOOP or CAPSTAN MECHANICAL
BAD	GOOD	BAD	GOOD	GOOD	CAPSTAN PHASE LOOP or CAPSTAN MECHANICAL
BAD	GOOD	GOOD	BAD	N/A	DRUM SPEED LOOP or DRUM MECHANICAL
BAD	GOOD	GOOD	GOOD	BAD	DRUM PHASE LOOP or DRUM MECHANICAL
BAD	BAD	N/A	BAD	N/A	REFERENCE FREQUENCY
BAD	BAD	N/A	GOOD	BAD	REFERENCE FREQUENCY

*NOTE: A noise bar that occurs periodically at a rate of one minute or greater could be a capstan or drum phase problem.

■ 3-32 *Use the results of the five servo tests to determine the most likely cause of a servo problem.*

The "good" and "bad" servos locked readings are based on extensive research that shows that bad capstan or phase servo circuits produce errors greater than two percent. No "good/bad" reading is displayed if the servos are varying widely. If you see a "bad" or no indication, suspect either a bad capstan or drum phase circuit. A "good" indication means that the phase circuits are operating properly.

Regardless of the servos locked test results, do all the servo analyzer tests before you begin to troubleshoot the problem. The results of the other tests might further isolate the problem to the capstan or drum phase loop.

Servo locked test notes

There are two servo locked test notes:

Note 1: Some newer machines have very tight capstan drum speed loops that operate marginally even with a bad phase loop. The servos locked test has a minimum resolution and will not identify a problem in which the reference signals change phase slower than one cycle per minute. If you observe a noise bar that comes and goes slower than once a minute, suspect a servo phase loop problem.

Note 2: Some machines use a combined PG/FG drum signal. The servos in these machines can lock incorrectly if the drum PG signal is missing, locking instead to the FG signal. Because the FG signal is a higher frequency, the servos can lock to one of several transitions. Depending on which transition it locks to, the drum will lock, but noise or a noise bar occurs in the picture. The VC93 servos locked test will indicate that the servos are locked. This is a correct diagnosis as the servo phase loops are locked. Refer to waveform drawing in Fig. 3-33.

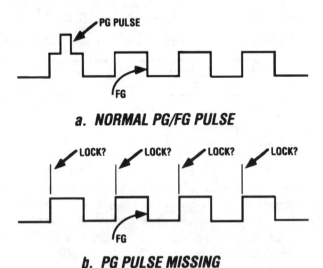

a. NORMAL PG/FG PULSE

b. PG PULSE MISSING

■ **3-33** *Some VCRs use a combined FG/PG pulse (a). If the PG pulse is missing, the VCR will lock to any cycle of the FG pulse.*

The best way to tackle this problem is to stop and start the VCR several times while observing the position of the noise bar. Because the servos can lock on any FG transition, the noise bar will change position each time the machine is placed into *play*. If you observe this symptom, check the combined FG/PG pulse for a missing PG pulse.

Understanding the capstan speed error test

The capstan speed error test checks how fast the tape is being pulled through the machine. The VC93 analyzes the frequency of the CLT pulse when using the troubleshooting test lead, or the playback audio frequency when using the performance test lead and performance test tape, and compares it to an internal 29.97-Hz reference. This test checks the actual tape speed, not just the capstan motor speed. This dynamic test locates mechanical problems such as tape drag and slippage, in addition to capstan circuit problems. Refer to diagram in (Fig. 3-34).

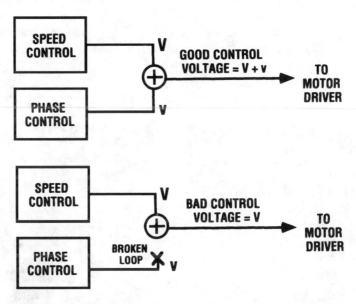

■ **3-34** *The phase control loop adds a small amount of correction voltage to the speed control voltage. A defective phase loop will cause the speed to be off slightly.*

Readings that are off more than 10 percent suggest a problem in the capstan speed loop. Readings that are within 10 percent are caused by either a defective capstan speed loop or a defective capstan phase loop. (A defective phase loop will produce a wrong correction voltage. This voltage adds to the speed loop correction voltage to control the capstan motor. Because the capstan motor control voltage is off slightly, the motor speed is also off.)

To determine which loop is defective, play back a tape that contains segments of different tape speeds. If the speed circuits are working properly, you will obtain low capstan speed error percentages at all speeds. But if the speed circuits are not working properly, you will obtain large percentage errors at some tape speeds.

Note: The slowest tape speed will likely have a larger error than faster tape speeds, even in a correctly functioning machine.

If the capstan speed error percentages are only slightly greater than two percent for all tape speeds, the problem is likely to be a mechanical problem, such as excessive oxide buildup on the capstan, a hard capstan idler or a loose capstan motor belt. If these check out good, the problem is likely a defective capstan phase loop.

The "good/bad" readings are based on extensive research that shows that capstan speed errors greater than +/− 0.5 percent usually cannot be corrected by the phase loop circuitry. No reading will be displayed if the tapes speed is varying widely. If you see a "bad" indicator, or no indication at all, use the percent readings to help determine what the capstan problem is. A technician is performing the capstan speed test with the VC93 in the Fig. 3-35 photo.

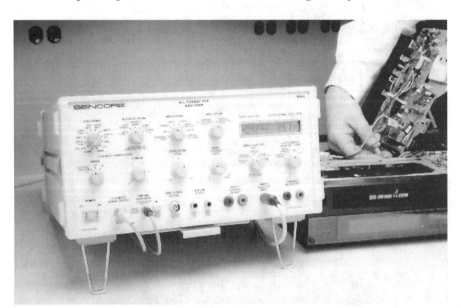

■ **3-35** *A technician is shown using the Sencore VC93 analyzer to check out a VCR loop control system.*

Understanding the capstan jitter test

The capstan jitter test analyzes the CLT pulse when using the troubleshooting test lead, or the playback audio when using the performance test lead and performance test tape, for minute speed variations. This test actually checks tape speed variations and not just capstan motor speed variations. Thus, it checks for mechanical problems such as tape drag and slippage, as well as for failures in the capstan control circuits.

The capstan jitter percentage readings indicate the amount of tape speed variation. All machines have some speed variation due to tape stretch, tightness of the capstan loops and other mechanical problems. Excessive speed variation causes the linear audio to vary excessively in pitch.

The "good/bad" indications are based on research that shows that capstan speed variations greater than 0.5 percent give unacceptable audio performance and can affect the video performance.

Likely causes of "bad" capstan jitter are: a defective capstan phase loop, a slipping capstan belt or idler, or a bent or dirty capstan shaft. Also suspect a possible bad capstan motor. A bad motor bearing or a motor winding can cause the capstan to catch as it turns. The lack of a "good/bad" reading indicates that one of the above potential defects is causing severe capstan jitter.

Understanding the drum speed error test

The drum error speed test analyzes the frequency of the SW30 pulse when using the troubleshooting test leads or the playback video vertical sync pulse when using the servo performance test lead and performance test tape. It compares the signal to an internal 29.97-Hz reference. The SW30 signal is universal between most VCRs and is derived from the drum PG signal used by the drum servo circuits.

The drum speed error percentage reading displays how far the drum speed deviates from the desired speed. Wrong drum speed will cause the horizontal sync pulses in the playback video to occur at the wrong time. If the drum speed is close to correct, the symptom appears as a misadjusted horizontal hold control on the playback monitor. A large drum speed error will cause the playback video to be so bad that no conclusion can be made by observing the playback monitor. Newer VCRs will mute the playback video when the drum speed is far off.

The "good/bad" readings are based on the amount of frequency offset that can be tolerated by most TV receivers. A speed error more than 0.1 percent will produce a "bad" indication. If the drum speed is varying widely, no "good/bad" indication is displayed. The most likely reason for a "bad" drum speed error reading is a missing FG pulse from the drum motor.

Understanding the drum jitter test

The drum jitter test analyzes the SW30 signal, or the playback video vertical sync pulse when you use the servo performance test lead, for minute speed variations. The percentage reading indicates how much the speed of the revolving drum varies. Excessive drum speed variation causes the picture to appear to "breathe" in and out.

The "good/bad" indication is based on research showing that drum speed variations greater than 0.1 percent give unacceptable picture quality. No "good/bad" reading is displayed if the speed is varying widely. This indicates a problem in the drum servo. A "bad" indication also means the drum servo is bad. The most likely reason is a problem in the drum phase circuit or a mechanical defect, such as a bad motor bearing, a drum that is out of balance, or a spot of dirt or oxide on the drum.

Making up a work tape

For the highest quality work tape, always use a direct connection to the video and audio inputs of the VCR instead of using the antenna input. Always record your work tape using a high-quality tape and the highest quality VCR available. A four-head or five-head VCR is better than a two-head VCR because it uses the ideal head width for each tape speed.

For fast troubleshooting, record three separate work tapes. Record the first tape with the performance and troubleshooting tests running at the SP speed. Record a second tape at the SLP speed and the third tape at the LP speed. This will speed troubleshooting of speed-related VCR problems.

The following sequence uses the VA62A video analyzer as the source for the audio and video signals. Connect the VA62A to the VC93 as shown in Fig. 3-36. Now perform the following steps:

1. Set VA62A controls:
   - *Video pattern* switch to *multiburst bar sweep.*
   - Push all *multiburst bar sweep interrupt* buttons to the ON position.
   - Set Audio to 1 kHz.

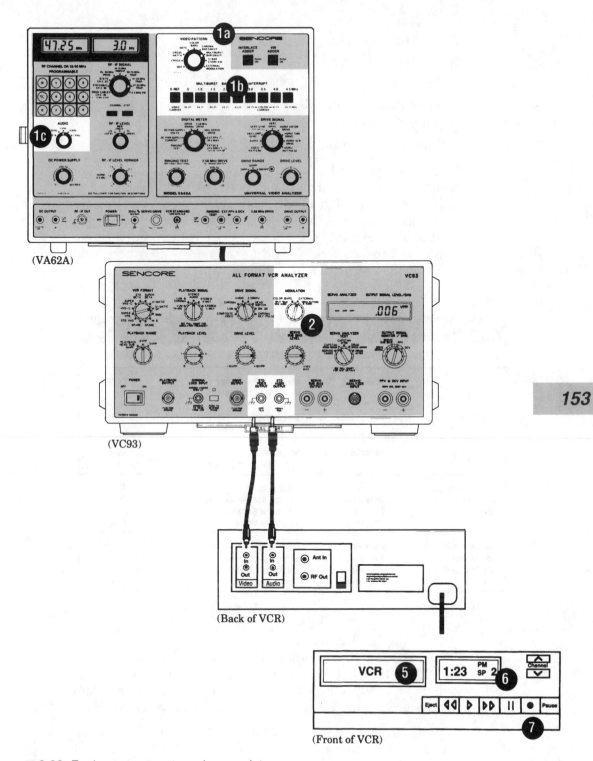

(VA62A)

(VC93)

(Back of VCR)

(Front of VCR)

153

■ **3-36** *Equipment set-up to make a work tape.*

2. Set VC93 *modulation* switch to external.

3. Connect *Std video output* jack to *video in* jack on VCR.

4. Connect *Std audio output* jack to *audio in* jack on VCR.

5. Place blank tape in VCR.

6. Set VCR speed to fast speed.

7. Press *record* button on VCR.

Record the speed select test sequence on the tape in the order shown in Fig. 3-37, changing tape speed on the VCR each 10 seconds. Switch the VCR to the SP mode and record the performance test sequence shown in Fig. 3-37, changing the video pattern and the audio on the VA62A every 20 seconds. Finally, record the troubleshooting test sequence, changing the video pattern and audio signal on the VA62A each two minutes.

TYPICAL WORK TAPE

PURPOSE	LENGTH	VIDEO PATTERN	AUDIO
Speed Select Test Sequence	10 seconds	Multiburst Bar Sweep (Fast Speed-SP)	1KHz
	10 seconds	Multiburst Bar Sweep (Medium Speed-LP)	1KHz
	10 seconds	Multiburst Bar Sweep (Slowest Sweep-SLP)	1KHz
Performance Test Sequence (single speed)	20 seconds	Color Bars	333Hz
	20 seconds	Chroma Bar Sweep	1KHz
	20 seconds	Multiburst Bar Sweep	5KHz
	20 seconds	10-Bar Staircase	7KHz
Troubleshooting Test Sequence (single speed)	2 minutes	Color Bars	333Hz
	2 minutes	Chroma Bar Sweep	1KHz
	2 minutes	Multiburst Bar Sweep	5KHz
	2 minutes	10-Bar Staircase	7KHz

■ **3-37** *Information that should be formed on a typical work tape.*

Make a separate tape following the preceding sequence, but place the VCR in the medium speed for the performance test sequence and the troubleshooting test sequence. Do the same with a third tape, but record the last two sequences at the slowest tape speed.

Note: If you do not have a VA62A, then use the color bars pattern and 1-kHz tone contained in the VC93. Record the speed select test sequence using the color bars pattern and switching the speed on the VCR every 10 seconds. Record the rest of the tape by switching the speed every 20 seconds.

The key VCR format characteristics are shown in Fig. 3-38. Refer to chapter 9 for more VCR troubleshooting procedures using the Sencore VC93 all-format VCR analyzer.

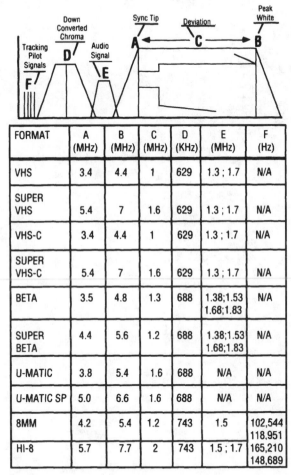

FORMAT	A (MHz)	B (MHz)	C (MHz)	D (KHz)	E (MHz)	F (Hz)
VHS	3.4	4.4	1	629	1.3 ; 1.7	N/A
SUPER VHS	5.4	7	1.6	629	1.3 ; 1.7	N/A
VHS-C	3.4	4.4	1	629	1.3 ; 1.7	N/A
SUPER VHS-C	5.4	7	1.6	629	1.3 ; 1.7	N/A
BETA	3.5	4.8	1.3	688	1.38;1.53 1.68;1.83	N/A
SUPER BETA	4.4	5.6	1.2	688	1.38;1.53 1.68;1.83	N/A
U-MATIC	3.8	5.4	1.6	688	N/A	N/A
U-MATIC SP	5.0	6.6	1.6	688	N/A	N/A
8MM	4.2	5.4	1.2	743	1.5	102,544 118,951
HI-8	5.7	7.7	2	743	1.5 ; 1.7	165,210 148,689

■ 3-38 *Key VCR format characteristics.*

155

Servo and
control systems

THE SERVO CIRCUITRY COMPARES THE PHASE OF THE 30 PG signal to that of the vertical sync separated from the video input signal in the record mode so as to control the drum rotation and keep its phase constant. A signal is twice as long as the cycle of the VD signal (vertical sync) fed to the CTL (control) head as the servo reference signal during playback, and is recorded onto the tape.

Control functions include autostop at the end of the tape, control of tape threading and unthreading, generation of the audio muting and video blanking signals, control of the pause circuit, and protection operations for such cases as tape slack and rotational failure of the drum.

Servo troubleshooting

VCR servo systems keep the cylinder head, capstan, and videotape moving at the correct speed. The servo must lock up in order to record and play a tape properly.

Some of the common servo problems are incorrect free-running speed, defective motor, incorrect voltage to the motor, and loose or worn drive belts. A new set of belts has solved many servo problems. Thus, servo problems can be electronic or mechanical. A good way to zero in on a servo problem is to go through all of the adjustment procedures. As an example, all that might be required is to properly set the free-running speed of the servo motor.

The picture symptom for a servo that is not locked up is the appearance of dark horizontal bands (see Fig. 4-1) that roll vertically up the screen. These bands might come and go, deceiving you into suspecting another type of intermittent problem.

157

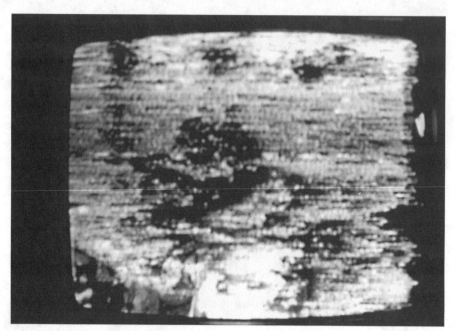

■ **4-1** *Horizontal bands across a screen indicate the servo is not locked up.*

The servo circuits form closed loops to control the speed of motors. Unfortunately, each of the two sets of servo circuits produces nearly the same symptoms when defective, making it difficult to discern which loop is causing the problem. To make matters worse, each set of servos usually contains two loops—one inside the other. The inside loop, or speed loop, causes the motor to turn at nearly the correct speed. The outside phase loop provides additional control over the first one. A problem in either loop causes all the signals in both loops to shift away from the normal values as the automatic circuits try to correct for the defective stage. This corrective action causes every component in the loop to appear defective. See the servo block diagram in Fig. 4-2.

Additionally, the servos have both electrical and mechanical components. Mechanical defects produce identical symptoms and measurements as electrical defects. The VA62 video analyzer helps break the loops to separate these confusing conditions.

Separating cylinder servo problems from capstan servo problems is the first step in troubleshooting. A simple look-and-listen test

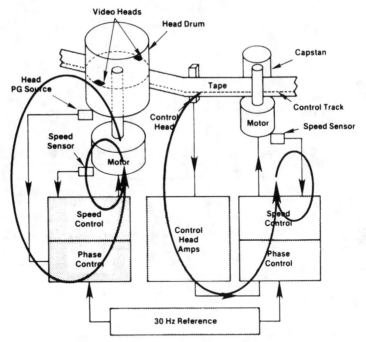

■ 4-2 *A pictorial of a VCR head drum and servo drive.*

tells which servo is causing the problem. What should you look and listen for?

Cylinder servo

The cylinder motor drives the video heads at the proper speed and position with respect to the tape. Except for the new hi-fi machines that use spinning audio heads, cylinder servo problems affect only the picture. Refer to the divide-and-conquer VCR block diagram in Fig. 4-3.

One IV frame consists of two fields of video information, or 525 lines. These frames of information are repeated 30 times each second. The VCR upper cylinder contains two video head tabs mounted 180 degrees apart. Each head tab records one field of video information each revolution of the cylinder to form a complete frame. The cylinder motor speed is approximately 1800 rpm or 30 rps. This speed must be maintained even during periods when the tape is not moving around the cylinder. This is the job of the cylinder servo.

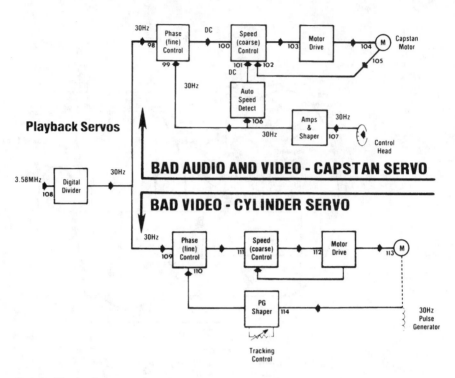

■ 4-3 *Block diagram of servo system divide-and-conquer technique.*

Capstan servo

The capstan motor controls the speed of the tape moving past the heads. Because most VCRs use stationary audio heads, capstan servo problems affect the sound and picture. If the picture is affected and the sound is not, you have a cylinder servo problem. But if the sound is also affected, you have a capstan servo problem.

Before you even open a schematic or reach for a test probe, divide and conquer the problem by listening and looking for the symptom. *Listen:* If the audio sounds too fast or slow, the problem is in the capstan servo circuits. *Look:* If the sound is okay but the picture is bad, put the VCR into the PAUSE mode. If the picture does not stay stationary and upright, the problem is in the cylinder circuits.

Sencore SC61 waveform analyzer

Now that you know which VCR servo is at fault, approach the circuits with an oscilloscope. I find that the Sencore SC61 Waveform Analyzer is *the* answer to VCR waveform troubleshooting. It is faster and more accurate than any scope I have ever used. With the SC61, I can conquer servo problems much quicker. Here are just a few of SC61's features:

- ☐ 60 MHz usable to 100 MHz gives you more confidence in waveform analyzing.
- ☐ Input protection to 3000 volts protects your investment and lets you make measurements other scopes can't.
- ☐ 100 percent automatic "auto-tracking" digital readout lets you make error-free measurements.
- ☐ Gives error-free readings of dc volts, peak-to-peak volts, and frequency at the push of a button. Measures any increment of a waveform in amplitude, time, or frequency and calculates frequency ratio.
- ☐ Rock-solid sync circuits latch onto the most elusive signal.
- ☐ Faster and more accurate than a conventional oscilloscope.

The block diagram of Fig. 4-4 shows the speed and phase relationships that servos control.

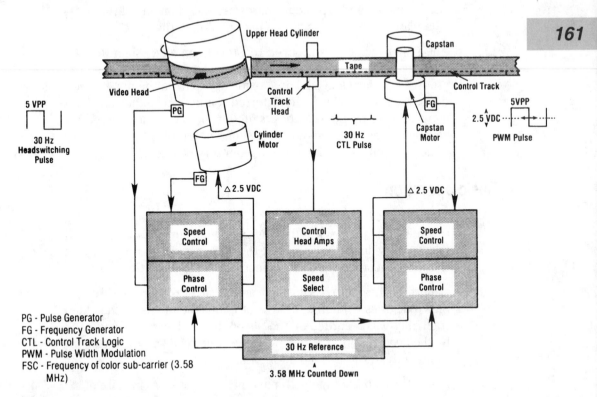

PG - Pulse Generator
FG - Frequency Generator
CTL - Control Track Logic
PWM - Pulse Width Modulation
FSC - Frequency of color sub-carrier (3.58 MHz)

■ 4-4 *Block diagram of VCR servo system.*

Oxide buildup

Accurate servo control is only possible in a clean, well-maintained machine. The capstan motor is responsible for moving the tape (in record and normal play) from left to right through the tape path. The tape path includes the following critical components:

- [ ] Back tension.
- [ ] Tape path entrance guide.
- [ ] Full erase head.
- [ ] Buffer roller.
- [ ] Cylinder entrance guide.
- [ ] Video head cylinder.
- [ ] Cylinder exit guide.
- [ ] Audio dub erase head.
- [ ] Audio and CTL heads.
- [ ] Capstan and pinch roller.
- [ ] Tape path exit guide.

Oxide buildup and tape edge shards deposited on any of these devices can cause some degree of malfunction in the VCR.

The upper cylinder is grooved to provide an air cushion for the tape that wraps around it. Greatly increased drag results when the grooves become clogged. Large oxide deposits on the upper cylinder cause poor tracking and distorted sound. You might think it's a capstan servo problem. Deposits on the audio/CTL head assembly will cause decreased audio or no sound at all, plus the possibility of no CTL pulses reaching the capstan servo circuit.

Basic operation

The servo system maintains the proper capstan speed during recording and playback and at the same time ensures that the head tabs on the video cylinder retrace the proper tracks during playback that were recorded during the record mode. Refer to Fig. 4-5.

Each video head tab has its gap azimuth set at 6 degrees. One is at plus 6 degrees while the other is at minus 6 degrees. If the video track that was recorded by the plus-6 degree head is played back by the head that is at minus 6 degrees, the result will be noise instead of the desired video information.

Speed through the tape path must be kept constant in spite of increases or decreases in drag caused by dirty surfaces and varying

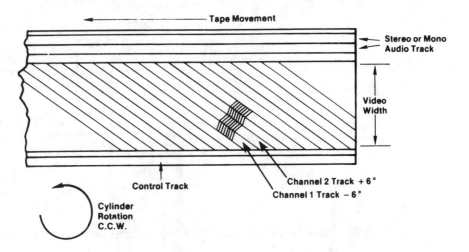

Tape Movement

Stereo or Mono
Audio Track

Video
Width

Control Track

Channel 2 Track + 6°
Channel 1 Track − 6°

Cylinder
Rotation
C.C.W.

■ **4-5** *Using control track signals from the tape, the servo system synchronizes the video heads to play back the proper track.*

back tension. The motors that operate both the cylinder and the capstan are controlled in speed and phase. Speed control takes care of large rotational changes while phase control takes care of the fine adjustments. The drive to each motor is the result of mixing or adding the speed and phase correction signals from the servo circuit.

Servo signals

Let's see what signals are used to control the speed of the capstan and cylinder motors.

Capstan frequency generator (FG) pulse

Note: Except for the 3.58-MHz and 30-Hz signals, all other servo frequencies might vary from one type of VCR to the next. Those provided in this text are typical for VHS machines.

The capstan motor (Fig. 4-6A) generates FG pulses of a certain frequency determined by the flywheel speed (720 Hz in SP, 360 Hz in LP, or 240 Hz in the EP speed). This is accomplished either by toothed magnetic elements (one fixed, the other moving within it), Hall effect devices, or fixed coils located near the outside diameter of the motor's magnetic flywheel. These frequency generator pulses are used to measure and maintain correct capstan operating speed in both record and playback. They are compared with the divided 3.58 MHz (see point B of Fig. 4-6). The result of

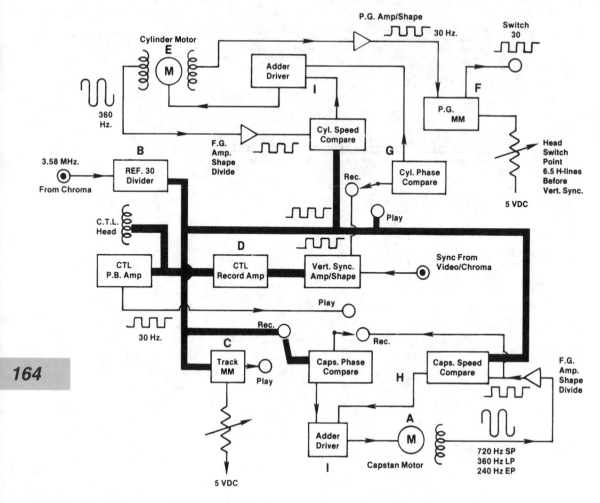

■ **4-6** *Block diagram of VCR servo circuits with test points (A) through (I).*

this comparison is the speed error signal and goes to the motor drive circuit.

Capstan speed and phase control

Capstan phase control is accomplished during the recording mode by comparing two 30-Hz reference signals. The first is obtained by dividing down the 3.58 MHz that comes from the video/chroma section of the machine. The second is a divided sample of the 720-Hz capstan FG signal (720 Hz divided by 24 equals 30 Hz) in the SP speed. The result of this comparison becomes the phase error signal that is used to keep the motor turning at a constant speed.

164

In the record mode, a 30-Hz pulse is developed from the vertical sync and is recorded on the control track.

During playback, the machine compares CTL and FG timing in a speed detect circuit to automatically select the speed the tape was recorded in. The divided 3.58-MHz reference signal (30 Hz) goes to the tracking control circuit monostable multivibrator, which allows for adjustment when playing tapes that were recorded in other machines (see Fig. 4-6C). The 30-Hz reference is compared with the 30-Hz pulse (Fig. 4-6D) to provide the phase error signal to the capstan motor drive circuit.

Cylinder motor control

The cylinder motor (Fig. 4-6E), like the capstan motor, makes use of a frequency generator (FG) device to produce a 360-Hz signal when it is operating at its correct speed. This signal is divided by 12 (to give 30 Hz) and is then compared to a divided vertical sync frequency during record and to a divided 3.58-MHz during playback.

The resulting error signal is sent to the cylinder motor drive circuit to maintain the 30 RPS rate required of the motor in both record and playback functions, regardless of capstan speed.

The PG pulse

The PG pulse indicates the position of the heads and is used in head switching. A magnet located on the cylinder motor flywheel corresponds to the position of one on the video head tabs. The magnet produces a pulse in a pickup coil or Hall-effect device with each revolution. This is the PG signal source.

From the PG pulses, another waveform is developed, known as the RF switch (or head switching) pulse. During formation of the RF switch pulse, the negative PG pulse is ignored and only the positive portion of the pulse is used. The RF switch pulse switches the video head during playback.

When this 33.3-ms (30-Hz) waveform is in its negative half cycle (during playback), the track 1 head switches on and the track 2 head is on during the other half cycle. By adjusting the PG MM control, the position of the switch point where each head is turned on can be placed at the required 6 1/2-horizontal fines before vertical sync. The RF switch pulse is also used in other areas of the machine for control purposes. Refer to waveform drawings in Fig. 4-7.

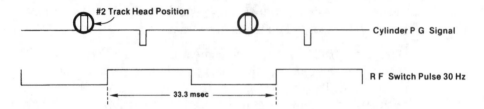

■ 4-7 *Comparison of cylinder PG signal and RF switching pulse signals.*

In the record mode, cylinder motor phase control is accomplished by comparison of the 30-Hz pulse generator (PG) signal with one-half of the divided vertical sync (30 Hz), which has been separated from the composite sync of the signal being recorded.

In playback, the 30-Hz PG signal is again compared with the 30-Hz reference signal obtained from the divided 3.58 MHz. The result of these comparisons, in both record and play, is the phase error signal output to the cylinder motor drive circuit for phase correction.

Cylinder phase in some machines is accomplished by adjusting it during forward and reverse search modes to lock the noise bar in the picture in one position. This prevents it from entering the vertical interval and provides a stable picture during search. This is done by adjusting the reference 30-Hz signal to 29.49 Hz in forward search and 30.51 Hz in reverse search.

For the cylinder motor, digital counters, clocked by the divided 3.58 MHz (cylinder speed Fsc/2 and cylinder phase by Fsc/8), develop the error information that is then used to pulse-width-modulate a 1.75-MHz carrier. Pulse-width modulation refers to the ability to control the duty cycle or "on" time of a rectangular waveform.

In the case of capstan digital speed error development (Fig. 4-6H), the master clock is again a divided 3.58 MHz (see Fig. 4-8). The divide rate is controlled by the speed select circuit. The digital phase error clock frequency is 447 kHz. The resulting error information is used to pulse-width-modulate a 3.5-kHz carrier for speed control. A 44-Hz carrier is used for phase correction.

These pulse-width-modulated signals, two from the cylinder circuit and two from the capstan circuit, are routed through lowpass filters and become dc levels that are then applied to adder circuits as shown in Fig. 4-6I. The results are the cylinder and capstan motor servo drive voltages.

Mode	Speed Error Development	Resulting Error Information
SP	3.58 MHz divided 4 times	895 kHz
LP	3.58 MHz divided 8 times	447 kHz
EP	3.58 MHz divided 12 times	298 kHz

■ **4-8** *Errors that develop for various VCR speed modes.*

Checking servo IC inputs

All of the circuitry used to perform these functions (with the exception of the lowpass filter) is usually contained in one large-scale integrated (LSI) chip. Because so many inputs and outputs are inaccessible within the body of the IC, you must be able to make troubleshooting determinations from the external information available to you. First, check the control signals from each assembly for proper level at the inputs of this IC.

Because digital counters continue to operate in the absence of reset pulses but provide nonsynchronized results, it is very important to confirm the presence of all primary and reference information at the appropriate servo IC inputs. This is best accomplished with a frequency counter and an oscilloscope.

If all necessary primary and reference information is getting to the servo IC and proper adjustment signals are not being produced, the IC itself is defective and must be changed. However, be aware that mechanical malfunctions in the motors themselves can cause erroneous primary information at the servo IC.

Many times, head switching problems are mistaken for servo problems when in fact they are the results of distortion of the RF switch pulse (SW30). The importance of RF switch pulse distortion cannot be stressed enough. This pulse is fed to several areas in the machine including system control. It can become distorted by component failures that have nothing to do with servo problems and can cause erratic head switching and noise bands in the video output.

Servo and pulse systems

The position relationship of the video heads and the 30 PG coils is shown in Fig. 4-9. A block diagram of the servo and pulse system circuits is shown in Fig. 4-10. The servo timing chart is shown in Fig. 4-11.

The drum servo system of this machine is a common magnetic brake servo system. The rotary drum is belt-driven by an ac hysteresis motor and controlled by a brake coil. The tape is also driven by the ac hysteresis motor.

The rotary head drum assembly is a stacked array consisting of the upper drum (fixed), the video head (rotating) and the lower drum (fixed). The two video heads are mounted on the periphery of the head disc. Two magnetic pole pieces, used in conjunction with the two 30 PG coils, are also mounted on the disc. The two PG coils (30 PG coils A and B) are mounted on the lower drum. The 30 PG (A) signal is used in the drum servo system. The 30 PG (A) and (B) signals are used to produce the RF switching pulses.

The drum servo circuit uses two ICs, CX138A and CX139A. The 30 PG (A) pulse triggers the lock PG delay multivibrator MMV (1) to obtain a 30-Hz rectangular wave. The output from MMV (1) toggles a second one-shot, MMV (4), which squares the signal into a 50 percent duty cycle waveform. The output from MMV (4) passes

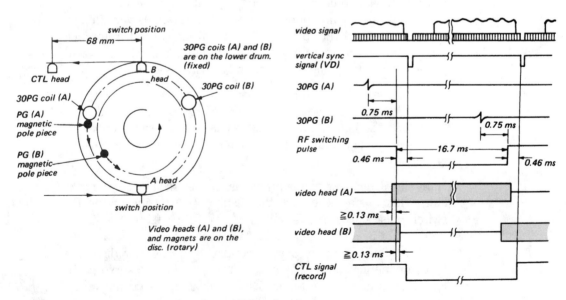

■ **4-9** *Relationship between heads and PG coils.*

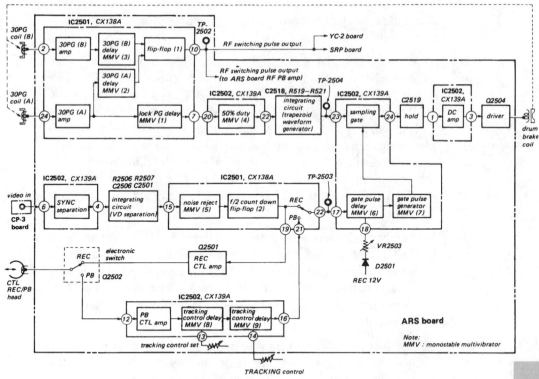

■ **4-10** *Block diagram of servo and pulse circuits.*

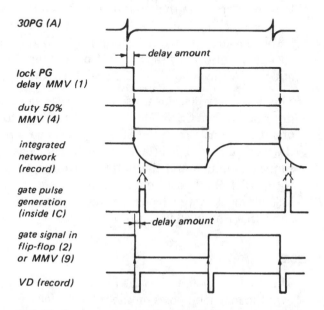

■ **4-11** *Servo waveform timing chart.*

through an integrator network that converts the square wave into a trapezoidal waveform.

The 60-Hz VD signal, separated from the video input signal, is fed to MMV (5), which eliminates noise. The output from MMV (5) toggles a flip-flop (2), which divides and shapes the signal into a 30-Hz square wave. This signal is used in the record mode as the CTL signal. It is delayed by the gate pulse delay MMV (6) and samples the trailing edge of the trapezoidal waveform in the gate circuit. The result is a voltage corresponding to the time from the PG pulse generation to the VD signal, i.e., the phase relationship between the generated PG pulse and the sampled vertical sync signal.

In the playback mode, the gate pulse delay MMV (6) is triggered by the CTL signal. The sampled voltage is stored in the hold circuit until the next sampling. The stored voltage, amplified by a dc amplifier and driver that also functions as a compensator, controls the drum brake coil. The compensator is used for stabilizing the operation of the servo system.

The correct phase relationship between the drum and the vertical sync signal is obtained by changing the delay time of the gate pulse delay MMV (6). The CTL signal triggers the tracking control MMV (8) and (9) in the playback mode. The output of MMV (9) is applied to the gate pulse delay MW (6).

Positional errors in the mounting of the 30 PG (A) and (B) units are corrected by the delay MMV (2) and (3). The outputs of the MMV (2) and (3) trigger flip-flop (1) to obtain the 30-Hz RF switching pulse. Only the output from the flip-flop (1) at pin 10 is affected by the delay time of the MMV (2) and (3) chip.

The drum servo circuit uses two ICs that were designed for the drum servo. The CS138A chip contains a block for obtaining various pulse signals from the 30 PG pulses and a block for dividing the vertical sync signal (60 Hz) in half. The phase comparison and sampling circuits and the output drive amplifiers are contained within CX139A. It also contains the sync separator circuit and the CTL playback system block. CX138A has a section for the capstan servo that is unused in this machine.

PG pulses

The PG coils, mounted on the lower drum, generate a pulse whenever the magnetic poles on the rotary video head disc pass above them (Fig. 4-12). As the magnet approaches a PG coil, a positive-

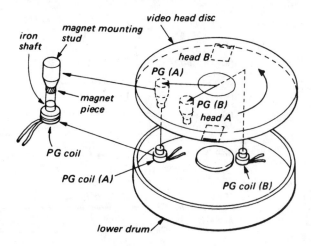

■ **4-12** *PG pulse generator mechanism.*

Labels in figure: iron shaft; magnet mounting stud; video head disc; head B; PG (A); magnet piece; PG (B); head A; PG coil; PG coil (A); PG coil (B); lower drum

going pulse is generated and when it recedes, a negative-going pulse is generated. There are two sets, (A) and (B), of PG coils and their magnetic pole pieces.

The distance of PG coils (A) and (B) from the center of the drum are different and each generates one pulse for each rotation of the drum. The negative-going pulse of the PG coil output is used to indicate the rotating phase of the drum. The pole pieces, which are strong permanent magnets, pass across the coils and induce a pulse by electromagnetic induction. The small head drum diameter used in this machine requires the use of a permanent magnetic system. The pole pieces cut across the coils at a slower relative speed than in the older model VCR machines. A timing chart of the 30 PG pulse circuit is shown in Fig. 4-13 and its block diagram in Fig. 4-14.

Both positive and negative pulses are obtained across R2501 and R2502 by the electric current induced in the PG coils. PG amplifiers amplify and shape the negative-going pulses to the PG coil outputs and trigger PG delay one-shots MMV (2) and MMV (3). The PG amplifier output waveform cannot be observed because it is inside the IC. The output of the PG (A) amplifier directly triggers the lock PG delay one-shot MMV (1). The lock PG delay MMV (1) is triggered by the signal passed through the PB (A) delay MMV (2) and the flip-flop (1). The lock PG delay MMV (1) is triggered by the phase of the PG amplifier output without regarding the delay time of the MMV (2).

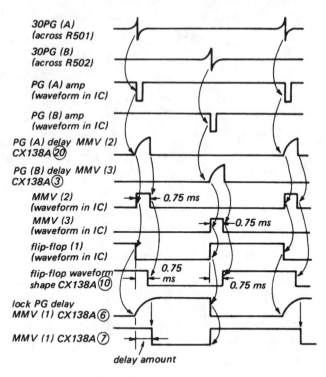

30PG (A)
(across R501)

30PG (B)
(across R502)

PG (A) amp
(waveform in IC)

PG (B) amp
(waveform in IC)

PG (A) delay MMV (2)
CX138A ⑳

PG (B) delay MMV (3)
CX138A ③

MMV (2)
(waveform in IC) ← 0.75 ms

MMV (3)
(waveform in IC) ← 0.75 ms

flip-flop (1)
(waveform in IC)

flip-flop waveform
shape CX138A ⑩ 0.75 ms ← 0.75 ms

lock PG delay
MMV (1) CX138A ⑥

MMV (1) CX138A ⑦

delay amount

■ **4-13** *Timing chart for the 30 PG pulse circuit.*

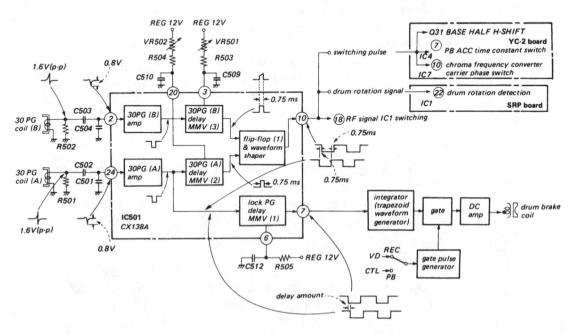

■ **4-14** *30 PG pulse circuit.*

Servo and control systems

The PG pulses (A) and (B) trigger the flip-flop (1) after mounting position errors in the two PG coils are compensated by the MMV (2) and MMV (3) and a 30-Hz square wave is produced. This delay mono multivibrator is formed by a combination of an RC integrated network and a Schmitt trigger circuit. The external circuit connected to pin 3 or pin 20 is the time constant circuit. All other multivibrators in CX138A and CX139A have the same arrangement.

Flip-flop (1) is triggered on the positive-going transitions of the output waveforms of MMV (2) and MMV (3), producing a 30-Hz square wave that is not affected by the delay time of the delay one-shots.

The output of the flip-flop (1) triggers the lock PG delay MMV (1) to obtain the output whose trailing edge is delayed for a constant width. The time constant circuit for the MMV (1) is connected to pin 6. The output of flip-flop (1) is wave shaped by the outputs of the PG delay MMV (2) and MMV (3) to obtain the same square wave output as the one obtained when the flip-flop is triggered by the delay phase of the MMV (2) and (3). This square wave is the RF switching pulse obtained from pin 10. This pulse goes to the playback RF amplifier, to the drum as a signal for indicating drum rotation, and for various switching signals.

Vertical sync and CTL signal generator

The block diagram and timing chart of the vertical sync signal and CTL signal systems are shown in Figs. 4-15 and 4-16, respectively.

The vertical sync signal, separated from the video input signal, becomes the servo reference signal in the record mode. The video input connection is provided on the CP-3 board. R5004 and C5001 on the CP-3 board form a lowpass filter. The filter rejects the chroma burst signal and high-frequency noise. The video signal is sync-separated in the circuit between pins 6 and 4 of IC502, CX139A. The sync separation circuit consists of a feedback clamp circuit for sag correction and a switching amplifier. The external circuit connected to pin 5 is the clamp time constant. An RC integrator circuit separates the V sync from the sync separator output and triggers the noise elimination one-shot, MMV (5), via SW (1) of IC2501. SW (1) switches to the pin 15 side for a 0-volt input at pin 16 and to the pin 14 side for 12 volts. The switch is kept permanently switched to the pin 15 side in this unit and is not used as a switch. The noise elimination one-shot, MMV (5), eliminates noise by means of the fact that a one-shot, once toggled, cannot be toggled again until after it has reset itself. The external circuit at pin 17 is the time constant network for MMV (5). The 60 Hz-VD vertical sync signal is divided into a 30-Hz square wave in the divide-by-two flip-flop. The 30-Hz square wave passes through SW (3) and appears as the gate signal output at pin 22. The negative-going transition of the flip-flop (2) output becomes the servo reference phase.

The flip-flop (2) output is inverted once in Q501, inverted again and amplified in Q502, and recorded on the tape as the CTL signal. The negative-going transition of the CTL amplifier output is used as the playback servo reference phase. The REC 12 V is supplied as the power supply for the record CTL amplifier.

SW (3) is the servo reference signal switching circuit between the record and playback modes. It is switched to the flip-flop (2) side when pin 23 is 12 volts and to the pin 21 side when pin 23 is 0 volts. In the playback mode, the playback CTL signal is used as the servo

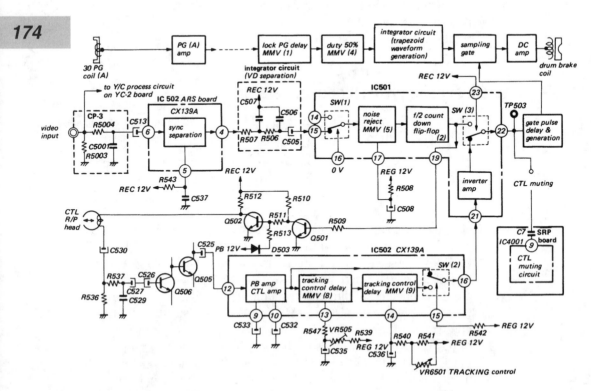

■ 4-15 *Vertical sync and CTL signal circuits.*

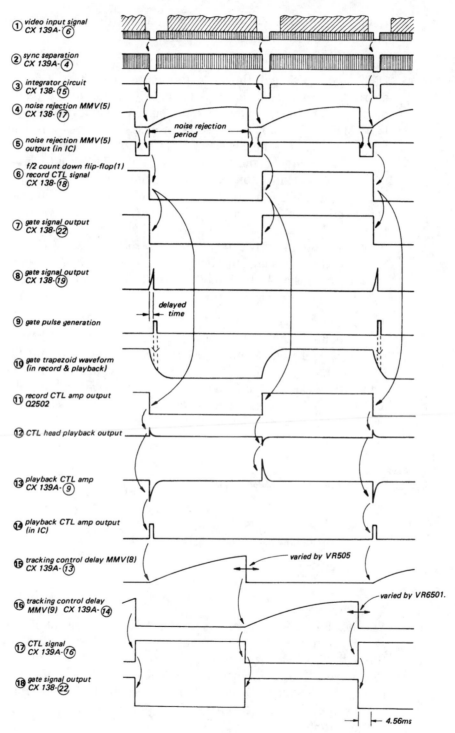

① video input signal
CX 139A- ⑥

② sync separation
CX 139A- ④

③ integrator circuit
CX 138- ⑮

④ noise rejection MMV(5)
CX 138- ⑰

⑤ noise rejection MMV(5)
output (in IC)

noise rejection
period

⑥ f/2 count down flip-flop(1)
record CTL signal
CX 138- ⑱

⑦ gate signal output
CX 138- ㉒

⑧ gate signal output
CX 138- ⑲

⑨ gate pulse generation

delayed
time

⑩ gate trapezoid waveform
(in record & playback)

⑪ record CTL amp output
Q2502

⑫ CTL head playback output

⑬ playback CTL amp
CX 139A- ⑨

⑭ playback CTL amp output
(in IC)

⑮ tracking control delay MMV(8)
CX 139A- ⑬

varied by VR505

⑯ tracking control delay
MMV(9) CX 139A- ⑭

varied by VR6501.

⑰ CTL signal
CX 139A- ⑯

⑱ gate signal output
CX 138- ㉒

4.56ms

■ 4-16 *Timing chart of vertical and CTL signal system.*

175

Vertical sync and CTL signal generator

reference signal. The positive polarity pulse of the CTL head output becomes the servo reference phase.

The PB 12 volts is fed to the base of Q502 in the playback mode via D503 and R513 and the collector-emitter circuit of Q502 is shorted. The terminal that was in the signal side in the record mode turns to the ground side in the playback mode, and the polarity of the CTL head is inverted. A lowpass filter rejects the high-frequency noise. The playback CTL amplifier amplifies the positive-going input signal in Q502 and Q506 and feeds it to pin 12 of IC502 for amplification and shape so as to obtain the positive pulse output. C533 (connected to pin 9) determines the frequency characteristics of the linear amplifier stage, which is a lowpass filter for 1 kHz, -3 dB. C532 (connected to pin 10) is a decoupling capacitor (ac grounded). The tracking control delay MMV (8) and (9) delay the CTL signal by about 2/3 cycle for the manual tracking control in the playback mode. The MMV (9) delays the CTL signal by about 1/3 cycle.

Thus, the output delayed 2/3 cycle is obtained. VR505 adjusts a delay amount of the MMV (8) so that the output phase of the MMV (9) has the relationship of the output phase of the CTL amplifier (as shown in Fig. 4-17) when the tracking control knob is at the center detent position. The tracking control varies the delay amount of MMV (9) to obtain a delay or advanced CTL signal reference to the playback output of the CTL head.

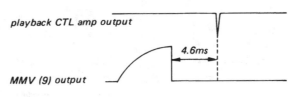

playback CTL amp output

4.6ms

MMV (9) output

■ **4-17** *Delay waveform.*

The positive-going transition of the MMV (9) output is the servo reference phase. The output of the MMV (9) goes through SW (2) and is fed to an inverter amplifier in CX138A. The negative-going transition becomes the servo reference phase as well as the VD signal in the record mode. The CTL signal at pin 22 of CX138A is fed to a gate pulse circuit and a CTL muting circuit.

Phase comparison gate

In this circuit, the trapezoidal waveform produced from the 30-PG pulse is sampled using a gate pulse produced from either the VD signal or the playback CTL signal. The gate output is amplified in a dc amplifier to drive the head drum brake coil. The sampling was done by the 30-PG pulse in older units, but in the Zenith KR9000, the VD/CTL signals are used to form the sampling pulses. Because the CTL head output of this video recorder is low (positive pulse of 1 mV), the CTL signal might drop out sometimes due to clogging or dust on the heads. In the CTL/VD gate system, the hold voltage is supplied to the dc amplifier until the CTL signal returns to the normal state. In this way, the influence of the CTL dropout on the reproduced picture is minimized. The schematic diagram of the phase comparison gate circuit is shown in Fig. 4-18, the timing chart of the trapezoidal wave former is in Fig. 4-19, and the timing chart of the gate pulse former is in Fig. 4-20.

The lock PG delay MMV (1) output, triggered by the 30 PG (A) pulse, is shaped into a 50 percent duty cycle square wave by the duty MMV (4) of IC502, CX139A. The MMV (4) supplies a comparison waveform to a gate circuit and produces the 50 percent duty cycle square wave output without regard to a pulse width of the trigger input. The falling slope of the input waveform is used as the reference phase. R2518 and C2517 (connected to pin 21) determine the time constant of MMV (4).

The response time of the servo is shortened by making the comparison waveform 50 percent duty. The trigger input waveform for MMV (4) is almost 50 percent duty cycle in this unit, but MMV (4) is utilized because it is already available in the chip.

The integrator circuit converts the rectangular wave into a trapezoidal wave by charge and discharge of an RC network. In the record mode, the trapezoidal wave is obtained by charge and discharge of C518. When pin 22 goes high, Q1 and Q2 in the IC are off, the 12-volt supply turns on Q3, which passes current through a 50-ohm resistor to R520 via pin 22. This charges C518. When pin 22 goes low, C518 discharges through R521. Q1 and Q2 in the IC are then ON and Q3 is OFF. The discharge current from C518 feeds through R520 to pin 22, the 50-ohm resistor, base-emitter circuit of Q2, and the collector to emitter of Q1. R519 is a biasing resistor for the dc amplifier in the later stage and raises the integrator output waveform about 6 volts dc above ground.

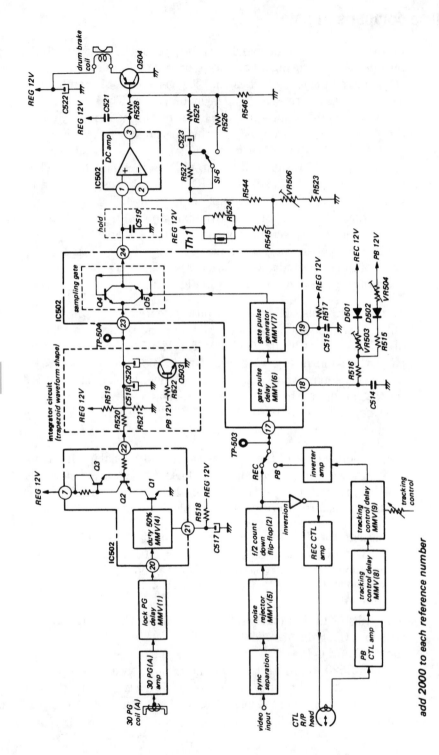

178

add 2000 to each reference number

■ **4-18** *Phase comparison gate circuit.*

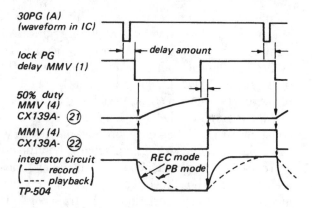

30PG (A)
(waveform in IC)

lock PG
delay MMV (1)

← delay amount →

50% duty
MMV (4)
CX139A- ㉑

MMV (4)
CX139A- ㉒

integrator circuit
(—— record)
(---- playback)
TP-504

REC mode
PB mode

■ **4-19** *Timing chart of trapezoidal waveform (comparison waveform) former.*

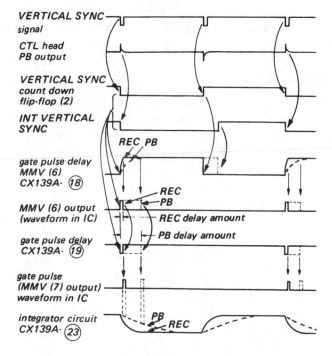

VERTICAL SYNC
signal

CTL head
PB output

VERTICAL SYNC
count down
flip-flop (2)

INT VERTICAL
SYNC

REC PB

gate pulse delay
MMV (6)
CX139A- ⑱

REC
PB

MMV (6) output
(waveform in IC)

REC delay amount
PB delay amount

gate pulse delay
CX139A- ⑲

gate pulse
(MMV (7) output)
waveform in IC

integrator circuit
CX139A- ㉓

PB
REC

■ **4-20** *Timing chart of gate pulse former.*

In the playback mode, the PB 12 V is fed through R522 to the base of Q503 and the collector to emitter of Q503 is shorted. This adds C520 to the integrator time constant circuit. The slope of the trapezoidal waveform is reduced, and the servo loop gain is reduced so that the integrator circuit does not respond to the high frequency variation contained in the playback CTL signal. The output trapezoidal waveform of the integrator circuit is fed to a sampling gate. The gate pulse is produced in the MMV (6) and MMV (7). The gate pulse delay MMV (6) uses the negative-going phase of the VD/CTL signal supplied to pin 17 as the reference phase and produces a constant delayed output. The RC network at pin 18 determines the time constant of the MMV (6). The time constants are switched in record and playback by D501 and D502. The delay amount is larger in playback than the one in record. This corrects the lock phase, because if the delay times in the MMV (6) are the same in both record and playback, the lock phase becomes incorrect due to the variation of the trapezoidal wave slope of the gate comparison waveform. The servo lock phase is adjusted by varying the delay time in MMV (6). VR503 is used for adjusting the lock phase in recording and VR504 in playback.

The gate-pulse-former one-shot, MMV (7), is triggered by the output of MMV (6) and generates a gate pulse of constant width. R517 and C515 at pin 19 determine the time constant of the MMV (7) (cannot be observed—part of the chip) and is supplied to the gate-sampling circuit.

The voltage held in C519 is amplified in the dc amplifier and supplied to the brake coil driver. This dc amplifier is a type of operational amplifier (op amp). It amplifies the potential difference between pins 1 and 2, using the potential at pin 2 as the reference. This voltage is applied to pin 2 via R544. VR506 adjusts the bias. The output, which has the same phase as that of the input to pin 1, is obtained at pin 3. This amplifier functions as a phase compensator by inserting an RC network into the feedback loop.

R525 and C523 form a feedback circuit for negative feedback of the output from pin 3 to pin 2. R527 and R526 are added to the circuit in the playback mode to reduce the loop gain. C522 is a high-frequency-bypass filter to prevent the playback RF amplifier from being interfered with by any high-frequency signal flowing to the brake coil. The dc amplifier output is applied to Q504 to drive the brake coil.

Control and pause circuits

The main function of the control circuit is to guide the mechanism during tape threading and unthreading operations, together with the subsequent control of the signal system. In the Zenith KR9000 VCR, the switching between function modes is mainly done by manually depressing the function buttons. Therefore, automatic controls of the mechanism by the system control are very few. What is controlled automatically by the system control circuit is the auto-stop solenoid. Figure 4-21 shows the system and pause control block diagram. There are five major operations provided by the system control circuit:

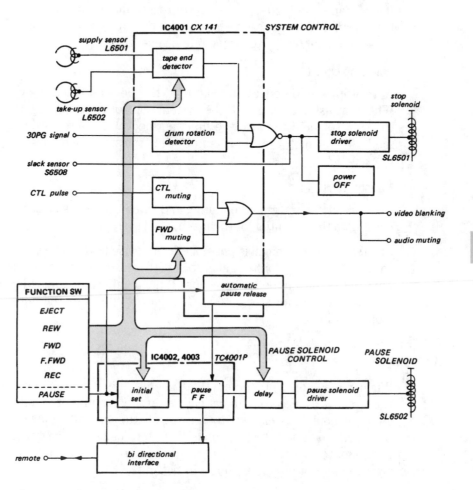

■ **4-21** *System control and pause control block diagram.*

Tape end sensor

The metallic foil attached to both the tape start and end are detected. An electrical circuit is closed, driving the auto-stop circuit.

Head drum rotation detector

When any of the function buttons are pressed, the system control circuit detects the head drum rotation by detecting the 30-PG pulses. If for any reason the head drum rotation should stop, the auto-stop circuit is energized.

Tape slack sensor

If slack tape is detected in the play or record mode, the tape must be rewound. When the tape slack is detected by the tape slack sensor element, the auto-stop circuit is energized.

Auto-stop circuit

Auto-stop is initiated by actuation of the devices listed above. When auto-stop is initiated, the system control circuit energizes the stop solenoid to place the function mechanism into the stop mode. The ac motor power is cut off when the stop solenoid is energized.

Muting circuit

This circuit generates audio muting and video blanking signals according to the operation of the mechanical system.

The take-up sensing coil is positioned very close to the tape in the cassette tape guide on the take-up side. It senses the tape beginning, in the rewind mode, by means of the metallic foil attached to the tape at the tape's beginning. The sensor energizes the auto-stop function.

Auto-stop

Whenever any of the detectors that generate the auto-stop operations senses a need to automatically stop the machine, the auto-stop solenoid is energized to release all the function buttons, putting the unit into the stop mode. The stop solenoid drive circuit is controlled by the signals generated in the respective auto-stop signal generator circuit as shown in Fig. 4-22.

The power supply for the ac motor is cut off when the auto-stop is energized. When any one of the function buttons is depressed, the button is latched mechanically by the lock slider of the function button. Release of the function button can occur in two ways: (1) manually by depressing the stop button, or (2) electrically by energizing of the auto-stop solenoid by an output from the system control (Fig. 4-23). The lock slider releases the function button in each case.

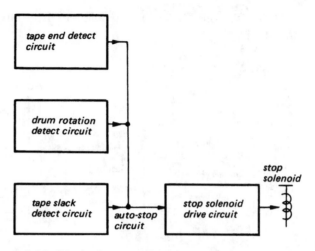

■ 4-22 *Block diagram of auto-stop circuit.*

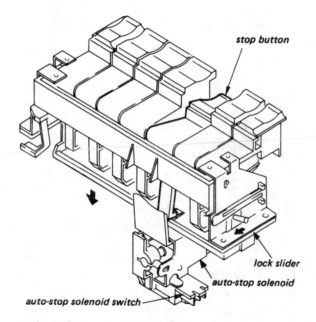

■ 4-23 *Stop solenoid location.*

When the auto-stop solenoid is energized, the lock slider is pulled in the direction shown by the arrow in Fig. 4-23 to release the function button lock. If the autostop solenoid is still energized, the lock slider will remain pulled even if any one of the buttons is pushed. The ac motor won't start, even if the function button is kept depressed after the auto-stop, because the stop solenoid switch is still on. The auto-stop solenoid drive circuit is shown in Fig. 4-24.

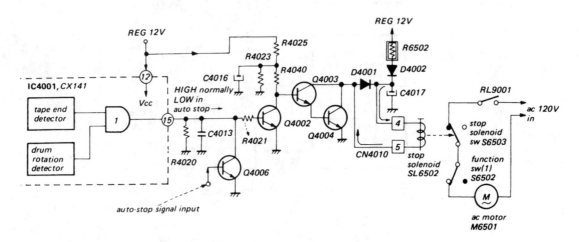

■ **4-24** *Stop solenoid drive circuit.*

184

Counter memory

The system control of the recorder during the rewind mode is automatically placed into the stop mode by the counter memory circuit when the tape counter reaches a 9999 indication. The counter memory circuit shown in Fig. 4-25 is included in the stop-mode circuit of the tape-end detector circuit. The bias voltage at pin 11 of the tape-end sensor oscillator circuit is made to increase the supply voltage instantaneously when the tape counter indicates 9999. This stops the oscillation for a moment and the machine is automatically stopped. Capacitor C4018, the counter switch, and the memory ON/OFF switch are connected in parallel with R4030, bias resistor of pin 11. The counter switch located in the tape counter turns on when the tape counter reaches its (0000) position. The oscillation is stopped for the C4018 charging period when the memory ON/OFF switch is on and the counter indication is 9999. If the rewind button is depressed once more after the machine has been automatically put into the stop mode, the rewind operation takes place because C4018 has been charged. Resistor R4601 prevents the machine from being placed in the stop mode automatically when the memory ON/OFF switch is turned on after the tape counter reaches 9999.

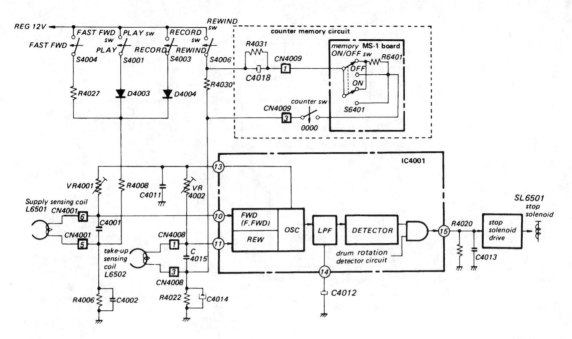

■ 4-25 *Tape end detector and memory circuits.*

Drum rotation detector

When normal operation cannot be performed due to motor trouble or overload, the recorder must be put into the stop mode. This is also required in the case where the head drum won't rotate due to a broken head drum belt. The drum rotation detector circuit uses the 30-PG signal to detect drum rotation. When the rotation stops, it activates the stop solenoid circuit. The detector circuit is activated when the ac motor rotates, i.e., when one of the function keys is pressed. The schematic diagram for this circuit is shown in Fig. 4-26.

Pause circuits

The pause circuits include a number of functions including memory, solenoid drive, auto release, and remote control.

Memory

The pause memory circuit is located in IC4003 (Fig. 4-27). The four NOR gates of the IC comprise two latches for the initial set and two toggle flip-flops. The basic operation of the latch is to latch a certain state (tape running or tape stopped) and to lock the

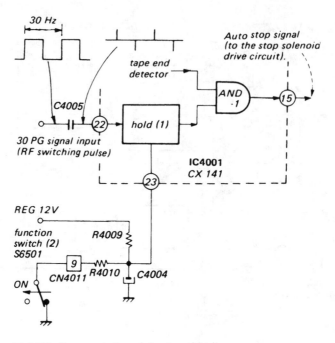

■ **4-26** *Drum rotation detector circuit.*

186

gate to ignore a change of the state. The latch operation is utilized for the initial set of the toggle flip-flop connected to the output of this machine. The input of gate 1 of IC4003 is connected to input P (of the pause solenoid), and its output is tied to the input of gate 2.

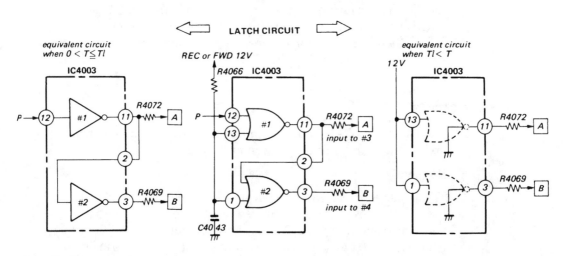

■ **4-27** *Pause memory circuit.*

Servo and control systems

One of the inputs of each of gate 1 and gate 2 is a low level for time (T). The output of gate 1 becomes P and the output of gate 2 P. The gates 1 and 2 work as inverters. Because one of the inputs of each gate becomes high when the latch time is (T), the inverter operation does not work, and the following flip-flop serves as an ordinary toggle-type flip-flop. The time is set for about two milliseconds by R4066 and C4043.

Solenoid drive

The pause solenoid drive circuit is formed by Q4016 and Q4017 as shown in Fig. 4-28. The pause solenoid is a two-winding three-terminal type with C (common), P (primary), and S (secondary) terminals. The winding between the C and P terminals serves as an initial pull-in while the one between the C and S terminals works for maintenance so that any temperature increase and the pull-in power requirement of the solenoid is reduced.

Transistor Q4016 turns on in the "run" state, causing current flow through the C-S winding and D4008. This turns on Q4017 and Q4018 (time constant determined by R4078 and C4046). At the same time, a current flows through the C-P winding, and the initial pull-in takes place. When Q4018 turns on, the potential of the P terminal and the current between C and S is cut off. At the same time, Q4017 and Q4018 turn off (by the same time constant determined by R4078 and C4046). Then current flows in the C-S winding again, and the pause solenoid energizing is maintained.

Transistor Q4019 (connected to the base of Q4016) is delayed about 350 ms after the play or record button is pressed by the time

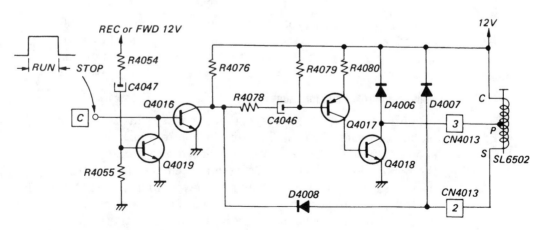

■ **4-28** *Pause solenoid drive circuit.*

constant circuit consisting of R4054, R4055, and C4047, and the run drive signal is bypassed during the delay (Q4019 is on for about 350 ms).

Auto-release

The pause auto-release circuit is one of the protection circuits for the tape. The pause auto-release circuit is shown in Fig. 4-29. The pause 12 V becomes a regulated voltage supply by the R4016 and R4014 resistor division. It is processed in the time constant circuit consisting of R4013 and C4006 to be a lamp voltage that increases with time in the pause mode. The lamp voltage is applied to the voltage comparator in IC4001 via pin 20. When it reaches the reference voltage in the IC, pin 21 of IC4001 goes high to set the pause flip-flop of IC4003 to the tape running state and release the pause state.

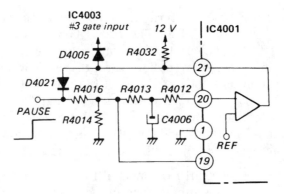

■ **4-29** *Pause auto-release circuit.*

Resistor R4012 is for protection of the internal circuit of IC4001 and D402 1. It serves to lower pin 21 to ground potential when the voltage is not applied to pin 19 in the stop mode or in the tape running state.

Remote control circuit for pause

A transmission signal and a reception signal are added in one cable in the remote control. The remote control block diagram is shown in Fig. 4-30. The reception signal is the control signal from the remote control to the videocassette recorder and the transmission signal is the pause signal from the recorder to the remote control. Because these two signals are transmitted with one signal line, the reception signal is a negative pulse signal and the pause signal is a positive pulse.

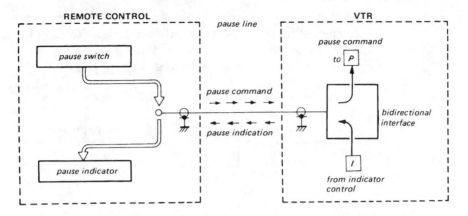

■ **4-30** *Remote control block diagram.*

Refer to Fig. 4-31 for the remote control circuit. The received remote control signal is divided by R4061 and R4060 and is applied to the base of Q4014. The threshold voltage of Q4014 is set to about 1.2 volts. When a pulse lower than the voltage comes in, Q4014 turns off and a positive trigger pulse appears at its collector. This pulse triggers the MMV formed by the two NOR gates (numbers 1 and 2 of IC4002) connected to the following stage. The MMV forms an OR circuit together with the pause switch of the recorder. Normally the output of "D" is a low level.

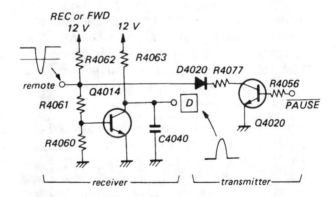

■ **4-31** *Remote control circuit.*

Basic drum servo

Servo is an abbreviation of the term *servo mechanism*. It is a self-correcting closedloop system, usually containing a mechanism, such as a motor controlling some other device until a feedback signal generated by the controlled device matches the reference signal. This definition of "Servo" is adhered to regardless of the simplicity or complexity of the design. Basically, every servo can be reduced to the diagram shown in Fig. 4-32.

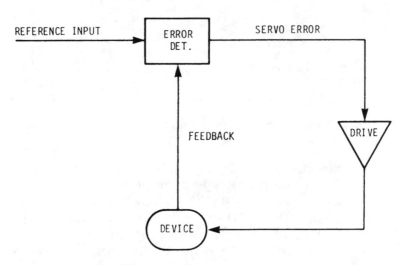

■ **4-32** *Block of basic servo system.*

The main purpose of the servo system is to operate the controlled device in a predetermined way. Should the operation pattern change for any reason, this change is sensed and relayed to the *error detector* (comparator), which compares the feedback signal with a stable reference. The output of the error detector is a signal representing the error necessary to bring the controlled device back to the correct operation.

In the Betamax, the servo system controls the speed and phase of the drum and capstan motors. In the helical scan system, the heads' path along the tape has to be tightly controlled so that they will write uniformly in the diagonal pattern along the tape. Also, the heads will read (scan) only along the tracks they have respectively recorded. (Track width is only 29.2 microns.) In addition to reading the correct track, the heads must maintain a speed of 30 rps and a phase or position relationship necessary to read and write sync pulses at the correct time for proper system operation. The motor used here is controlled by the *drum servo*. The *capstan servo* controls the capstan motor, which pulls the tape past the drum at a constant speed.

In many applications, these two servo systems operate independently, but they are linked together for the overall correction and control of the VCR. The servos are speed controllers that control the drum and capstan motors to place the tape in the proper position to match the required phase relationship between the video signal and the position of the video head along the track.

The diagram in Fig. 4-33 shows the basic drum servo. The free-running speed of the motor is set by the voltage across R1 and R2. A dc error voltage from the phase comparator will change the free-running voltage level, thus changing the motor speed.

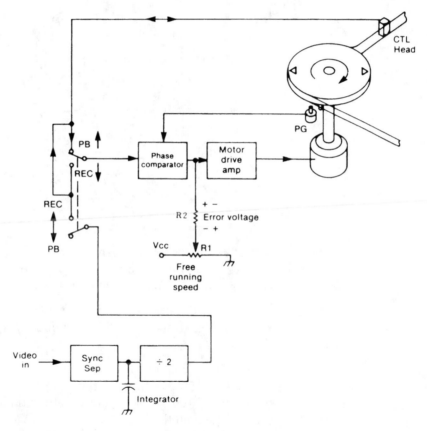

■ **4-33** *Basic drum servo block diagram.*

The phase comparator compares two inputs. One is a pulse from the PG coils called the PG pulse. As the drum turns, two magnets strategically placed on it cause sine wave-shaped pulses to be induced into two pick-up coils in the drum assembly.

The diagram in Fig. 4-34 shows the locations of the coils and magnets in the drum assembly. As the magnet approaches the coil, flux builds up quickly due to the expanding field induced into the coil's core. When the magnet and the core piece of the coil are aligned, the flux is no longer changing and the induced voltage drops to zero. As the magnet moves away from alignment, the induced core flux collapses and produces the negative part of the sine wave-shaped pulse. Two sets of pulses are developed: PGA and PGB pulses.

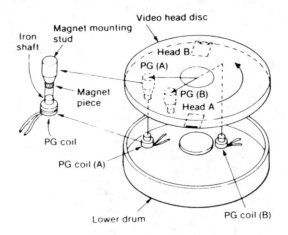

■ **4-34** *PM-type pulse generator.*

The arrival of the PGA pulse signals the arrival of the A head at the point of the track where vertical sync is to be recorded or read. The PG pulses are present both in record and playback.

The reference pulse to the phase comparator comes from the video in the record mode and from the CTL head in the playback mode.

In the record mode, sync is separated from the video, divided by 2, and is used as the 30-Hz reference input to the comparator. Early or late arrival of PG pulses from the scanner indicates a timing error with the head's position on the tape. A dc error voltage generated at the output of the phase comparator will cause the scanner to speed up or slow down to correct the head's timing error.

In addition to being the reference to the phase comparator, the 30-Hz sync pulse is used to obtain a square wave that is recorded longitudinally on the bottom of the tape forming the CTL track. This recorded square wave is called a control pulse (CTL pulse), and is used as the phase comparator's reference input during playback. When the CTL pulse is read by the CTL head, this is a signal that a prerecorded track is in place around the scanner. The A head should then be crossing the point on that track where vertical sync has been recorded. Should PG timing be off, the comparator's output voltage will change to adjust the drum's speed accordingly.

To operate properly, other circuits are necessary in the PG pulse's route to the comparator and in the CTL route to the phase comparator. We shall first look at circuits in the PG pulse's route.

The pulse generated to signal the arrival of the A head is called the PGA pulse. In practice, this magnet is placed on the scanner in a position where the PGA pulse will occur before the A head actually contacts the tape. By doing this and using delay multivibrators (MMVs), adjustments can be made to the timing of the pulse for optimum circuit operation.

Between the PGA coil and the phase comparator are two delay MMVs. The first one's output is fed to the second MMV, and also to a flip-flop. The flip-flop has a second input coming from a delay MMV, which delays a PGB pulse developed in the same manner as the PGA.

These two signals are used to toggle the flip-flop, creating a 30 Hz square wave whose transitions are used to switch on and off the playback head's pre-amps, 7 horizontal lines before vertical sync occurs. This is done to ensure switching transients do not disrupt vertical sync. Adjusting the delay time of the PGA and PGB MMVs effectively adjusts the square wave's transitions, which affects the head pre-amp switching time. Adjustments to the delay MMVs are done during the playback mode using the alignment tape. Note block diagram in Fig. 4-35.

The second delay, a lock phase delay, in the path of the PGA pulse, delays the pulse enough to have it coincide with the 30-Hz sync pulse. This is because the phase comparator compares sync with the PGA pulse. Any timing error in the drum would still be present after the actions of the delays and would cause the phase comparator to generate a correction voltage. Adjustments for the lock phase delay are done in the record mode.

Also in the record mode, the 30-Hz sync signal is converted to a 30-Hz square wave and fed directly to the CTL head for recording. However, in playback it is delayed by MMVs to achieve optimum timing of these pulses for use as the reference input to the phase comparator.

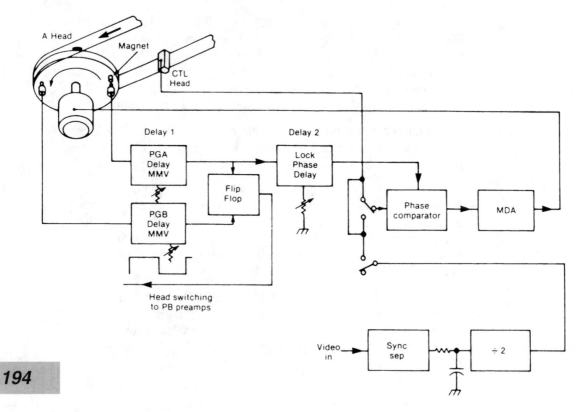

■ **4-35** *PGA, PGB, and lock phase delay added.*

The CTL head is placed a specific distance away from the drum along the tape path. This distance can vary minutely between machines and, if not adjusted, will cause tracking problems when a tape recorded on one machine is played back on another. Using the alignment tape, the CTL head is moved along the tape path as necessary to achieve maximum RF output. When there is an error, however, a customer tracking control can be used. The effect of the tracking control is the same as moving the CTL head along the tape path, but without physically moving it.

The diagram in Fig. 4-36 shows two tracking delay MMVs being used. The first MMV inserts a delay that is determined by a service adjustment. The delay added to the second MMV is controlled by the front panel tracking control. When the second MMV is set at mid range, the total delay inserted by both MMVs is equal to the time interval between control track pulses, which is approximately 1/30 second.

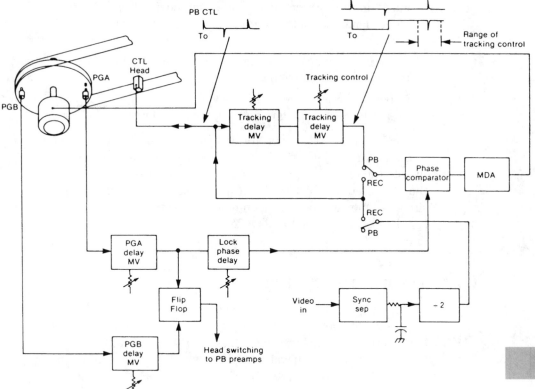

■ **4-36** *Addition of tracking delay multivibrators.*

Under these conditions, the phase comparator's reference input is the control track pulse of the previous frame of the recording (one CTL pulse per frame). By varying the tracking control, the delay of the MMV is changed and thereby changes the timing of the CTL pulse to the phase comparator. The effect is the same as physically shifting the CTL head along the tape path.

In older machines, the output of the phase comparator was applied to a motor drive amp, which in turn varied the current through a coil to vary the effect of a two-pole electromagnet around a squirrel cage rotor. This system slowed down or speeded up the drum, which was belt driven at a range of 30.3 to 30.5 rps by an ac hysteresis motor spinning at 30 rps. The system was rugged and reliable, but impractical for use in portable machines. Modern machines use dc motors. The three main drum servo functions are:

☐ Maintain the correct video head position with respect to the video signals so that the signal is recorded in a definite pattern.

☐ The correct video head position for proper tracking of the tape in the playback mode.

☐ Generate a pulse identifying the video head position for proper head switching in both record and playback modes.

These functions are only performed when the position and speed of the heads are known. PG pulses are used for this purpose. Although the error-correcting signals affect the speed of the heads, the position of the heads is of primary concern. If the relative position of a moving object is maintained correctly, its speed must be correct. A correct speed, however, does not ensure correct position.

Contemporary servo systems

Because of great advances in semiconductor technology, the main circuits of the servo once made up of discrete components are now contained in integrated circuits. These ICs perform the same functions, in some applications with greater accuracy. Of course, they must have the sync, PG, FG, CTL pulses, etc, to function properly.

Let's take a quick look at how the servo control system functions in a modern Betamax machine. The whole drum/capstan servo control circuit is contained in one IC chip in four sections as follows: drum speed servo, drum phase servo, capstan speed servo, and capstan phase servo.

Each servo control circuit measures the phase difference between two input signals, thus creating an error signal that is pulse width modulated, filtered and used to control the motor drive.

The drum/capstan servo IC is a digital IC in this application. It therefore must have a reference clock. The 3.58 MHz from the color crystal oscillator is input to the IC at pin 28 and is internally divided. Various divisions of this clock are used for all operations of the IC, such as measuring the time between input pulses or the duty cycle of the output pulses.

The four servo outputs (because this is a digital IC) are pulse width modulated (PWM) rather than dc analog voltages. They must therefore be lowpass filtered (or digital to analog converted) to provide the necessary dc drive voltage for the dc motors. By

196

changing (or modulating) the pulse width of a fixed frequency signal, the filtered output voltage can be made to change accordingly. For example, if the duty cycle (the amount of time the signal is high vs. low) is increased, the filtered dc voltage output will increase. When amplified and applied to a dc motor, the speed of the motor will increase.

On some Betamax machines, the frequency of both the drum and capstan speed PWM outputs is approximately 13.8 kHz. The frequency of both phase PWM outputs is approximately 833 Hz.

Drum speed servo

In Fig. 4-37 waveforms, three PG pulses are shown developed in the drum assembly. They are PGA, PGB, and PGS pulses. PGA and PGS are used for speed servo, and PGB is used in the phase servo. PGA and PGB are both used to develop the head switching (RF switching) pulse. Adjustments to the RF switching pulse are made with RV5 and RV6 at IC1 pins 2 and 41.

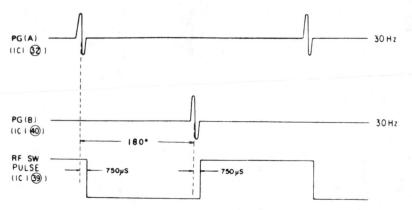

■ **4-37** *Timing chart for the RF SW pulse signal.*

Routine VCR maintenance

5

THIS CHAPTER COVERS VCR CLEANING AS WELL AS OTHER maintenance procedures. All of these tips can be used by professional electronics technicians or VCR owners. These maintenance points include how to properly clean the VCR heads and tape guides. The care and feeding of both the Betamax and VHS videotape systems is covered. Other points of VCR service are degaussing the video heads and cleaning and checking the drive belt. The chapter covers what to do when video heads have been worn smooth and develop "stiction." The chapter continues with information on VCR head cleaning tapes, which was adapted from a paper compiled by the 3M Co.

In many ways, the VCR is not too much more complicated for routine cleaning than some of the more sophisticated audio tape machines. However, some of the electronic and mechanical tape transport devices are considerably more sophisticated.

In an audio machine, the magnetic tape passes over stationary heads to record and reproduce sound. The VCR, by comparison, has rotating video heads along with fixed audio, control track, and erase heads. The rotating heads are needed because of the higher frequencies. The tape must travel across the heads at a much faster rate than for audio recording. This is accomplished by rotating the video heads at a very high speed past the slower moving tape.

The original video taping technique was perfected for the professional broadcast industry by Amprex. For home video recorders, the operation has been refined for use in a cassette format, eliminating the need to handle large rolls of tape. The chapter concludes with an odd infrared remote TV/VCR control interference problem. Then some information on the Hi-Tachi flash memory camcorder. And to round out this chapter a set of drawings are included to show you how to connect the VCR/TV converted box, etc.

Demagnetizing heads

Every metal part of the VCR that comes into contact with the tape will gradually become slightly magnetized. This especially pertains to the video and audio heads, which should be demagnetized regularly. Residual magnetism affects all types of magnetic heads. Heads and other metal contact points in the tape path will become partially magnetized from such sources as the normal on-and-off surges from the recorder's electronics. Items such as a faulty bias oscillator, the use of an ohmmeter for VCR troubleshooting, or the use of some magnetized tools near the heads could easily cause unwanted magnetism.

A magnetized tape head (and other magnetized metal parts) will erase some of the high frequencies on prerecorded tapes or will cause a hiss or visible background noise. Also, magnetization of the videotape from these spurious sources can cause loss of color and partial erasure of the tape.

It is simple to perform VCR machine demagnetization. Just plug in the demagnetizing tool, turn it on, and then slowly bring it near the part to be demagnetized. Now move it slowly up and down a few times, and then slowly retract the tool to about three feet

■ **5-1** *The head demagnetizer in use.*

away. Bring the tip of the demagnetizer as close as possible to the head tip without actually contacting it. Should the tool be turned off while close to the head, it will have the reverse effect and leave the head magnetized. If you are careful to avoid this, the head and other machine parts will be completely demagnetized. The Nortronics VCR-205 head demagnetizer is shown in action in Fig. 5-1.

Please note that you should only use an approved VCR head demagnetizer for this operation. The video head demagnetizers produce a weaker flux than most of those used for audio tape machines. Using a demagnetizer made for an audio tape player can actually shatter the video head chips.

Removing VCR insides from the case enclosure

Before you remove the VCR case to clean the heads, make adjustments or perform minor repairs be sure the VCR is unplugged from the ac socket.

To remove the case use a screwdriver to take out the screws. These will usually be located on the rear of the case and/or on the bottom. Now put the screws in a small plastic bag or cup so they will not be lost. Some of the screws might be of different sizes so remember to put them back in the same place so as not to strip out the holes. Also, do not drop any screws into the VCR mechanism.

Some VCRs might have a circuit board or cover over the cylinder head and transport. This must be removed before you can clean the heads and tape guide tracks. You are now ready to clean the heads and other mechanism such as tape guides and pressure and pinch rollers.

Why clean VCRs?

Even with normal operation, contaminants in the atmosphere can cause a VCR to perform poorly. Particles of dust, smoke, loose magnetic tape oxide, and oxide binder combine over a period of time to form a hard buildup on the head surfaces. This buildup causes the tape to be physically spaced away from the proper firm contact with the face of the head, which reduces the amount of signal being recorded and played back. Normal spacing between the tape and head face is 0.000020 inch (for comparison, the thickness of a human hair is 0.004 inch). When this extremely close contact is lost due to dirt buildup on the heads, picture playback performance will be diminished. This signal loss can cause mushy, distorted sound with a noticeable lack of high frequencies. Visually, it appears as a lack of overall picture clarity or as a snowy picture. It is also necessary to clean tape oxide buildup from all the contact points in the tape path, like rollers and guides.

In extreme cases where tape oxides are allowed to build up over a long period, the accumulation can become large and ragged enough to scratch or tear tapes or interfere with the precise tape speed required for VCR operation. Regular VCR cleaning is the only way to prevent oxide buildup and other major machine problems. Professional broadcasters and recording studios know this well and clean all heads daily. It is recommended that the home model VCRs be cleaned and the heads demagnetized every 100 hours of operation, or two to four times a year.

The rotating video heads are the most delicate and expensive parts of the VCR. They must be treated with the utmost care and be kept very clean.

The video heads actually penetrate into the tape oxide when recorded and played back; thus, there will always be a small amount of oxide shedding from the tape onto the heads and guides in the normal VCR running process. A good VCR cleaning procedure is required on a regular basis.

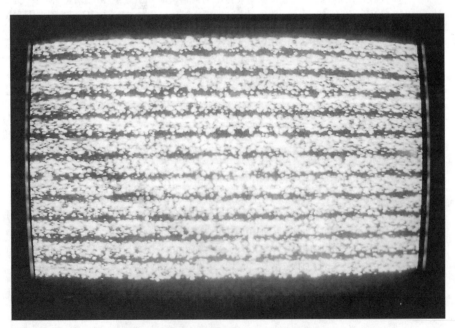

■ **5-2** *Poor tape playback due to oxide build-up problem.*

An oxide buildup on guides and heads will cause very poor record and playback of the videotapes. This shows up as streaks, noise (Fig. 5-2), or if built up enough, no picture at all. You might even suspect a faulty head drum. In fact, dirty heads will show up first in the playback mode because of the relatively weak magnetic field from the tape that must be transferred to the head drum. Another streak condition you might think is due to dirty heads is tape dropout. This is when a few horizontal lines across the picture are missing (Fig. 5-3) and appear as a line of interference. This dropout is caused by a streak of oxide missing from the tape or a faulty badly worn tape.

■ **5-3** *Streaks through the picture due to dropout.*

How to clean VCRs

To clean the heads, use a special cleaning pad such as chamois cloth or cellular foam swabs. If these are not available, lintless cloth or muslin can be used. Cotton-tipped swabs should not be used for cleaning the video heads. The cotton strands can catch on the edges of the video heads and pull the small ferrite chip away from its mounting and ruin the head. However, cotton swabs soaked in cleaning fluid can be used to clean tape guides, control track, audio, and erase heads. To clean the heads and guides, you can use methanol or surgical isopropyl alcohol.

The cleaning pad should be liberally soaked in the cleaning fluid (alcohol) and then gently and firmly rubbed sideways across the heads. Never rub it up and down, as this might damage the heads. Clean the whole head in this same sideways motion. Make sure you then clean all places that the tape touches. Do not touch these parts with your fingers, as the oil will attract dust and dirt. But, it's a good idea to hold the head on top so it will not rotate as you rub it clean (in the direction of head rotation). This correct cleaning technique is illustrated in Fig. 5-4. The main thing to remember in head cleaning is to be very careful. The more often you perform this task the easier it will become.

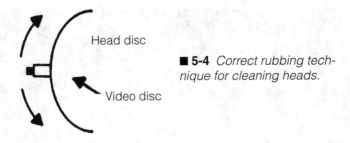

Head disc

Video disc

■ **5-4** *Correct rubbing technique for cleaning heads.*

Should you find a head that is very dirty and the head chip is plugged up with oxide, try the following tip for cleaning. The first thing to do is soak the area around the head chip with alcohol two or three times. Then, as shown in Fig. 5-5, use an old toothbrush also soaked in alcohol to clean out around the head chip. Do this very carefully, as the head can be damaged. (If it cannot be cleaned, replace the heads.) you might want to cut the bristles of the toothbrush shorter. A spray-type video head cleaner is being used in Fig. 5-6.

Cleaning Beta VCRs

Let's now go through the actual steps for cleaning a Beta machine. Caution: Make sure the ac power cord is disconnected before cleaning the machine.

1. Remove the screws that hold the top cover of the VCR in place. These usually require a Phillips screwdriver. Be careful not to damage the screw heads. If this is an older model Zenith or Sony Betamax, remove the tracking knob in the lower, left-hand corner by grasping it and pulling it up and off the shaft. Now, press the eject button to raise the cassette carrier platform, lift off the top cover, then push the cassette carrier down again.

204

■ **5-5** *A toothbrush can be used for cleaning clogged heads.*

■ **5-6** *Spray type head cleaner being used.*

2. The two video heads are located within the rotating center ring. Rotate this ring to bring each head (gap) into a convenient position by turning the black motor fan blower shroud shown by the arrow in the upper-right corner. You might also want to use an inspection mirror to get a better look at the heads. Do not actually touch the head face with your fingers.

In order to clean these heads, first saturate one of the cellular foam cleaning swabs with a good tape head spray cleaner. Clean the heads using only a horizontal (side-to-side) motion. To ensure that cleaning is done in a side-to-side fashion, hold the swab stationary against the head and use the fan motor shroud to rotate the head back and forth. Caution: Do not clean heads with a vertical (up-and-down) motion. This can easily damage the very delicate VCR heads.

3. The control track and audio heads are cleaned next. The control track and audio heads are located near the tip of the swab. Clean these heads in the same way you cleaned the video heads (using only a horizontal scrubbing motion).

4. Now the erase head should be cleaned. Remember, use only a horizontal cleaning motion.

5. After all the heads have been cleaned, perform the same cleaning function on all contact points (rollers, guides, etc.) that are in the tape path.

6. Push the eject button to raise the cassette carrier and reassemble the unit by placing the top cover in position. Install and tighten all cover screws with the Phillips screwdriver. Replace the tracking control if there is one. The VCR should now be wiped clean with an antistatic dust cloth.

Cleaning VHS VCRs

Let's go through the actual steps for cleaning a VHS-format machine. Caution: Make certain that the ac power cord is disconnected from the machine before removing the cover and cleaning.

1. First, remove the screws that hold the top cover of the machine in place so that it can be removed. These are usually Phillips screws. Be careful not to damage the screw heads.

2. The two very delicate video heads are located on the rotating head disc. Gently rotate this wheel disc to bring each head into a convenient position for cleaning. You might want to use an inspection mirror to get a better look at the head faces. Do not actually touch the highly polished face of the disc wheel

with your fingers. The control track, audio heads, and erase head are located on each side of the head cylinder.

To clean these heads, first saturate one of the cellular foam cleaning swabs with a good tape head cleaner fluid. Clean the heads using only a horizontal (side-to-side) motion. To ensure that cleaning is done in a side-to-side fashion, hold the swab stationary against the head and use the head wheel to rotate it back and forth. Caution: Do not clean with a vertical (up-and-down) motion. This could damage the fragile heads.

3. The control track and audio heads should be cleaned in the same way you cleaned the video heads (using only a horizontal scrubbing motion).

4. After all the heads have been cleaned, perform the same cleaning function on all contact points (rollers, guides, etc.) that are in the tape path.

5. Now replace the top cover and install all screws. And if you wish to do a class-A job, wipe the machine clean with an anti-static cloth. Caution: Do not use cotton swabs or other lint-producing materials, as the lint might be left during the cleaning process and could damage delicate VCR machine components.

Make sure the tape heads in the VCR unit are totally dry after the cleaning procedure before the machine is put into operation. Otherwise, the tape might not thread properly and jam up the machine. Nortronics VCR-103 spray tape head cleaner ensures completely dry heads, because this cleaner evaporates completely, leaving no oil or other residue.

Notes on tape tension

Remember that the tape tension is very important in the proper operation of a VCR. Too much tension causes excessive head and tape guide wear or tracking errors due to tape stretch that can permanently stretch the tape. Too much tension in the threading loop can actually stop the tape from traversing its path and even pull it out of its path. Too little tension allows the tape to fall out of its true path, preventing proper contact with the heads, which causes misalignment with the video tracks and results in picture dropouts. Many tape transport problems are due to the wrong tension on the tape, and checks should be performed on a regular basis.

Belts and drive wheels

All of the drive belts and wheels should also be checked and cleaned when the VCR machine is apart for head cleaning. Check for loose drive belts and worn rubber drive wheels. If only the rewind and fast forward operate properly, then suspect worn drive wheels. The drive belts and wheels can be cleaned with isopropyl alcohol or any other cleaner made for this purpose. If new belts are installed, make sure they are put on properly and check them for proper tension. Always make a careful check when drive belts, wheels, or other mechanical parts are changed or adjusted.

Much of the VCR's mechanical alignment and parts replacement require using special jigs, gauges, and fixtures for correct operation. Make sure you are properly trained before attempting these procedures.

Video head "stiction"

Video cassette recorders that have logged many hours of use might begin to exhibit a condition described as *stiction*. The word, a combination of the words *sticking* and *friction*, indicates a condition of the video head drum assembly that causes the tape to stop moving during record or playback. If this occurs and continues unchecked (as in recording with the timer mode), severe clogging of the video heads and tape damage could result.

The apparent cause of stiction is the loss of an air cushion between the tape and the record drum head. As the VCR is used, friction from tape travel polishes the drum surface smooth. This prevents the required air buildup, and the tape adheres to the drum. Obviously, replacement of the head drum assembly is the ultimate solution, but this is very costly.

The following alternative has been tested and recommended whenever the drum exhibits highly polished areas and is a last resort before the video head disc is changed. This procedure requires the use of a specially designed brush tool (Zenith part no. DS-21179) that can be obtained from your local Zenith distributor. Caution: Do not use sandpaper, emery cloth or other similar abrasives to perform this repair.

If the head disc is not to be replaced after this procedure, use extreme caution when working around the video heads, as they can be easily damaged. Proceed as follows:

1. Remove the cassette lift assembly and threading arm guides (for Zenith and Sony machines).

2. Remove the arm lock bracket assembly and the tape retainer spring assembly.

3. Remove the rear panel assembly and perform the threading operation.

4. Fold the narrow side of a calling card (see Fig. 5-7) and insert this between the drum and head bracket assembly.

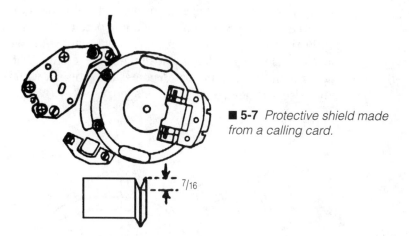

■ **5-7** *Protective shield made from a calling card.*

5. Position the video heads as for dihedral adjustment and lock the head in place with partially inserted dihedral screws, (if dihedral screws are not available, hold the rotating disc firmly with thumb and forefinger, taking care to not touch the heads).

6. Use the brushing tool (part no. SD-21179), with bristles set to approximately 1/8 inch. Brush horizontally across the upper (stationary) head disc, center (rotating) head disc, and lower (stationary) head disc. *Do not* come closer than approximately 3/8 inch to the head disc coil.

7. Continue in this way from the front of the head disc assembly to within 3/8 inch of the coil at the rear. Then rotate the video head disc clockwise until the rear coil is on line with the rewind sensing coil.

8. Hold the rotating disc in place and continued brushing toward the rear of the head disc assembly.

9. When the tape path around the drum has been brushed to a dull finish, blow any possible remaining aluminum particles out the rear of the VCR machine with a compressed air gun or

use the canned air that can be found in some VCR head cleaning and maintenance kits. This brushing procedure should give many more taping hours without head replacement.

Head cleaning tips

To clean the VCR head you might first try a cleaning tape. If you cannot clean the head this way, then use a cleaning patch. Coat the cleaning patch with alcohol or head-cleaning fluid to the point indicated in Fig. 5-8. Touch the cleaning patch to the head tip and gently turn the head (rotating cylinder) right and left. Do not move the cleaning patch vertically and make sure that only the buckskin on the cleaning patch comes into contact with the head. Otherwise, the head might be damaged.

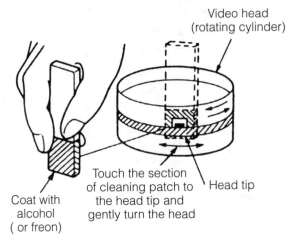

Video head
(rotating cylinder)

Coat with
alcohol
(or freon)

Touch the section
of cleaning patch to
the head tip and
gently turn the head

Head tip

■ **5-8** *Technique for using the cleaning patch.*

Thoroughly dry the head. Then test tape-running. If alcohol or head-cleaning fluid remains on the video head, the tape might be damaged when it comes into contact with the head surface.

Next, clean the tape transport system and drive system, etc., by wiping with a cleaning patch wetted with alcohol or head-cleaning fluid.

Note: It is the tape transport system that comes into contact with the running tape. The drive system consists of those parts that move the tape. Make sure that during cleaning you do not touch the tape transport system with the tip of a screwdriver and that no force is applied to the system that would cause deforming.

Required maintenance

The recording density of a VCR is much higher than that of an audio tape recorder. VCR components must be very precise, at tolerances of 1/1000 mm, to ensure compatibility with other VCRs. If any of these components are worn or dirty, the symptoms will be the same as if the part is defective. To ensure a good picture, periodic inspection and maintenance, including replacement of worn out parts and lubrication, are necessary.

Scheduled maintenance

Schedules for maintenance and inspection are not fixed because they vary greatly according to the way the customer uses the VCR and the environment in which the VCR is used. But, in general home use, a good picture will be maintained if the inspection and maintenance is made every 1,000 hours. The drawing in Fig. 5-9 shows the relation between time used and inspection period.

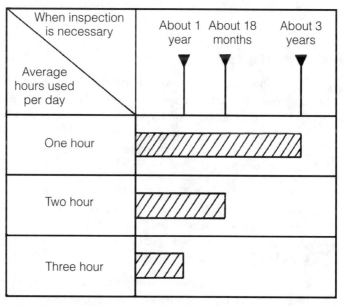

■ **5-9** *The scheduled maintenance chart.*

Check before repairs are performed

Many faults can be remedied by cleaning and oiling. Check the needed lubrication and the conditions of cleanliness in the unit. Check with the customer to find out how often the unit is used, and then determine that the unit is ready for inspection and maintenance. Check the following parts that are listed in Fig. 5-10.

Poor S/N, no color	Dirt on video head or worn video head
Tape does not run or tape is slack	Dirt on pressure roller, belt or flywheel belt
Vertical jitter, horizontal jitter	Dirt on video head or in tape transport system
Color beats	Dirt on full-erase head
Low volume or rewind is not done or rotation is slow	Dirt on audio/control head
Fast forward or rewind is not done or rotation is slow	Dirt on belt

■ **5-10** *Various picture symptoms caused by dirty video heads.*

VCR head cleaning tapes

The following information is adapted from a paper compiled by 3M Co. Wear and tear causes VCR problems ranging from poor video and sound quality to the inability to operate because the machine is dirty. As with any sophisticated machine, the VCR requires proper care and periodic maintenance to continue operating at maximum effectiveness.

According to VCR makers, as often as 90 percent of the time the primary cause of problems in VCRs returned for repairs is simply a dirty machine in need of thorough cleaning. Of particular concern is the proper maintenance of the delicate record and playback heads in the machine.

In response, a number of different head-cleaning systems have been introduced for consumer use. Some use liquid solvents. Others use a dry abrasion method.

Each promises to do an effective job of keeping the VCR's record and playback heads clean and in proper condition. Unfortunately, not all deliver on that promise. Some, in fact, can cause more harm than good, as witnessed by widespread consumer complaints about head cleaning systems that might have caused damage and needless repairs.

So serious has the problem become that the California Bureau of Electronics and Appliance Repair recently held hearings on a proposal to ban the marketing of video head-cleaning systems intended for home use.

The video recording process

Recording and playing back video information involves placing and sensing magnetic signals on a thin polyester film coated with magnetic oxide particles suspended in a binder. To record the signals, a magnetic field is created in the VCR's recording heads. An extremely small space, or gap, in each head provides a place where the magnetic signal can be transferred to the videotape as the tape is pulled past the head by the VCR's transport mechanism. In playback mode, the process is reversed, allowing the gaps in the playback heads to detect the magnetic signals and convert them back into the electronic signals that recreate the sound and picture information stored on the tape.

The video heads are mounted in a rotating drum assembly. The heads protrude slightly from the drum, which allows them to push into the tape coating. This design provides the extremely intimate contact between head and tape that is essential to recording and resolving the high-frequency video information. Refer to Fig. 5-11.

A video is much more complex than the audio-only signal carried on typical audio cassettes. For each second of taped video picture viewed on a TV monitor, the videotape carries 30 frames of magnetic information, each frame consisting of two lines—one line recorded in alternating sequence by each of the two heads. In addition, separate synchronization and audio tracks are also recorded on the tape as it passes over the rotating heads at the equivalent of up to 80 mph.

So small are the distances and tolerances in video recording that the microscopic gap in the heads measures approximately one-half micron across. A human hair, by comparison, has a thickness of about 100 microns—200 times the size of the head gap. It is also in this tiny gap that the magnetic flux lines, which are representations of the video images, are either generated or sensed. Any in-

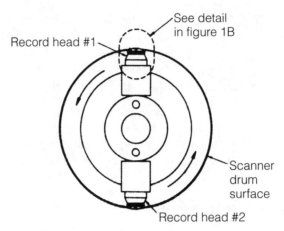

See detail
in figure 1B

Record head #1

Scanner
drum
surface

Record head #2

Top view of ratating record-head assembly

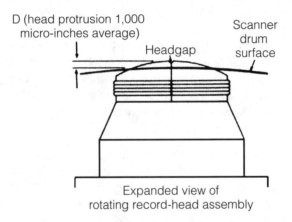

D (head protrusion 1,000
micro-inches average)

Headgap

Scanner
drum
surface

Expanded view of
rotating record-head assembly

■ **5-11** *The video heads are mounted in a rotating drum assembly. The heads protrude slightly from the drum, which allows them to push into the tape coating.*

terference with the surface of these tiny record heads, or the even smaller head gap, will cause problems ranging from a noisy picture to complete loss of the video image.

Because of the complexity of this recording process, keeping the video heads, and particularly the head gaps, clean and free from contaminants is essential for proper video reproduction. Contaminants as tiny as microscopic dust or airborne smoke particles can cause interference with magnetic signals and result in unacceptable video recordings or noisy playback images. See Fig. 5-12.

214

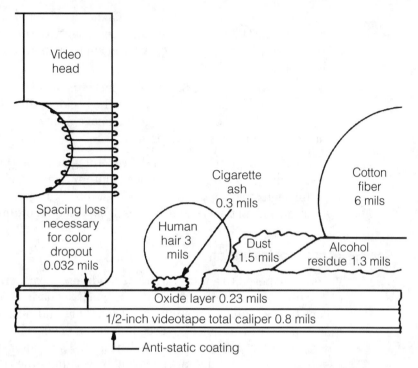

5-12 *Contaminants as tiny as microscopic dust or airborne smoke particles can cause interference with magnetic signals and result in unacceptable video recordings or noisy playback images.*

In addition, it is important to keep the entire path taken by the magnetic tape clean, and to properly maintain the VCR's tape guide mechanisms and tension adjustments because these components can also adversely affect video performance.

Videotape quality

The tape that carriers the picture and sound information is an important factor in proper VCR operation. Videotape consists of a thin, tough, flexible polyester film coated with microscopic particles of magnetic oxide suspended in a binder. Three important components affect the quality of the tape's performance and ultimately the cleanliness of the VCR's heads:

☐ The oxide particles are magnetically imprinted by the heads with the video, audio and synchronization signals. The greater the number and the more uniform the shape of the particles, the better the quality of the video image.

☐ The binder holds the microscopic magnetic oxide particles in suspension. The better the formulation quality of the binder, the better the oxide particles will remain in suspension against the backing.

☐ The tape backing must be strong, nonstretchable, and environmentally stable. The weaker the backing, the more subject it will be to stretching and damage from continued use and environmental conditions such as heat.

Because of the physical nature of the recording process—involving the penetration of the recorder's heads into the tape itself during stressful, high-speed operation—only high-quality, rugged, technologically advanced tapes can ensure reliable video recording.

Sources of head contamination

VCRs are exposed to a wide variety of conditions, a vast cross section of users, a selection of tapes whose quality can range from excellent to poor, and an abundance of handling and mishandling. Videotapes are also subject to broad ranges of environmental conditions, users and forms of use and misuse. The combination of users, use conditions and tapes inevitably will affect the condition of the VCR's delicate recording and playback heads.

Contaminants come in many shapes and sizes, and from a wide variety of sources. Dust is present in every home environment. Smoke particles from cigarettes are many times larger than the head gaps. Some debris can be generated by the machine itself. Particles of the magnetic oxide from videotapes can be worn off by repeated passes over the heads. The way in which both recorder and tapes are stored and used is also a key element in the generation of tape and head contaminants.

A major, often unsuspected source of head contamination is rented tapes. These cassettes are used by consumers of all ages and with varying levels of awareness of proper tape care and handling. They are played in a variety of machines, many of them not maintained very well, and critical adjustments to the tape—guiding mechanisms—are rarely made. All of these factors can lead to undue wear, stress, and damage to the thin, delicate magnetic tape and can increase the deposit of debris from the tape to the VCR's heads.

Another source of head contamination is the use of unlicensed and cheaper quality videotapes. The VHS format is a licensed product of JVC; Beta is licensed from Sony. Bargain tapes might not be produced to the standards specified in these formats, and there is an unavoidable tradeoff between quality and price. The result can be a greater potential for tape pigments and binders to be transferred from tape to the recorder's heads, especially as the poorer quality cassette's performance deteriorates with repeated use.

Over time, contaminants from a combination of all these sources can build up on the heads and across the head gaps. The intrusion of these contaminants between the heads and the videotape impairs the heads' ability to properly record and play back signals. As the heads become progressively dirtier, picture quality deteriorates, leading ultimately to a complete clogging of the gap, at which point a complete loss of the video image occurs.

Preventive maintenance

Basic VCR maintenance includes tape path alignment, tension adjustment, and general cleaning. Without proper maintenance and cleaning, performance quality can be expected to diminish over time, resulting not only in poorer video but also possible damage to the videotapes played in the machine. This level of maintenance is not something that can or should be done by the consumer. A consumer's VCR should be periodically reconditioned by experienced service technicians to maintain machine and tape integrity. The frequency of maintenance is naturally dictated by the frequency and type of use.

Video head cleaning

In addition to a regular schedule of professional machine maintenance, the video heads require regular attention to prevent deterioration or loss of the video signal.

It is important to note that recording tapes and video recorders are generally self-cleaning. The manufacturers of good- and premium-quality videotapes add trace amounts of cleaning agents to their oxide formulations to help prevent the buildup of debris on the VCR's heads. In addition, the recorder's own threading and tension adjustment mechanisms are designed to help keep debris buildup to a minimum.

Predictably, though, contaminants and other debris will accumulate on the video heads over time. This problem can be treated by the consumer. Head-cleaning videocassettes can provide an effective and inexpensive means of removing foreign matter from the VCR's heads and getting the equipment safely back into operation.

Three basic types of head-cleaning systems designed specifically for use on VCR heads are currently available to the consumer: wet systems, dry abrasive systems, and a magnetic tape-based system. This selection of VCR head-cleaning products is available in a marketplace characterized by conflicting claims, variations in product quality, and aggressive pricing and marketing strategies.

Wet/fabric systems

Wet systems rely on a liquid solvent applied to a fabric material housed in a videocassette casing. The cassette is then in the VCR, allowing the solvent to be transferred to the heads. This form of cleaning depends on the ability of the solvent to safely and effectively dissolve all the various types of debris from tapes and other sources present in the VCR. Furthermore, solvent cleaners cannot refurbish the ferrite surfaces of the video heads, which is a necessary part of restoring the VCR's performance and video signal.

Even when properly formulated, wet systems present a number of problems, not the least of which is the fact that the solutions used might damage plastic and rubber components inside the VCR. To effectively dissolve debris left on the heads, the wet system's fluid must be an aggressive solvent. Unfortunately, the process of transporting the solvent-soaked fabric past the VCR heads requires contact with other internal parts, which the solvent might also attack.

The primary danger is to the rubber pinch rollers of the VCR's tape transport system. These rollers are a vital part of the main tape drive system, and even the least deterioration can cause failure, typically resulting in a tape being "eaten" by the machine.

The fabric media used as a carrier are also a potential danger to proper VCR operation. Video recording equipment is specifically designed and manufactured to safely and reliably handle an extremely thin, polished magnetic tape. The introduction of thicker and relatively coarse fibrous materials employed by some wet cleaning systems can put unintended stress on the tape transport mechanism and conceivably the fragile video heads.

Compared to the small dimensions of the video heads (and especially the microscopic gaps in the heads), the fiber materials used in many wet cleaning systems are relatively large. Given these sizes, it is easy to understand how physical damage can be sustained by the delicate recording heads.

Finally, professional-quality camcorders and most home VCRs have a dew sensor to detect the presence of excess moisture and shut off the machine when conditions are too moist for safe operation. In some older VCRs, the dew sensor must be manually reset by a service technician, but even in those that automatically reset the introduction of unnecessary moisture is not recommended. Liquid cleaning solutions are best left to qualified service technicians.

Dry/abrasive systems

Dry/abrasive systems offer a second alternative to the consumer. These systems avoid the problem associated with the use of a wet solvent on a fabric and do not attack the rubber pinch rollers or plastic internal parts of the VCR.

However, although many dry/abrasive systems are capable of doing an effective job of contaminant removal, they are often too aggressive and cause premature video head wear. Some of the materials used in dry/abrasive cleaning systems are equivalent to very fine sandpapers. The cleaning effectiveness of abrasion is clear, but the abrasive nature of the system itself creates the danger of improper use or overuse.

In most typical dry/abrasive systems, there is no reliable way for the consumer to judge when the VCR's heads have been properly cleaned and conditioned. As a result, the consumer must make a best guess regarding how long to leave the dry/abrasive cassette in the VCR.

Magnetic-tape-based systems

The third consumer alternative is a magnetic tape-based head-cleaning system. The chief advantage of such a system is that magnetic tape is the medium for which the VCR was designed. It will not damage the VCR's delicate heads and internal mechanisms.

In a magnetic tape-based video head cleaning system, the level of cleaning agents in the tape formulation is increased to produce a cleaning cassette. Because the system is based on magnetic recording tape, the tape can include instructions for the user. As the system is played in the VCR and the heads become clean, a prerecorded video message tells the user when the recorder heads are clean and the machine should be stopped.

In tests conducted by 3M, testers found that with the magnetic tape head cleaner, most cleaning can be accomplished within 30 seconds or less. On this basis, the consumer can expect 240 or more cleanings. If all 240 cleanings were made, head life would be reduced by an average of only 2.2 percent on clean heads.

Because new machines with clean heads were tested, the tape would appear to do less damage in everyday use because the tape is not in intimate contact with the record head until the head is clean, at which time the tape's prerecorded video message alerts the user to stop the machine.

As noted, premium-quality recording tapes from most major manufacturers already contain trace amounts of head-cleaning agents in their formulations, along with lubricants and other necessary additives. This small amount of cleaning agent helps keep the video heads free of debris during operation and under normal conditions. Without it, the extremely small head gaps of the typical VCR would be subject to frequent clogging.

The user's responsibility

In a best case scenario, a customer using exclusively premium tapes in a clean, smoke-free home environment could expect a low incidence of dirty or clogged heads. In reality, this is almost never the case. More typically, video recorders are used with a wide variety of recording tapes of vastly varying quality and condition, including tapes that are rented, borrowed, and less-than-good quality.

Also, most home video equipment is not periodically adjusted or maintained in accordance with the manufacturer's recommendations. The result can be undue wear, stress, and damage to the thin magnetic tapes used in the VCR and, almost inevitably, an occasional head clog that renders the video heads partially or totally inoperable.

The amount of cleaning agent in premium-tape formulations does provide a basic form of preventive maintenance. It is not sufficient, however, to remove more than a small amount of contaminants under normal operating conditions because too much cleaning agent compromises the tape's recording quality.

When problems occur, consumer head-cleaning tapes can help keep the VCR in operation. However, periodic cleaning and adjustment by an electronic servicer should be part of the preventive maintenance for all VCRs and can prevent more costly damage.

Unlike dry abrasive and wet/fabric head cleaners, Scotch head cleaning videocassettes are magnetic tape, combined with a carefully controlled cleaning agent. They do the best job of clog removal and head reconditioning without risk of damage or undue wear. And Scotch head cleaners are cleaning up the market. Customers are learning that the dry abrasive head cleaners can cause premature video head wear by abrading the surface of the heads. And some have found that wet/fabric type cleaners are not only messy, but some solvents can attack plastic and rubber parts, speeding up wear and deterioration.

Intermittent remote control operation

Should you have a strange VCR or TV remote control intermittent operation problem and you have some of these new power saver fluorescent tubes that screw into a standard light socket (see Fig. 5-13), this might be the problem. These lights have been known to interfere with the infrared pulses that are emitted from the remote unit.

■ **5-13** *Power saver fluorescent light.*

Most remote control units transmit pulses of infrared energy. These coded pulses can vary, but the remote control circuits are much the same. They usually transmit pulses of infrared energy at a frequency of 56 KHz. This happens each time a button is pushed, thus producing a group of coded pulses.

A code usually contains 24 bits of binary data. The first four bits determines which one of 16 types of equipment is to be controlled. Generally, the 0 is for a TV set, 1 is for number 1 VCR, 2 is for the number 2 VCR, and 3 for a compact disk player. The next eight bits in this sequence are used to identify the pressed key. Now these eight bits permit a total of 256 key control codes. The last 12 bits are check sum bits that are compared to the first 12 bits to detect any errors.

When the TV/VCR remote receiver detects an infrared pulse train, it compares it with a clock pulse operating at the same frequency. If the signal is high, a binary one is produced. If not a binary zero

■ **5-14** *Compact fluorescent screw-in tube light.*

is produced. After the code pulses are processed, the VCR/TV equipment responds for the proper action.

These compact fluorescent tubes as shown in Fig. 5-14, operate at high frequencies. These high frequencies modulate the tube's infrared output and cause interference for remote controls based on the infrared emission.

The big and heavy ballast used with standard fluorescent tubes limits the flow of 60 Hz current through the tube. This ballast produces a flicker that can be annoying when read by these older fluorescent lights. This flicker rate is 120 Hz, which is twice of the power line power frequency.

Now it is this nonvisible flicker of a switched infrared pulse signal that will interfere with the infrared remote control units. The remote control units for controlling VCRs, TV sets, stereos, and cable boxes also switch in the 50- to 100-KHz range to transmit this binary data to the equipments receiver remote circuits. So, if you are having remote control problems and have these type of fluorescent lights, this might be the cause. If you turn off these lights and your remotes work OK, that's the problem.

Flash memory camcorder

In the very near future a video recording system with no moving parts is possible, according to Hitachi, which reports it expects to market a consumer camcorder that uses flash memory instead of tape as its recording medium within a few years. Hitachi even demonstrated an early prototype that played back full-motion color and stereo sound. As Hitachi currently plans it, the camcorder will weigh about 10 ounces and will provide 30 minutes of color recording using as the storage device a 400-megabit multilayered flash memory about the size of a sugar cube. According to Hitachi's preliminary design, the camcorder would use a single-chip MPEG-1 encoder/decoder and have an electronic zoom system, further eliminating moving parts. Hitachi believes that because the MPEG-1 encoder might not supply adequate picture quality, it is developing a proprietary compression algorithm. In addition to the solid-state video memory, Hitachi says it is developing a hard disk as a possible camcorder storage medium.

Step 1: Disconnect the Antenna or Cable from Your TV.

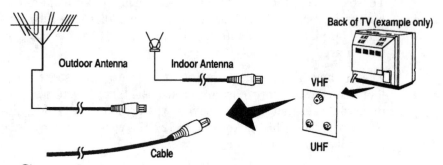

Step 2: Connect Your Antenna or Cable to Your VCR.

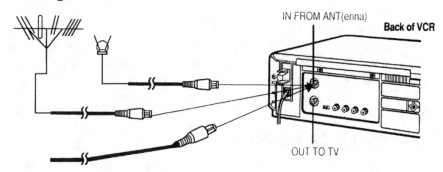

Step 3: Connect Your VCR to Your TV.

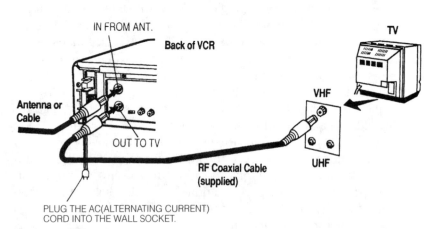

■ 5-15 *Drawing of various ways to connect your TV set and VCR to cable and antenna systems.*

Step 1: Disconnect The Antenna or Cable from Your TV.

Your antenna or cable should have one of the ends shown below. To avoid confusion, label all terminals (on TV) and ends before disconnecting.

Step 2: Connect Your Antenna or Cable to Your VCR.

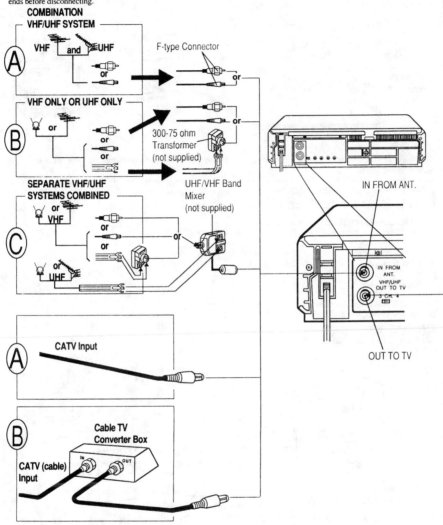

COMBINATION
VHF/UHF SYSTEM

Ⓐ VHF and UHF

or

F-type Connector

or

VHF ONLY OR UHF ONLY

Ⓑ or

or

or

300-75 ohm
Transformer
(not supplied)

or

SEPARATE VHF/UHF
SYSTEMS COMBINED

Ⓒ or
VHF

or

or

or

or
UHF

UHF/VHF Band
Mixer
(not supplied)

IN FROM ANT.

IN FROM
ANT.

VHF/UHF
OUT TO TV

3 CH. 4

OUT TO TV

Ⓐ CATV Input

Ⓑ Cable TV
Converter Box

CATV (cable)
Input

IN

OUT

■ **5-16** *Various other ways to connect your TV and VCR.*

225

If your cable system uses a converter box, you may use this connection instead of the basic cable connection shown on the previous page.

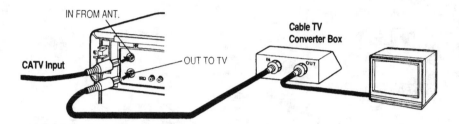

With this connection you can 1) program unattended recording of more than one unscrambled channel, or 2) view one channel while recording another.

| Note: | • | Use the VCR CHANNEL UP/DOWN buttons to select a channel to be recorded. |
| | • | You cannot record scrambled channels with this connection. |

To Record One Channel While Viewing Another
1. Select the channel to be recorded with the VCR CHANNEL UP/DOWN buttons.
2. On the VCR set the VCR/TV selector to "TV".
3. Set the channel selector on the TV to the output channel of the Converter Box.
4. Select the channel to be viewed on the Converter Box.

To Play Back a Tape
1. Set the VCR/TV selector to "VCR".
2. Set the TV channel selector to the output channel of the Converter Box.
3. Set the Converter Box channel selector to the VCR output channel (channel 3 or 4).

VCR to Stereo TV Hook-Ups

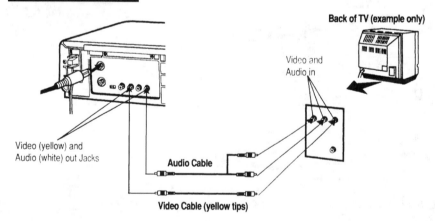

For any additional adjustments you may need to make to your TV, please refer to your TV owner's manual. Cables not included.

■ **5-17** *Connections of VCR/TV to the converter/descrambler box.*

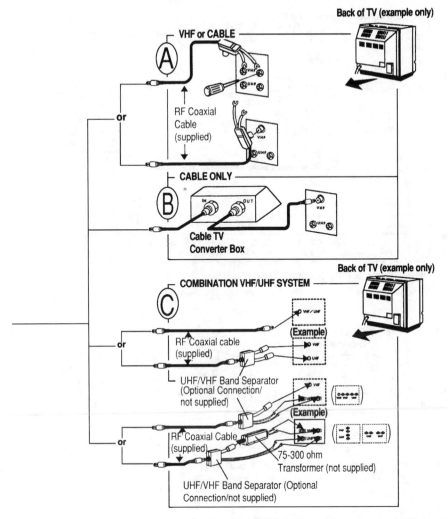

5-18 *More connection configurations for VHF/UHF hook-ups to your VCR and TV sets.*

Illustrated VCR and TV set hook-ups

The drawings in Fig. 5-15 shows the various ways to connect your VCR to the TV receiver. The drawings in Fig. 5-16 show various other hooks-ups you can try. The illustrations in Fig. 5-17 show connection of the VCR/TV to the converter/descramble box. Also, VCR to Stereo TV hook-ups. Figure 5-18 illustrates of various connections for VHF and UHF antenna connections to your VCR and TV sets. The drawings in Fig. 5-19 show hook-ups without a converter/descramble box and a cable to VCR connections with a converter/descrambler box.

Without a Converter/Descrambler Box

Use this connection if your cable system connects directly to your TV without a converter box.

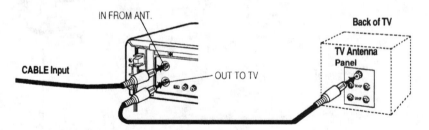

With this connection, you can 1) use your VCR Remote Control to select channels, or 2) program one or more unscrambled channels for unattended recording, or 3) view one channel while recording another.

> **Note:** • You can also use the VCR CHANNEL UP/DOWN buttons to select the desired channel.
> • <u>You cannot record or view</u> scrambled channels with this connection.

With a (Cable to VCR) Converter/Descrambler Box

If your cable service supplies you with a converter box, you can use the basic connection shown here.

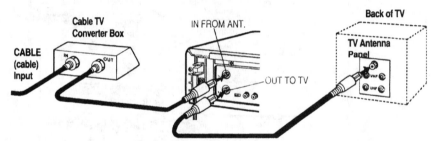

■ **5-19** *Some hook-ups without a converter/descrambler box and a cable to VCR connection with a converter/descrambler box.*

228

Troubleshooting VCR tuners and IF circuits

6

THIS CHAPTER DETAILS HOW TO ANALYZE AND TROU-bleshoot VCR/TV tuner and IF control problems, and how to check sensitivity and tuner/IF gain. Also included is information on some tuner/IF alignment techniques, as well as AGC and AFT operations. The last part of this chapter is devoted to tuner and IF circuit troubleshooting techniques.

For troubleshooting and repairing these sophisticated systems we will be featuring the Sencore VG91 Universal Video Generator.[1]

Today's sophisticated VCR/TV tuning systems are more reliable than the mechanical tuners of yesteryear, yet they still develop problems. No reception, missing channel(s), poor color, poor audio, snowy picture, or herringbone interference can be caused by either the tuning system or a problem in the VCR machine. Sometimes the tuner is sent out for repairs, only to discover the same problem still exists after it is reinstalled.

The Sencore VG91 Video Generator (Fig. 6-1) provides all the accurate RF-IF TV signals and tests needed to fully analyze and troubleshoot them with the VG91.

The modern VCR tuner

Today's tuning systems bring together three technologies: varactor tuners, frequency channel synthesis (phase lock loop-PLL), and microprocessor control. The block diagram in Fig. 6-2 illustrates the three different technologies that come together in a modern tuning system.

Some photos and drawings are courtesy of Sencore, inc.

[1]Information in this chapter courtesy of Sencore

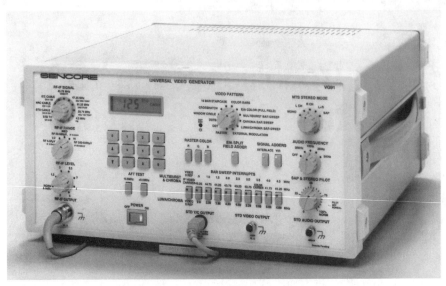

■ **6-1** *The Sencore VG91 Universal Video Generator.*

The varactor tuner selects and amplifies the specified channel and converts it to a lower IF carrier frequency (41.25 to 45.75 MHz). The varactor tuner contains a VHF and UHF section each consisting of an RF amp, a mixer, and an oscillator. Tuned LC circuits before and after the RF amp pass the selected channel. The oscillator mixes with the selected TV channel converting the channel to the IF carrier frequencies.

Today's varactor tuners no longer have moving parts or switch contacts to cause problems. Tuned circuits are formed with coils and varactor diodes. With specific dc voltages, the tuned circuits resonate to select the desired TV channel. "Switching diodes" transfer various coils in-circuit extending the frequency range of the tuned circuits. This enables it to tune through various bands to select VHF and cable channels.

Channel frequency synthesis (PLL)

The tuner's oscillator frequency determines which TV channel is received. The oscillator is set 45.75 MHz above the desired channel, ranging from 100 to 850 MHz. Modern tuning systems use frequency synthesis (PLL) for precise control of the tuner's oscillator.

Frequency synthesis consists of the tuner's oscillator, a crystal reference oscillator, frequency dividers, and a frequency comparator.

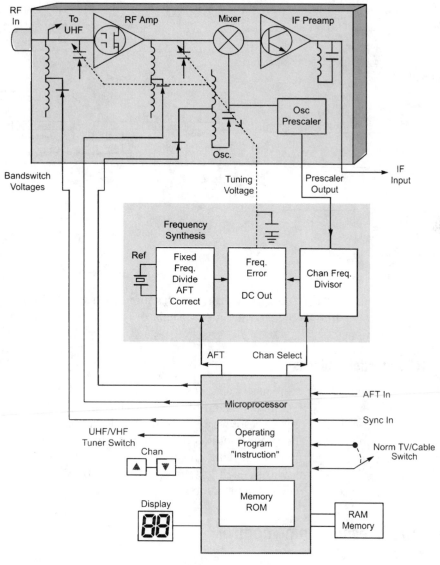

■ 6-2 *A modern TV tuning system includes a varactor tuner, frequency synthesis (PLL), and microprocessor.*

These stages can be found in the varactor tuner or as part of the microprocessor IC.

In a frequency synthesis system, the tuner's oscillator is divided down (prescaled) and then divided again by a factor for that specific channel. The resultant frequency is compared to the fre-

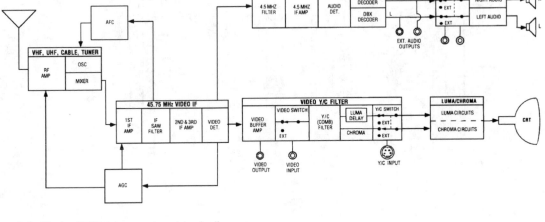

quency of a reference oscillator. Any frequency error results in a dc correction voltage to the tuner's oscillator to restore the correct frequency.

VCR/TV functional blocks

A basic video system has two stages: video and audio. The VCR RF system adds three stages ahead of these stages: Video IF, Tuner, and Audio IF. Figure 6-3 shows a functional block diagram of a TV-Video system. The tuner contains only a few of these operational sections or all five. Identifying these sections and the common functional blocks within each greatly simplifies testing and troubleshooting of VCR-video products.

Microprocessor tuning control

A microprocessor controls the frequency synthesis, AFT, and all tuning functions. It also executes action directed by a program (step-by-step instructions) stored in permanent "read only memory" (ROM). These memories provide instructions and data needed to control frequency synthesis.

When a channel is selected, the microprocessor retrieves and decodes channel information from memory. Then the microprocessor outputs the proper PLL divisors and bandswitching voltages. Random-Access-Memory (RAM) is used to store changing tuning information such as channel tuning offsets, channel scan memory, and last channel tuning information.

The receiver's Normal TV/Cable TV selection changes the microprocessor's tuning instructions. Off-air TV channels are always within 10 KHz of allocated frequencies, while cable channels can be shifted as far as 2 MHz according to HRC or ICC cable formats. To tune-off-air TV channels, the microprocessor instructs the PLL to set the tuner's oscillator 45.75 MHz above the channel frequency. AFT voltage provides the precise tuning correction.

To tune cable channels, the microprocessor increments the tuner's oscillator through several searches 2 MHz above and below the cable channel's allocated carrier frequency. These searches are done when a channel is selected, or when AFT indicates the loss of an IF carrier (on some chassis). AFT and sync inputs to the microprocessor identify a video channel during the search.

Performance tests with the VG91

Many technicians are quick to blame problem symptoms on the varactor tuner. But problems in the IF, video detector, sync, AGC, AFC, and frequency synthesis circuits can cause the same symptoms as a bad tuner. Before you begin troubleshooting, it's best to do a full tuning system performance test. A comprehensive performance test provides you with a clear understanding of all the symptoms so you can troubleshoot the real defect—saving you time and money in the long run.

The first step in troubleshooting any tuner-RF-video system is to confirm symptoms or customer complaints by performance testing. Performance testing also should be the final step after you repair the chassis to ensure total operation. The VG91 provides modulated reference signals so that you can observe the operation at the outputs using the system's monitor or oscilloscope.

To reliably performance test the system you need to apply a known good signal at a known level. Over-the-air and cable signals vary in signal level, video content, and signal quality. These variations make it extremely difficult to properly evaluate the system's performance. The special pattern's adjustable RF-IF levels, AFT test frequencies, and standard outputs supplied by the VG91 allows you to evaluate the operation of each functional block common to TV/VCR video systems.

The VG91 performance tests fall into two general categories: testing receiver tuner and IF circuits and testing video and audio processing circuits. The connections between the VG91 and the video system will differ depending on which category you are testing.

For example, to test the receiver portion (tuner/IFs) of a VCR or TV you would apply RF signals to the tuner input. To test the audio and video processing circuits of a video monitor or VCR you would apply a signal to the VIDEO and AUDIO LINE IN jacks of the system. A full performance test confirms:

☐ VHF/UHF reception.

☐ Cable reception.

☐ RF-IF sensitivity or gain.

☐ AGC operation.

☐ AFT operation.

The VG91 universal video generator gives you all the signals you need for completely performance testing tuners. To see how it is done, refer to Fig. 6-4 for a run-down of these tests.

TUNER/IF PERFORMANCE TESTS			
TEST	**INSTRUCTIONS**	**EXPECTED RESULTS**	**CIRCUITS TESTED**
Standard TV Tuner Test	Input Each VHF/UHF Channel at 1000 μV	Good Video Output On Each Channel	Tuner, Tuner Control
Cable Tuner Test	Input Each Cable Channel At 1000 μV	Good Video Output On Each Channel	Tuner (Cable Channel Search Control)
RF-IF Sensitivity	Set RF Range To "MED." Reduce Level From 5-2.	Video Becomes Noisy, Sync & Color Locked	RF/IF Gain, Sync, & Color
AGC	Set Range To "HI". Increase Level To 5.	Little Or No Change To Video Level	AGC Circuits
AFT	Set Level To 1000 μV, Select EIA Color, Perform AFT Tests	AFT Restores Proper Video To + Or – Shift	AFT Circuits
IF Video Frequency Response	Select Multiburst Bar Sweep	Stripes To 3.0 Bar, Stripes Fade In 3.5-4.0 Bars. Roll-Off In 3.0-4.5 Bars.	IF Alignment
IF Color Bandpass Response	Select Chroma Bar Sweep	Stripes Fade In 4.0 Bar. Roll-Off In 3.5 & 4.0 Bars	IF Alignment
Video Detector	Select 10 Bar Staircase	10 Linear Steps, No Compression	Video Detector Alignment
Audio	Select 300 Hz Audio, Select Mono Or L&R Stereo Mode	Good Audio Output On Both Channels, Stereo "ON" (MTS Receiver Only)	Audio IF, Det., MTS/Sap Decoder

■ **6-4** *VCR Tuner/IF performance tests information.*

The first performance test checks the tuner's ability to receive standard off-air VHF and UHF TV channels. The set-ups for the tuner/IF VCR performance tests are shown in Fig. 6-5. To test VHF/UHF reception using the VG91, hook the RF-IF cable to the

234

antenna input, select the VG91 channel, select video and audio test signals, and adjust the RF output level to 1,000 mV. Now set the tuner channel to match the VG91 (select the "Normal TV" mode), observe the picture on the CRT, and listen to the audio. Do a quick check of channels in the low, middle, and top end of the VHF-low, VHF-high, and UHF bands. Switch through channels 2, 6, 7, 13, 45, and 69 for a quick test. If you see problems with any of these channels, check all of the VHF and UHF channels.

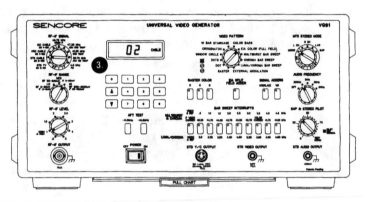

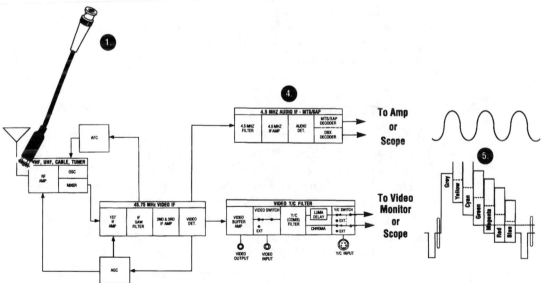

■ **6-5** *Test set-up for tuner/IF performance tests.*

Cable reception

If the tuner is capable of receiving "all cable" channels, you should make several additional performance tests. Varactor tuners equipped to receive cable channels have several additional tuning bands, and need to locate and lock onto offset cable channels. It is possible for the tuner to receive VHF and UHF channels, but not receive some cable channels. You should always test these tuner capabilities to avoid overlooking hidden problems or unseen symptoms.

To test cable channel reception, set your VG91 to "STD CABLE" mode, and match the tuner channel to the VG91. Then observe the video output again to confirm reception. You will need to test cable channels in the middle and extreme edges within each cable tuning band. If you see tuning problems in the receiver, check all the channels and look for similar symptoms.

To check the tuner's ability to search and lock to cable channels, offset your VG91 to "HRC CABLE" or "ICC CABLE." When performing these tests, tune the channel on the VG91 before you select the receiver channel. This sequence is important because many tuning search routines return to the cable channel center frequency when no signal is detected. This causes improper tuning to HRC or ICC carriers.

The receiver should produce a good picture for each of the channels tested. Test all channels noting which ones are bad. Check the service literature to see if channels relate to a particular band or location in each band.

RF-IF sensitivity or gain

An important, but often overlooked receiver performance test is RF-IF sensitivity. The tuner must select and amplify TV signals to provide clear video pictures. Problems in the antenna input, RF amplifier, or AGC circuits can reduce tuner gain or cause excessive signal loss. Severe problems cause very snowy reception. Marginal problems can provide good pictures at higher-than-normal signal levels but cause snowy pictures at the customer's home with typical signal levels of 1,000 mV.

To test RF-IF sensitivity, set your VG91 and tuner to the same channel. Starting with the VG91 output at 1,000 mV, decrease the RF output to 500 mV and observe the CRT. All receivers should produce an acceptable video picture at the initial 1,000-mV level. Most receivers will begin to show snow at 500 mV but should maintain proper sync and color. If the video is snowy at the 1,000-mV level, the receiver has low gain or excessive signal loss in the RF or IF signal path.

Acceptable tuner input signal levels can vary from less than 5,000 mV to over 5,000 mV. Many receivers use automatic gain control (AGC) to maintain a relatively constant video detector output and prevent overdriving of RF and IF stages. A strong input signal produces a dc voltage from the AGC circuits to reduce the gain of the RF and/or IF amplifier stages. Problems with the AGC circuits can improperly reduce gain causing snowy reception or fail to reduce gain with strong signals causing overloading problems.

To test AGC operation, once again set your VG91 and VCR tuner to the same channel. Starting with the VG91 output at 1,000 mV, increase the RF output to 5,000 mV and observe the CRT. If the receiver's AGC is functioning properly, little or no change should be seen in the video display. A large change in picture brightness, contrast, or unstable sync indicates an AGC circuit problem.

AFT operation and troubleshooting

Proper alignment and operation of the AFT circuits are vital for proper tuning and operation of the receiver. AFT circuits monitor the 45.75-MHz video-IF carrier frequency. When the tuner's oscillator is at the correct frequency, the incoming TV channel mixes and produces a 45.75-MHz carrier.

Problems in the AFT circuits can cause improper tuning, or AFT lock-in range problems. If the AFT detector is not tuned to the 45.75-MHz carrier the AFT will pull the fine tuning off frequency instead of on to the correct frequency. If the AFT tuning is not properly centered the AFT lock-in range might be inadequate.

Troubleshoot and align the AFT circuits with the VG91's MHz video-IF signal connected to the input of the IF stages. Use a DVM to measure the AFT voltage returned to the tuner. The VG91's accurate 45.75-MHz IF carrier centers the tuner AFT voltage within the AFT correction voltage range while the AFT TEST shifts the VG91's IF carrier to simulate a drift in the local oscillator. The AFT circuit should respond to the AFT TEST shifts by a change in the AFT voltage returned to the tuner.

The AFT circuit produces a specific dc voltage when the tuner's oscillator is properly set. If the oscillator drifts in either direction, so does the video-IF carrier frequency. The frequency change causes an AFT voltage change back to the PLL control. Based on the AFT voltage, the microprocessor moves the PLL (tuner's oscillator) accordingly to recover the 45.75-MHz IF frequency and normal AFT voltage. If the AFT circuit is defective or improperly aligned, tuning errors result.

The incoming RF signal mixes with the local oscillator signal inside the tuner to produce a 45.75-MHz IF carrier. If the local oscillator frequency drifts the IF carrier frequency will be incorrect and the signal will not pass through the IF stages properly. VCR tuners use an automatic fine-tuning (AFT) system to maintain the correct IF carrier frequency. The AFT circuit senses a frequency change in the IF carrier and applies a correction voltage to the local oscillator.

The AFT circuits in digital tuners are always active and cannot be switched off. The AFT works with the digital tuner to center the 45.75-MHz IF carrier in the IF passband. The AFT circuits must be properly aligned and working for a digital tuner to tune correctly.

The AFT voltage should change approximately equal amounts but opposite polarity to the "+" and "−" carrier shift. Unequal voltage

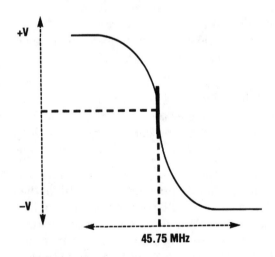

■ 6-6 *The AFT voltage should be centered at 45.75 MHz and change equally to minimum and maximum levels with the VG91 AFT tests.*

changes to the AFT TEST shifts indicates improper AFT centering and the need for alignment. No change in the AFT voltage indicates an AFT problem. Note Fig. 6-6.

If the AFT detector is severely misaligned, inject the VG91's 45.75-MHz Video IF carrier and monitor the DMV while adjusting the AFT detector coil. Determine the coil setting that produces the fastest voltage change between minimum and maximum. Then adjust the coil to center the output within this maximum and minimum voltage change.

As a final test of complete AFT operation, apply an RF signal to the antenna and use the AFT TEST to shift the RF carrier. Again monitor the AFT voltage with a DVM. The AFT voltages should be approximately the same as those observed when the IF 45.75 MHz was used. A large difference indicates a problem with the tuner.

"Time out" to analyze symptoms

As previously noted, modern VCR channel varactor tuners are influenced by the AGC, IF, AFC, video detector, and frequency synthesis (PLL) circuits. A problem in any of these circuits can cause reception or tuning problems. An important first step in troubleshooting these problems is performance testing with an accurate all-channel TV generator. This lets you clearly understand the symptoms, saving you time and money in the long run. You should performance test VHF/UHF reception, cable reception, tuner sensitivity or gain, AFT operation, and AGC operation. Note the VCR/TV troubleshooting tuner system guide in Fig. 6-7.

Once you have fully analyzed all symptoms, you are ready to begin troubleshooting. The symptoms will place the problem in one of three categories:

☐ System control problems (microprocessor or RAM memory)

☐ RF-IF gain problems (signal loss or reduced gain in RF-IF path)

☐ Improper tuning (tuner's oscillator frequency error)

Troubleshooting system control problems

Problems with the system control (microprocessor) usually alter many control functions. To identify system control problems, first analyze all the control functions. Check volume, video adjust-

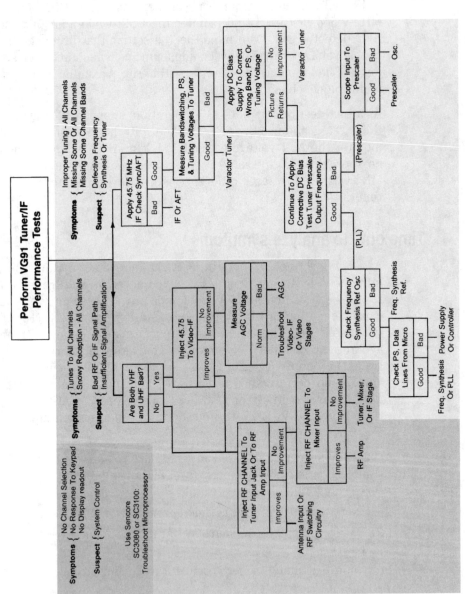

6-7 Guide to troubleshoot TV tuning systems with the VG91 Universal Video Generator.

ments, interface selections, and all other operations first. If the microprocessor selects channels, displays channel, performs other operations, and has digital activity to tuning control/synthesis circuits, the microprocessor and its memory are likely to be okay.

Finally, you can look for bandswitching voltage changes between VHF channels 6 to 7 and channel 14. An abrupt voltage change shows the microprocessor is likely providing the proper tuning instructions to the frequency synthesis circuits. If control problems are noted, use the SC3100 scope shown in Fig. 6-8. The "AUTO-TRACKER" and "Delta" features of this oscilloscope are excellent for troubleshooting microprocessors.

241

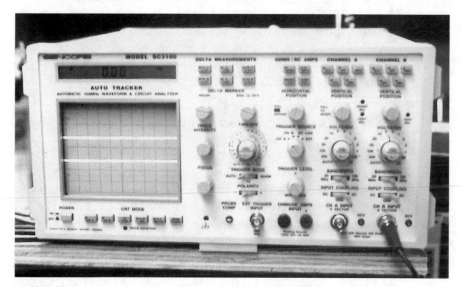

■ **6-8** *Photo of the Sencore SC3100 "AUTO TRACKER" scope.*

Gain problems in the RF and IF stages should be distinguished from improper tuning problems. If the receiver properly tunes to each channel, but the picture is noisy, the microprocessor, frequency synthesis system (PLL), and AFT circuits are working. There is no need to test tuning control circuits or voltages. The problem is a lack of signal gain or excessive signal loss in the RF and/or IF signal paths.

The inability to tune to all or some of the station channels indicates a problem in the tuner, frequency synthesis, or video IF/AFT

Troubleshooting system control problems

stages. These problems are the most difficult to isolate because you must check IF, sync, and AFT circuits. You will need to substitute accurate RF-IF signals with the VG91 universal video generator and use conventional troubleshooting techniques to isolate the defect. RF-IF troubleshooting has never been easy.

Troubleshooting RF-IF gain/problems

The VHF and UHF sections of the varactor tuner use separate RF amplifiers, oscillators, and mixer stages. Before troubleshooting RF-IF gain problems, consider if both VHF and UHF reception is being affected. Be sure to connect the RF cable to the proper input if the receiver has separate UHF and VHF inputs. If either the VHF or UHF reception is good while the other is poor, you have isolated the problem to one tuner section and at the same time determined that the IF and AGC circuits are good. If both have snowy reception, the problem is common to both sections. The problem could then be in the video IF, AGC, antenna input, or tuner-IF pre-amp stage.

To isolate gain problems, start by injecting your VG91's 45.75-MHz Video IF signal at the IF input of the tuner module. The normal IF level to the first IF amplifier transistor is approximately 1,500 mV and 20,000 mV to the SAW filter. Set the VG91's IF output to match these normal circuit levels when injecting into these circuit points. If good video returns, you have confirmed the problem is either the antenna input or varactor tuner. If there is no improvement, a gain problem exists in the video-IF or video stages.

The AGC voltage commonly controls the gain of an IF stage and the tuner's RF amplifier. While substituting the 45.75-MHz Video-IF, use a dc volt meter to check the AGC voltage and verify proper operation. Adjust the VG91's RF-IF LEVEL control just above the point where the picture becomes snowy. Slowly increase the RF-IF LEVEL control and observe the changing AGC voltage.

Compare the AGC voltage to the VCR schematic. Most tuning systems have AGC voltages of five to eight volts at low input signal levels (full gain). AGC voltages fall to approximately one volt to reduce gain of RF and IF amplifiers as signal levels increase. A voltage that does not change or is continually low indicates an AGC problem.

If the IF and AGC circuits test good, the next step is to determine if the problem is in the varactor tuner or antenna input signal paths. Apply the VG91's RF channel signal at 1,000 mV directly to the input connector on the varactor tuner. You might need to use the RCA type adapter cable supplied with the VG91 to adapt the RF-IF test cable to the tuner's RCA input jack. If a noise-free picture returns, the problem is in the antenna input connections, circuits, or cable.

If this injection does not restore good video, you have isolated the gain problem to the varactor tuner. You might now elect to have the tuner repaired by a tuner repair service or you can use the VG91 to further isolate the problem.

In many instances you can pinpoint the problem within the tuner and make the repair. To isolate the defective tuner stage, use the VG91's RF output and RF-IF troubleshooting balum to inject an RF channel signal at the mixer input. Use the "Lo" position of the RF-IF RANGE switch. If good video returns, the problem is in the RF amplifier or tuned circuits. If you see no improvement, the problem is in the mixer or IF amplifier stage.

Troubleshooting tuning problems

The first step in isolating tuning symptoms is to isolate the problem to the tuner/tuner control circuits or IF/AFT/sync stage. Because IF/AFT/sync circuit problems can alter tuning operation, you must first check these stages. Use the process of elimination to prove the IF circuits plus the AFT, AGC, and sync feedbacks to the tuning system are not causing the PLL to mistune. The last step is to use the accurate TV RF signals from the VG91 video analyzer along with conventional troubleshooting methods to isolate the defect.

Start by injecting the VG91's 45.75-MHz Video-IF signal at the video-IF output of the tuner module. Refer to test set-up in Fig. 6-9. If good video returns, you have confirmed the problem is tuning related. If the video does not improve, a problem exists to the video-IF, AFT, or video stages—but not the tuner.

Once you have confirmed the symptoms are tuning related, the next step is to test the sync and AFT circuits. By using the process of elimination, you can determine if any of these stages are the cause of tuning problems.

Continue to substitute the 45.75-MHz video IF signal with the VG91 while using an oscilloscope to check for proper vertical and horizontal sync to the tuner control circuits. Do not assume that because the receiver produces a good video picture, the tuner is receiving proper sync. Some receivers use separate sync detection circuits to feed the microprocessor. Refer to the basic test set-up in Fig. 6-10.

Next, connect a dc volt meter to measure the voltage present at the AFT test point. Compare the AFT voltage to the schematic information. The AFT voltage typically centers at 3.5 to 4.5 volts. Alternately push and hold the VG91's AFT TEST buttons and note the AFT voltage change. You should see near equal, but opposite changes in the AFT voltage as you alternate between the AFT TEST buttons. The AFT voltage typically falls to about 0.1 to 1.5 volts or rises to 6 to 8 volts as the AFT TEST buttons are pushed.

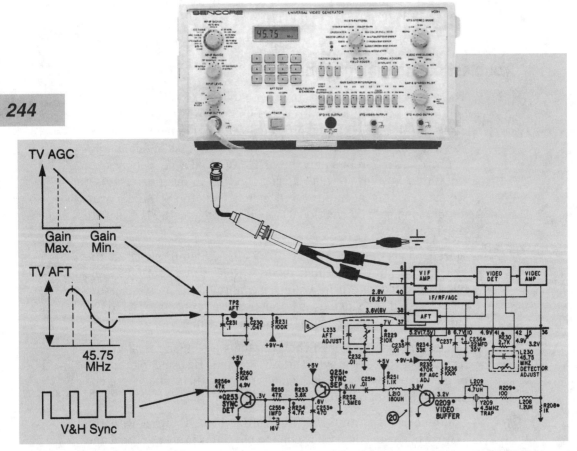

■ 6-9 To isolate tuning symptoms, substitute the accurate 45.75-MHz video-IF signal of the VG91 and test for proper IF, sync, AFT, and AGC operation.

Troubleshooting VCR tuners and IF circuits

If the voltage is not centered, does not change, or changes in only one direction, an AFT problem or misalignment is probable.

Many modern tuning systems use a comparator circuit to interface the AFT voltage to the tuning control microprocessor. You should make a final check of the AFT voltage change at the input to the tuner control microprocessor. This can find problems with AFT voltage paths or components that interface the AFT voltage to the tuner control microprocessor.

Once you have tested the video-IF stages and determined voltages and signals returned to the tuner control microprocessor are correct, you are ready to analyze the frequency synthesis and varactor tuner. To do this, you will need to measure power supply,

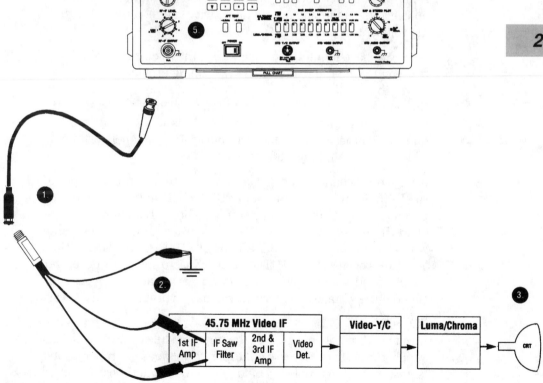

■ **6-10** *Basic test set-up for injecting into the IF stages.*

bandswitching, and tuning voltages applied to the varactor tuner and analyze the tuner's "prescaler" output.

You will find two common variations to the location of the frequency synthesis (PLL) circuits. Many tuner control microprocessors ICs have the frequency synthesis stages inside the IC. In these tuning systems, bandswitching and tuning voltages are available at the input pins to the varactor tuner. The varactor tuner contains a frequency divider prescaler that divides down the frequency of the tuner's oscillator. The prescaler output is returned to the frequency synthesis circuits and is available on an output of the varactor tuner.

In some tuning systems, frequency synthesis is contained inside the varactor tuner enclosure. The microprocessor or tuner control IC inputs data and control signals, which are decoded by the frequency synthesis IC inside the varactor tuner. In this system, bandswitching voltages are internal to the varactor tuner, but a test pin is provided for measuring the tuning voltage. To measure bandswitching and prescaler frequencies, you must remove the tuner's metal enclosure. If you do not, you are limited to checking enable and data inputs for digital activity and monitoring the tuning voltage test point.

Once you have found the frequency synthesis circuits and voltage test points, you are ready to analyze tuning problems. Set the receiver to the "Normal TV" tuning mode (if symptoms permit). In this mode, tuning systems synthesize the center channel frequency—then implement AFT action. This lets you analyze frequency synthesis operation without cable channel searching, which alters symptoms and measurements.

Apply an RF channel signal from the VG91 to the antenna input of the VCR. Set the VCR tuner and VG91 to channel 2 (if symptoms permit). This channel permits you to compare readings to voltage charts in the service literature and make waveform measurements with a scope such as the SENCORE SC3100 "AUTO TRACKER." You can also use the "AUTO TRACKER" to measure dc power supply, bandswitching, and tuning voltages applied to the varactor tuner. If the voltages are normal, you have a problem with the varactor tuner.

If voltages are not correct, test the tuner module by using a dc bias supply to substitute for any improper dc voltages. If proper tuning is restored, the problem could be in the tuner's oscillator or prescaler or in the frequency synthesis stages.

The first step in troubleshooting the frequency synthesis circuits is to test the tuner's oscillator and prescaler output. This determines if the PLL problem is in the varactor tuner. Because the PLL is a closed circuit loop, a problem here can cause all the frequencies and voltages in the frequency synthesis to be defective, including the tuner's oscillator and prescaler output. You will need to correct the defective tuning voltage with a dc bias supply to analyze the tuner's oscillator and prescaler.

To test the prescaler inside the varactor tuner, substitute with the dc bias supply to restore proper tuning. Use an oscilloscope to test for an output with the proper frequency from the prescaler. A missing or improper output indicates a problem with the tuner's prescaler or oscillator. A proper output indicates these stages are good and that the problem is elsewhere in the frequency synthesis.

Because the frequency synthesis circuits are usually part of a microprocessor or IC in the varactor tuner, the number of discrete stages, components, and test points are minimal. To isolate PLL problems, first check the presence and frequency of the reference PLL oscillator. This might be the main clock crystal of the microprocessor or tuner control IC. Also check for proper voltages to the PLL circuits and digital data or clock input.

If the varactor tuner signals are normal, or applying proper dc voltages to the varactor tuner fails to restore proper tuning, the tuner is defective. You will have to decide if you are going to attempt to repair the tuner, or send it to a tuner repair specialist.

IF response testing

Note that the bandpass requirements differ between VCR/TV-RF tuners and receivers. A large-screen television typically has a bandpass of 3.5–4.0 MHz to display good picture detail. Small-screen televisions usually have about 3 MHz of bandwidth, while a black-and-white receiver has only a 2–2.5-MHz bandwidth. (The narrow bandwidth in a black-and-white receiver is necessary to prevent the color signals from interfering with the luminance signals.)

IF bandwidth is determined by the alignment of LC tuned circuits or by the fixed response of a SAW filter. LC tuned circuits easily drift or change alignment due to thermal or mechanical stresses. SAW filters also become defective and drift.

The VG91's multiburst bar sweep video pattern quickly tests the complete response of video-IF stages. This pattern provides various video frequencies that allow you to test the IF response by visually observing the output of the video detector.

Special circuits and accessories

7

THIS CHAPTER FEATURES SPECIAL VCR CIRCUITS AND ACcessories that make your VCR become more versatile. One accessory is the VCR PLUS+, which is an instant programmer for your VCR. All you need to do with this unit is punch in your PLUS CODE and the VCR is programmed and ready to record.

Other features include automatic VCR head cleaners, freeze frame, speed search, and tape search. This chapter concludes with Mitsubishi's noiseless speed search system.

The VCR PLUS+ system

The VCR PLUS+ unit is used to program your VCR from the codes printed in most TV logs. The unit does it quickly and correctly the first time. It does away with all those program mistakes.

Using the VCR PLUS+

To tape the show of your choice, just enter the PLUS CODE printed in your TV listing next to that program. Then press either: *Once* or *Weekly* or *Daily M - F* to select how often you wish to tape that program. See Fig. 7-1. The display will confirm the date, channel, start time, and length of the program. (The recording length of the show is shown by time bars ——).

How to set the VCR

To set the VCR, first refer to Table 7-1, and look up the number corresponding to the model of VCR you are using. Turn the VCR to channel 3 or 4 and turn it OFF.

Open the VCR PLUS+ cover and follow the following sequence:
1. Press *VCR*.

Using VCR Plus+

1. Punch in the *PlusCode* for the show you want to record.

2. Choose how often you want the show recorded: *Press*.

 or or (DAILY M-F)

The display will confirm the date, channel, start time and length of the show you have just entered.

3. Leave the VCR Plus+ near your VCR and cable box.

At the right time, VCR Plus+ will turn on your VCR, change to the correct channel and tape your show. Just don't forget to insert the blank tape.

Leave your VCR OFF and your cable Box ON

■ **7-1** *How to use the VCR PLUS+.*

VCR brand list with codes.

VCR Brand Name	Code		
Akai	15	Panasonic	03
Audio Dynamics	11	Pentax	38
Broksonic	36	Philco	03
Canon	03	Philips	28
Citizen	22	Pioneer	21
		Quasar	03
Craig	32		
Curtis-Mathes	03	RCA	01
Daewoo (Daytron)	16	Realistic	34
DBX	11	Samsung	23
Emerson	05	Sanyo	20
		Scott	26
Fisher	04		
Funai	14	Sears	06
General Electric	02	Sharp	19
Goldstar	22	Shintom	17
Hitachi	21	Signature	03
Instant Replay	03	Sony (VHS, Beta)	13
J C Penney	25	Sound Design	14
JVC	12	Sylvania	28
Kenwood	33	Symphonic	14
Lloyds	14	Tashiko	22
Magnavox	07	Tatung	35
Magnin	23	TEAC	14
Marta	22	Teknika	18
Marantz	27	Thomas	14
Memorex	29	TMK	37
MGN	23	Toshiba	10
Minolta	21	Vector Research	27
Mitsubishi	08	Video Concepts	27
Montgomery Ward	19	XR-1000	14
MultiTech	14	Yamaha	31
NEC	11	Zenith	09

2. When the display shows *VCR* press the 2-digit code you looked up in Table 7-1 (e.g., 01 is for RCA). Point the VCR PLUS+ at your VCR and press *Enter*. VCR PLUS+'s red light will flash while it is sending a test signal to your VCR.

3. If your VCR turned ON and changed to channel 09 press *SAVE*. Set clock. If your VCR did not turn ON or turned ON but did not change to channel 09 press *ENTER* again. Wait until the red light stops flashing. VCR PLUS+ is sending the next possible VCR code. If your VCR turned ON and changed to channel 09 press *SAVE*. If your VCR did not turn ON or turned ON but did not change to channel 09 press *ENTER* again until you find the code that works for your VCR. Then

Step

2

Set Clock

···

1. Press CLOCK

2. When the Display shows:

YR	Press year (for example 90), then press **ENTER**.
MO	Press month (for example 07 is July), then press **ENTER**.
DA	Press date (for example 01 for the 1st) then press **ENTER**.
Hr	Press hour (for example 02 for 2 o'clock), then press **ENTER**.
Mn	Press minute (for example 05 for 5 minutes), then press **ENTER**.
AM/PM	Press 1 for AM or 2 for PM,

The display will show **SAVE**. Wait for a few seconds and the display will show the current time and date that you have just entered.

Note: You no longer need to set the clock on your VCR.

■ **7-2** *How to set the clock with VCR PLUS+.*

press *SAVE*. (The display shows *END* if you have tried all possible VCR codes for that model. If so, press VCR code 00 and then *ENTER*. Repeat steps in 3. above to try all possible codes, one at a time.)

To set the clock refer to Fig. 7-2. To set the cable mode refer to Fig. 7-3.

Setting cable channel numbers

For some viewers (probably because they have cable), the channels listed in their TV book are different from the channels on their TV at home. If you are one of these viewers, you need to tell VCR PLUS+ that your channels are different. You only need to tell VCR PLUS+ once. If your channel numbers change in the fu-

1. Press (CABLE)

2. When the Display shows:

CA B-   Press the two-digit code you looked up in Table 1, point VCR Plus+ at your Cable Box (converter) and press *(ENTER)*.
VCR Plus+ red light will flash while it is sending a test signal to your Cable Box.

3. **IF your Cable Box changed to Channel 09:** press *(SAVE)*. (Continue to Step 4 : Set Cable Channel Numbers)

IF your Cable Box did not change to Channel 09: Press *(ENTER)* again, wait until the red light stops flashing. VCR Plus+ is sending the next possible code.

IF your Cable Box changed to Channel 09: press *(SAVE)*.

IF your cable box did not change to Channel 09: press *(ENTER)* until you find the code that works for your Cable Box, then press *(SAVE)*.

(The display shows END if you have tried all possible Cable Box codes for that brand. If so, press Cable code 00 and then *(ENTER)* . Repeat steps in 3 above to try all possible brands' codes, one at a time.)

■ **7-3** *Setting VCR+ when using cable.*

ture, you only need to tell VCR PLUS+ the ones that have changed.

Additional explanation

Your TV guide will have a chart indicating the channel number assigned for VCR PLUS+ to each cable and broadcast channel (e.g., HBO, CNN, ABC, CBS, NBC, etc.) Refer to Fig. 7-4 for more examples.

For example: Suppose your TV book has assigned channel 14 to HBO but your cable company delivers HBO on channel 18. Since the channel numbers are different, you need to use the channel button. Do the following:

(1) Press ⬭ CH ⬭ (the two blank spaces under the display "Guide CH" will flash).

(2) Press *14*. (now the two blank spaces under the display "TV CH" will flash).

(3) Press *18*.

(4) Press ⬭ENTER⬭. Repeat Steps 2-4 for each channel that is different.

(5) Press ⬭SAVE⬭.

Note: After you have saved your setting, you may review your settings by pressing ⬭ CH ⬭ and then ⬭REVIEW⬭. Keep pressing ⬭REVIEW⬭ to move through the channel settings.

To change a channel setting repeat steps 1, 2, and 3 on the left hand page. (Refer to the example in the Supplemental Operating Instructions.)

■ 7-4 *Block diagram of index signal record circuits.*

253

Display messages

The following is an explanation of VCR PLUS+ display messages:

LOW BAT: It's time to change the batteries. All four AAA batteries must be replaced at the same time.

ERR: ENTRY: You have entered an invalid entry during setup.

ERR: CODE: The Plus Code number you have entered is not a valid number. Check your TV listing and reenter the number. The TV listing also could have an error.

ERR: DATE: There are several reasons why this message will appear:

☐ You might have tried to select a daily recording (Monday to Friday) or a Saturday or Sunday program.

☐ You might have tried to select weekly or daily recording for a show more than 7 days ahead. VCR PLUS+ only allows the weekly or daily recording option to be used for the current week's programs.

☐ You might have tried to enter a program that has already ended.

FULL: There are no programs entered in VCR PLUS+.

CLASH: The show you have selected overlaps with a previously entered show. If you want to tape the show you have selected, first cancel the previously entered show. Use the *REVIEW* button to locate and *CANCEL* the previously entered show or program.

Panasonic quick play

Some VCRs keep the tape loaded at all times. By keeping the tape fully loaded around the head cylinder and maintaining rotation speed in the stop mode, the Panasonic quick play mechanism helps reduce the response time from stop to play, and from play to rewind search. Transitions are quick (approximately two seconds) and very smooth.

Twin tuner systems

Some combination VCRs are equipped with a built-in twin tuner system featuring two separate full-channel digital quartz tuners. That means you can record one channel while watching something else on another channel. A push of the monitor key on the remote unit will show which channel is being viewed and which is being recorded on the screen, along with the time and tape counter display. Also, an auto set feature automatically tunes in only those channels being received in your area or on your antenna system. No manual tuning is necessary.

Auto head cleaners

A clean video head helps produce a clear picture and extends the life of the videotape and cylinder heads. You will find that some brands of VCRs have a built-in head cleaning system that maintains the video heads, automatically, upon insertion and ejection of a videotape.

Noiseless speed search systems

Home VCRs usually produce noise bars across the screen when in the forward and reverse search modes. The number of noise bars is directly proportional to the tape speed during search, with the number of bars increasing as the tape speed increases.

Mitsubishi has come up with a way to get rid of these noise bars. A set of video heads was added to the head drum that moves up and down relative to the circular face of the head drum as the drum rotates and the tape is pulled past the head drum.

How noise develops

Noise occurs because, as the tape travels horizontally, the head drum spins in a direction at an angle to the tape travel, tracing out the diagonal "stripes" of video information.

In the speed search mode, the tape travels at a much faster speed, so the video heads scribe a path across the tape that is much closer to vertical than diagonal. Thus, instead of tracing out the recorded stripe of information, the head traverses more than one video track.

Because a video head will only pick up signals from a video track that is at the same azimuth angle, the left azimuth head picks up a signal only when crossing a left azimuth track. When crossing a right azimuth track, only noise is produced. Conversely, the right azimuth head generates noise when crossing a left azimuth track. The absence of a signal from scanning opposing azimuth tracks produces the noise bars in the picture when the VCR is in the speed search mode.

Moving heads

The two speed-search movable heads are mounted on the rotating drum. In these VCRs a total of ten heads are mounted on the drum. The normal play/record two-hour and six-hour video heads are mounted in the conventional 4 × head configuration. The hi-fi audio heads are positioned 120 degrees ahead of the video heads.

In this VCR, the hi-fi head assemblies also house the flying erase heads, which minimize color distortion at the start of a recording and at points during the recording where record pause is activated.

The two new movable heads are 180 degrees apart and positioned midway between the hi-fi and video heads. Both heads are left azimuth and therefore will pick up signals only from a left azimuth video track.

The head is mounted on an actuator assembly within the drum. Current through the actuator coil moves the head vertically, controlling the height of the head. There is one actuator for each moving head.

The drive signal for the moving head actuators is coupled to the rotating drum through a brush contact assembly mounted on the top of the drum.

Both movable heads are left azimuth, and noise is produced when crossing right azimuth tracks. This results in noise bars similar to those encountered in conventional speed search.

To prevent the movable heads from crossing opposing azimuth tracks, the head path must be altered to correspond to the angle of the video tracks on the tape. This is accomplished by raising the head height at the start of each video track, decreasing the height to its normal position in the center of the track, and then lowering the height as the head moves toward the end of the track. Because the head no longer traverses right azimuth tracks, it does not produce any noise.

Tape search

Some VCRs include a feature that allows you to quickly locate the start of each prerecorded program. This is called *tape search* and is accomplished by recording an index signal for approximately one second at the start of each program. One such circuit is shown in Fig. 7-5. This index signal is a 30-Hz signal recorded across the full width of the tape for approximately one second by the full erase head. The index signal is recorded on the tape each time a recording is initiated, whether the recording is a manual start recording or was initiated electronically by the programmable timer.

Tape search (index signal record) is one of a sequence of events that takes place just prior to the start of each recording. The sequence (or countdown) is somewhat different for the first program (unloaded) than for subsequent (loaded machine) programs where the tape is already in place. With the first programs, the ma-

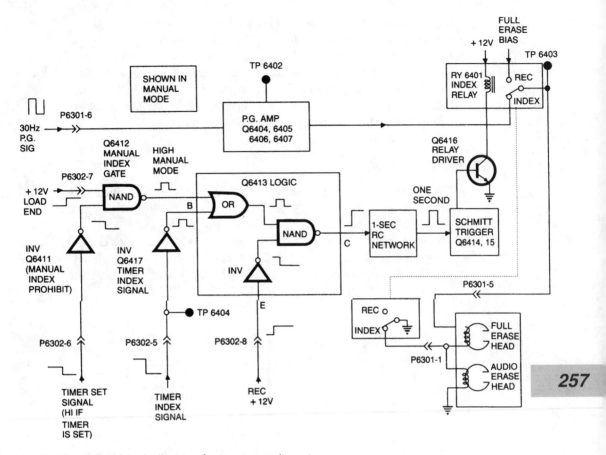

■ 7-5 *Simplified block diagram for tape search system.*

chine starts unloaded. As can be noted during operation of the machine, the timer actually starts the VCR 16 seconds prior to the actual desired starting time to allow the machine to start, load tape, and stabilize at the required constant speed.

This action can be observed at start time minus 16 seconds. The *timer-on* signal is sent to the transport control board. As a result, the VCR starts and loads, which requires about three seconds. Once the machine is loaded, the pressure roller engages and tape transportation stabilizes. At start time minus 12 seconds, the timer sends a signal to the tape search board to record the index signal, which is a 30-Hz signal recorded across the full width of the tape for approximately one second by the full erase head.

The second and subsequent programs start with the machine loaded, so there is a slight difference in the sequence of events.

The main thing to remember is that in order to allow the D-D motor to start without load and to prevent tape damage it is necessary that the operation of the pressure roller, which puts tension on the tape, be stopped for about two seconds to allow the D-D motor to come up to speed. Thus, a pressure roller prohibit circuit, which is timed, is included in the machine to allow about a two-second delay before the pressure roller engages.

Circuit description

The index signal, used to mark the tape in the record mode, is produced by amplifying the 30-Hz PG signal as shown in the signal record block diagram Fig. 7-6. The output of the PG amplifier is applied to a relay that, when energized (keyed on), applies the 30-Hz signal to the full erase head. When the relay is in its unenergized state, full erase bias is applied to the erase head and the series-connected audio erase head.

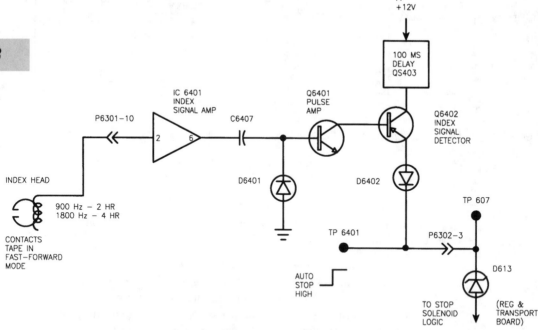

■ 7-6 *Recorded spectrum for original Betamax.*

The remainder of the circuitry controls the turn ON and turn OFF state of relay driver transistor Q6416. As previously noted, there are two modes of operation: one is the (first program on the tape) manual record mode; the second mode of operation is the timed-record mode.

Transistors Q6411 through Q6417, via various base and emitter connections, constitute a logic network that assists in determining when the index signal should be recorded. Whenever the collector voltage of Q6413 (output of NAND gate) goes high, a one-second pulse is produced by an RC network to switch a Schmitt trigger consisting of Q6414 and Q6415 and provide base drive to relay driver transistor Q6416.

The index signal is required only in the record mode, so one of the inputs to the logic circuit is record +12 volts, which is supplied via connector P6302 pin 8. The other input to the NAND gate is a signal from the output of the NOR that indicates that a recording (manual or timed) has been started and a pulse is necessary.

When a manual recording is made, the signal to start recording the index signal is the indication that loading is completed. This is provided by a logic +12-V (load completion) signal that enters the board via connector P6302 pin 5. When the timer set signal is HIGH, the inverted output of Q6411 inhibits the manual index signal, which is fed to the NOR gate part of the Q6413 logic circuit. Thus, manual indexing occurs when the +12-V load completion signal goes HIGH and the timer set signal is LOW so that the inverted output of Q6411 is HIGH.

When the timer is set, and the manual index is prohibited, the timer index signal fed via P6302 pin 5 goes HIGH when the microprocessor determines that it is time to record the index signal on the tape. As the signal at P6302-5 goes HIGH, the output of inverter transistor Q6417 goes LOW and the signal is passed through the NOR gate part of Q6413 logic circuit and into the NAND gate (Q6413 collector), which goes HIGH to initiate the one-second pulse into the Schmitt trigger.

The index signal that was recorded on the tape comes back at a higher frequency when the machine is operated in the fast forward tape search mode. As shown in Fig. 7-5 block diagram, the tape search signal (originally 30 Hz) is approximately a 900-Hz signal in the two-hour mode and approximately 1.8 kHz in the four-hour mode.

The index signal (recovered by the index head is in contact with the tape whenever the machine is in the fast-forward mode) enters the tape search board via connector P6301-10, where it is applied to index signal amplifier IC6401. The output signal of the index signal amplifier is detected by Q6401 and rectified into a dc signal sufficient to drive the base of Q6402. Presence of a signal at the base of Q6402 causes the device to conduct and send a pulse to the stop solenoid logic circuit located on the transport control board to stop the machine.

The emitter of Q6402 receives power via a 100-ms delay circuit. The purpose of the 100-ms delay is to ensure that the fast forward button latches at the start of a search to the next program. If this circuit was not incorporated, it is possible when the fast forward button is depressed to immediately detect the present index signal and trigger the stop solenoid, thereby making it impossible to fast forward to the next program.

Stereo and digital audio circuits

8

THIS CHAPTER DELVES INTO THE CONCEPT, OPERATION, and circuitry of the Sony Super-Beta VCR recording system and Sony's Audio-Beta hi-fi record/playback system. Then information is presented on the high-quality (HQ) and Super-VHS circuits found in some VHS machines. This information is followed by the details and principles of the multi-channel television sound (MTS) techniques now being used in some TV and VCR recorders. I also talk about how the TV stereo sound standard was established and the workings of the multichannel TV sound (MTS) system. The chapter concludes with Sony's digital audio for 8-mm PCM units.

Sony Beta hi-fi system

The original Betamax was a composite video that consisted of luminance and chroma. The chroma occupies a band of frequencies from 0 Hz to 4.5 MHz. Note recorded spectrum shown in Fig. 8-1. The chroma occupies an area within the luminance band centered around 3.58 MHz. To record this composite video information, the luminance and chroma must be separated and dealt with individually. The luminance signal is FM-modulated. This was done to reduce the number of octaves required to record the luminance signal. The luminance signal is recorded by causing an FM modulator to deviate from 3.6 MHz (for sync tips) to a high frequency of 4.8 MHz (for peak white levels). This 1.2 MHz deviation of the FM carrier produces upper and lower sidebands. The width of these sidebands determines the resolution of the signal recorded.

The chroma part of the composite video occupies an area within the frequency spectrum that overlaps the record luminance. To avoid these frequencies from interacting and causing chroma beats, the chroma signal was down-converted to 688 kHz. This combination of down-converted chroma and FM luminance comprises the RF signal recorded by the video heads.

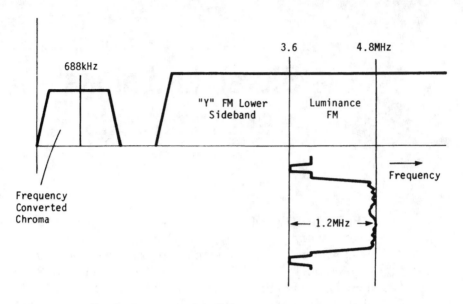

■ **8-1** *Recorded spectrum or original Betamax.*

Recorded spectrum Beta hi-fi

Beta hi-fi is a method that enables the videotape recorder to record left and right channel audio, separately using the video heads with a much higher fidelity than previously possible using the conventional longitudinal tracks. To do this, the left and right channel audio signals were FM-modulated and divided into four pilot audio carriers. Four carriers are required to maintain the separation between the left and right channel audio, as well as to reduce the crosstalk between adjacent tracks of the video information. These four audio carriers are centered about 1.5 MHz and are mixed with the luminance and chroma information to be recorded by the video heads on the tape. See waveform drawing of Fig. 8-2.

The addition of the four audio FM pilot carriers required the FM luminance signal to be shifted upward by 0.4 MHz to make room between the chroma and luminance information for these carriers. The high-frequency limitations of the video heads resulted in a loss of some of the FM sidebands due to this 0.42-MHz shift in the luminance frequency. This caused a slight reduction of the amount of resolution in the picture produced by Beta hi-fi units.

Recorded spectrum Super-Beta and Beta hi-fi

The Super-Beta system overcomes the limited resolution of not only Beta hi-fi but also of conventional Betamax units. This in-

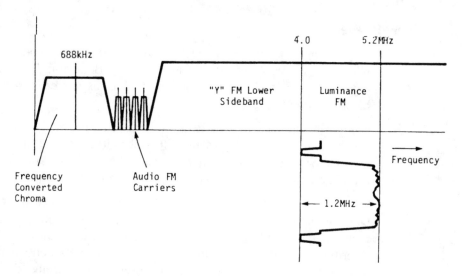

■ 8-2 *Recorded spectrum for Betamax hi-fi.*

creased resolution is achieved by using narrower gap heads that results in an improved high-frequency response, and a 0.8-MHz upward shift of the FM luminance carrier results in a larger lower sideband. The increased bandwidth of the total luminance signal results in resolution greater than achieved by both Beta hi-fi and the conventional Betamax system. Refer to waveform drawing of Fig. 8-3.

263

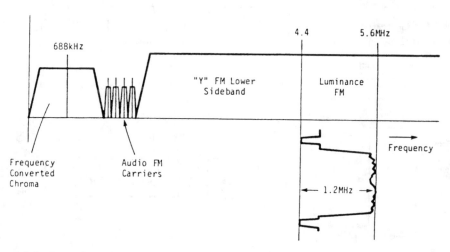

■ 8-3 *Recorded spectrum for Super-Beta and Beta hi-fi.*

Super-Beta record block

Few changes to the conventional video processing chain are needed to record Super-Beta (highband) or Beta hi-fi. They are highlighted as the overall record system is explained. Use the block diagram in Fig. 8-4.

Composite video from the VCR tuner or the rear panel line input is selected for the AGC stage. The AGC stage either increases or decreases its gain to produce a video signal with a constant amplitude horizontal sync pulse, and therefore a constant output level.

The composite video leaves the AGC stage and takes two paths. One path is through the E-E RFU video output line for an external monitor to display the recorded picture. The other is into a comb filter that separates the luminance and chroma components of the composite video signal and outputs them. The chroma down-converting process is conventional and is not explained here. When receiving composite video from another VCR, the EDIT switch can be engaged to negate the effects of the comb filter, permitting individual bandpass filters afterwards to separate the composite video for a clearer, edited recording.

264

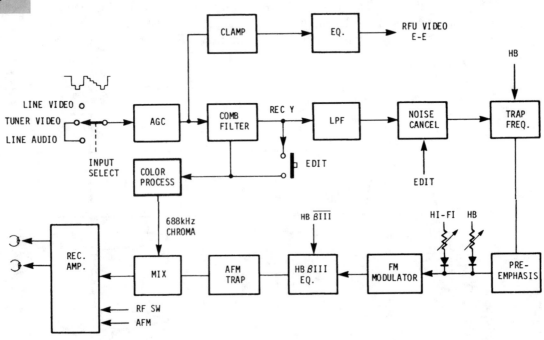

■ **8-4** *Super-Beta record block diagram.*

The lowpass filter restores an overall flat frequency response of the luminance signal that was altered by the previous comb filter stage. The luminance is acted upon by the noise cancel stage that follows. When the EDIT switch is engaged, less noise cancellation is achieved in this stage. This permits the full bandwidth of the playback signal to pass during the recording EDIT mode.

The luminance signal's upper frequency limit is fixed by a trap in the next processing stage. During high-band or Super-Beta recording, a wider bandwidth is desired. Therefore, a higher frequency trap to raise the upper frequency limit is selected with the front panel Super-Beta switch.

With the amplitude and frequency limits of the luminance signal now controlled and therefore known, an increase in gain at higher frequencies is brought about in the preemphasis stages. Upon playback, the high frequencies are returned to normal levels in the de-emphasis stages. This boost in record and reduction in playback reduces the loss of detail during the record/playback process.

The FM modulator changes the AM luminance signal to an FM signal containing upper and lower sidebands. Beta hi-fi and Super-Beta modes add a voltage to the incoming luminance signal to shift the carrier frequency of the AM modulator.

The equalization stage that follows not only rolls off the upper sideband that would contribute noise to the signal but also balances the level of lower sideband (lower frequencies) compared to the luminance FM carrier signal (higher frequencies). This prevents black streaks in the playback video commonly called over modulation noise. The Super-Beta switch and the BIII mode connect to this stage. The extended bandwidth in Super-Beta requires that the equalization emphasis be changed to maintain frequency spectrum balance. However, this is only necessary in the Super-Beta BII speed. The normal high-frequency losses at the slower Super-Beta BIII speed maintain proper balance without the Super-Beta equalization emphasis. Therefore, this Super-Beta emphasis is not used in Super-Beta BIII.

The AFM trap stage removes any noise in the frequency band between the chroma and the luminance's lower sideband, so the four pilot carriers used to record Beta hi-fi can be inserted here later on by the record amplifiers.

Luminance signal that has been controlled in amplitude and bandwidth, boosted, moderated, and cleaned is mixed with processed chroma in the next stage. The new luminance and chroma signal is mixed with the AFM audio if present, amplified, and stored on magnetic tape using video heads.

Super-Beta carrier shift

Of the three high-band changes to the record luminance stages, the most important is the FM modulation frequency shift. Super-Beta gains a larger sideband and therefore detail by shifting the record luminance signal +0.8 MHz. The record carrier shift diagram (Fig. 8-5), shows this is done by adding a small voltage to the input of the modulator at IC1 (pin 30) from the front panel Super-Beta switch through D212 and series resistors.

Beta hi-fi also shifts the modulator frequency, but only +0.4 MHz. This is also done by adding a small voltage to the input of the modulator from the front panel Beta hi-fi switch. In record, when the Beta hi-fi switch is turned on, 0 Vdc appears on the base of Q801, turning it off. The resultant HIGH at Q801/collector turns ON Q504 and Q505, which luminates the front panel Beta hi-fi light. When the Beta hi-fi lamp is lit, voltage is also applied to Q610 and Q611 turning them both on. The voltage from Q611's conduction is controlled by RV204 to set the maximum luminance carrier frequency in Beta hi-fi. This voltage forward biases a diode in the D212 package and is then applied to the modulator at IC1 (pin 30) to shift the modulator frequency to 0.4 MHz for Beta hi-fi.

Super-Beta playback block

The Super-Beta and Beta hi-fi modes make few changes to the playback signal path. They will be highlighted as the overall record system is explained. Refer to the block diagram in Fig. 8-6.

The signal from the tape is picked up by the video heads, amplified by the head amplifier, and mixed together. The luminance and chroma signals are separated using bandpass filters before taking different paths. (The chroma process path is not affected by Super-Beta nor Beta hi-fi modes and therefore is not discussed.)

The luminance path is through an equalization amplifier stage that provides a flat response from the higher frequency FM luminance carrier signal to the lower frequency FM luminance lower sideband signal. This stage is necessary because the video heads cannot play back with a linear (equal) output over the entire luminance spectrum.

During Super-Beta (high band), a wider luminance bandwidth is played back. The balance between the lower frequencies and ex-

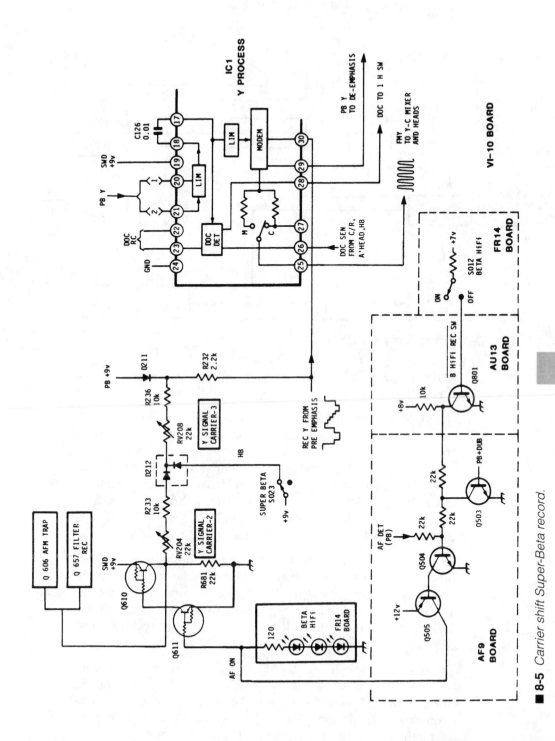

■ 8-5 Carrier shift Super-Beta record.

267

Sony Beta hi-fi system

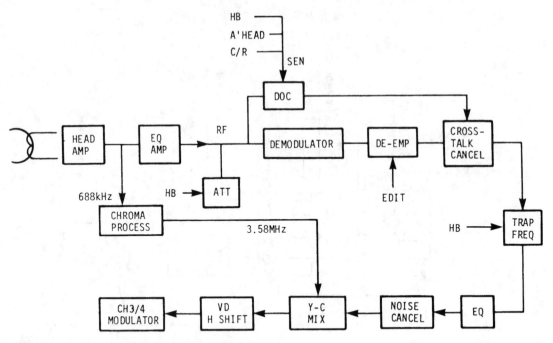

■ **8-6** *Super-Beta playback block diagram tie-in.*

tended higher frequencies must be maintained with a Super-Beta attenuation stage.

The RF luminance signal path divides after the equalization amplifier. One path is into the dropout compensator stage. This section produces an output pulse if the input RF luminance signal falls below threshold level. During Super-Beta mode, the greater amount of short-time-constant, high-frequency signals might not reach the DOC's threshold level and incorrectly signal a dropout. Therefore, the DOC sensitivity is reduced in Super-Beta. During pause, cue, or review, the DOC is disabled by further reducing DOC sensitivity.

The main path of the RF luminance signal is through the demodulator stage. During the recording process, the low-level, high-frequency components of the luminance signal were boosted, so they can be attenuated in the playback deemphasis stage. Thus, the low-level high-frequency losses that occur when recording and playing back on magnetic tape are reduced.

The crosstalk cancellation stage compares the past and present horizontal lines of signal, derives a difference of the two signals, and subtracts if from the present active horizontal line. High-frequency crosstalk cancellation can be done using this comparison method, because one horizontal line is similar to the next one.

The crosstalk cancel stage also receives signals from the dropout compensator stage. When the DOC signal is HIGH, a loss of RF from the tape has been detected. This DOS signal toggles a switch in the crosstalk cancel stage that selects the last active line of horizontal luminance from a 1H delay line and continues to insert it into the luminance chain until the DOC signal toggle returns LOW.

After the crosstalk cancel stage, a trap frequency stage is used to set a maximum frequency limit to the playback signal. This is necessary to eliminate high-frequency noise outside the luminance bandpass. In the Super-Beta mode, a higher frequency trap is chosen because of the wider bandwidth of the playback signal. An equalization circuit follows the trap to maintain a flat frequency response across the luminance band.

The noise-canceling stage removes high-frequency noise within the luminance spectrum. This completes the processing of the luminance signal from an RF signal played back from the video heads through demodulation, frequency response equalization, and noise reduction stages.

The processed luminance and chroma are combined in a Y-C mixer. The composite video output is acted upon by the next VD/H shift stage only during special effects modes to correct for possible distorted horizontal and vertical sync signals. The completed playback video signal is then converted by a modulator for viewing on the TV receiver.

Super-Beta playback circuit

Of the three Super-Beta changes to the playback luminance stages, the most important is the selection of trap frequencies. Each trap determines the upper frequency response limit and is set just above the luminance signal to eliminate noise above that point. Because the luminance signal is higher in frequency during Super-Beta, two limits must be set. One trap sets a frequency limit for normal playback and a second trap sets a frequency limit for Super-Beta playback. If the correct selection of these traps is not made, Super-Beta frequency response could be limited or noise could be added to regular playback. This is why this stage is most important.

The incoming luminance signal after demodulation and de-emphasis is applied through R102 to L-C traps. They establish an upper cutoff or termination frequency, but only one trap is used at a time. During conventional playback, when the Super-Beta switch is OFF, no voltage appears on the HB line, so Q108 and Q722 are off. When Q108 is OFF, the L-C trap at its collector is not used. When Q722 is OFF, its collector is HIGH, turning on Q107 and placing the low-frequency L102-C103 trap in the conventional playback luminance circuit.

In Super-Beta mode, when the front panel Super-Beta switch is closed, +9 Vdc appears on the HB line, turning on Q108 and Q722. Refer to Super-Beta circuit tie-in shown in Fig. 8-7. When Q722 is ON, its collector is LOW, turning off Q107, which disconnects its L-C trap from the luminance circuit. When Q108 is ON in Super-Beta, the high frequency L104-C161 trap is used. After the upper frequency limit has been set, noise above this limit has been eliminated, and signal below this limit is allowed to pass using a low-pass filter consisting of L101 and C162. The luminance signal is then passed through an equalization stage, a buffer transistor (Q106), and on to a mixer to combine with chroma to complete the return to a composite video signal.

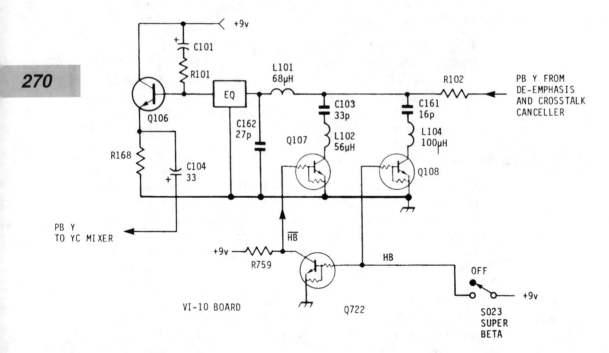

■ **8-7** *Circuit tie-in for Super-Beta playback.*

Capstan control for slow speed effect

Refer to Fig. 8-8. Speed and direction signals leave system control 3 and are made available to the main systems control 1 (IC401) for

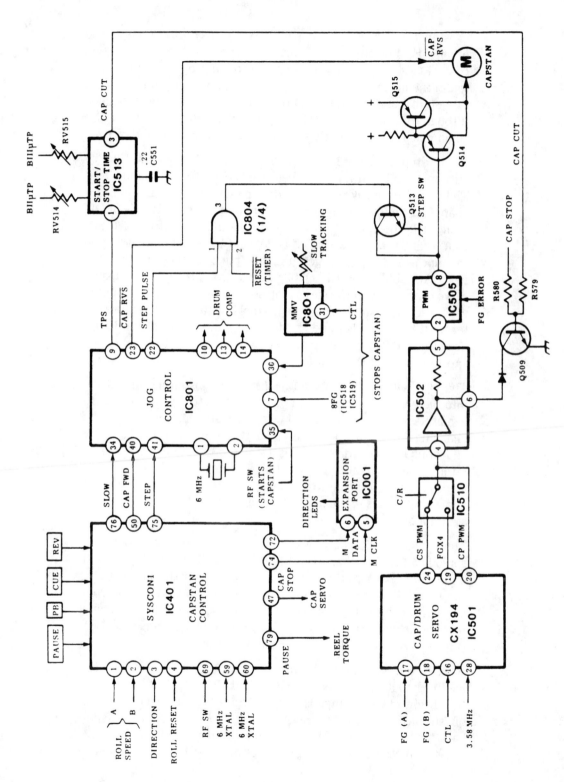

■ 8-8 Capstan control (slow) diagram.

capstan control. The front panel pause, play, cue, review, and other push buttons are also made available to the same system control (IC401) and are key-scanned to identify which push button in the matrix was pushed. When the pause button is pressed, the jog dial and shuttle ring are enabled. Information is clocked through system control 1(IC401) using an RF switching pulse coming in at pin 69 (there is a 6-MHz clock at pins 59 and 60).

Pause, step advance, 1/5 speed, and X1 speeds are controlled by outputs from IC401 and are labeled SLOW, CAPS FWD, and STEP. They are fed into IC801 where they come out with an important change in timing. During pause, step, and 1/5 speeds, the SLOW input at IC801 (pin 34) goes HIGH, causing the output of IC801 (pin 9, TPS) to go HIGH. TPS is delayed by IC513, depending on the tape speed, and emerges at IC513 (pin 3) as the CAP CUT signal. CAP CUT turns ON Q509, which grounds out the capstan servo drive, breaking its servo loop so step advance or slow motion can be accomplished. When the jog dial is rotated, a step pulse entering IC801 (pin 41) and the RF switching pulse at pin 35 are used to manufacture a step pulse that leaves pin 22 to advance the capstan motor to the next field.

The jog control IC801 not only starts the capstan motor moving into the next field, but it also must stop it on the A field to ensure a noiseless still-picture. Braking the capstan motor is done by changing the capstan direction signal CAP RVS from HIGH to LOW and applying step drive pulses (step pulse) from drive pulses (step pulse) of a controlled duration. In order for IC801 to develop step pulses to stop the tape exactly on the A field, it must first know how far it is to the new A field, and second, how fast the capstan is moving.

These two questions are best answered by referring to the slow servo/head timing chart, (Fig. 8-9), which shows the inputs and outputs of IC801 during a step advance. From the jog dial and through IC401, IC801 receives a step advance signal at IC801/pin 41 to advance the tape. A short time later when the RF switching pulse again makes a LOW-to-HIGH transition, a step pulse at IC801 (pin 22) is output to drive the capstan motor.

When the motor turns, the tape also moves. The tape provides feedback to IC801 (pin 30) and 31 called CTL. The CTL signal tells IC801 how close the tape is to the A track. This answers the first question (how far is it to the A field?). The motor provides feedback to the IC801 (pin 7), called 8FG. IC801 uses 8FG to tell how fast the motor's going by counting the time between the 8FG

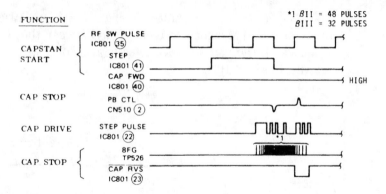

FUNCTION

| | | *1 8II = 48 PULSES |
| | | 8III = 32 PULSES |

CAPSTAN
START
{
RF SW PULSE
IC801 (35)

STEP
IC801 (41)

CAP FWD
IC801 (40)   HIGH

CAP STOP   PB CTL
CN510 (2)

CAP DRIVE   STEP PULSE
IC801 (22)

CAP STOP
{
8FG
TP526

CAP RVS
IC801 (23)

■ **8-9** *Slow servo/head control timing chart waveforms.*

pulses. This answers the question about how fast the capstan motor is turning. IC801 also knows the number of 8FG pulses from one A field to the next and can therefore gauge how close it is to the next A track, even if CTL is missing.

Now that both questions have been answered, pulses determined by 8FG and CTL can be manufactured to stop the tape on the A track. This signal is called the step pulse from IC801 (pin 22).

As the jog dial is rotated faster, normal PB (X1) speed is reached. At this time, the SLOW input to jog control IC801/pin 34 then goes LOW causing the TPS output to go LOW, which causes TPS IC801 (pin 9) to turn off Q509. This restores the capstan servo loop, and the tape is locked to X1 speed.

Audio-Beta hi-fi

In the Beta hi-fi system, the audio signals are converted to modulated FM carriers and recorded on the video track of the videotape using the two rotary video heads. The advantages to using the FM recording process are as follows:

☐ Excellent frequency response.

☐ Low noise and distortion.

☐ Wide dynamic range.

Using the rotary video heads results in a relative tape-to-head speed that is about 350 times that of a conventional audio recording system. Wow and flutter specifications are below the measurable limit.

273

The Y signals have been shifted up slightly from 0.4 MHz to make room for the AFM signals to be inserted between the chroma and Y signals on the video track. Refer to recorded spectrum signal drawing in Fig. 8-10.

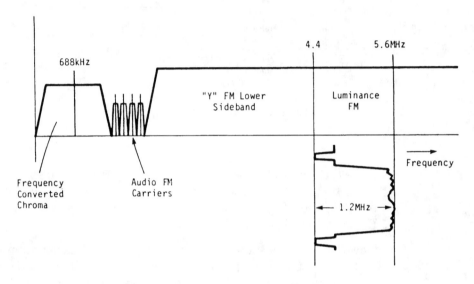

■ **8-10** *Recorded spectrum-Super-Beta and Beta hi-fi.*

The AFM signals consist of four different carriers—two for each channel—separated by 150 kHz. Four carriers are used to reduce crosstalk between the left and right channels and between A and B video tracks.

The frequency assignments of the four FM carriers are shown in the chart of Fig. 8-11 for the pilot frequencies. The F1 and F3 signals represent left and right channel audio and are recorded on the A video track.

Carrier	Frequency	Video Track	Audio Channel
f1	1.380682 MHz (87.75fH)	A	L
f2	1.530157 MHz (97.25fH)	B	L
f3	1.679633 MHz (106.75fH)	A	R
f4	1.829108 MHz (116.25fH)	B	R

■ **8-11** *Beta hi-fi pilot frequencies.*

In the playback process, the four AFM carriers are reconstructed into left and right channel audio by alternately switching the A and B tracks in synchronization with the video heads.

VHS high-quality video

The high-quality video system is a new technology that enables high picture quality in every part of the picture while maintaining superior VHS compatibility.

The two HQ improvements incorporated into the design of these VHS recorders are; WC-White Clip and DE-Detail Enhancement. The block diagram (without detail enhancement) is shown in Fig. 8-12.

Circuitry without detail enhancement

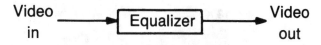

■ **8-12** *Circuitry without detail enhancement.*

275

The bandwidth of magnetically recordable frequencies is an important factor that determines the picture quality of VCRs. With technological advancements in the fields of material, device, and circuit engineering, it is now possible to raise the white clip level by 20 percent over previous recorders to accommodate recordings of higher frequencies.

By raising the white clip level, the rising flank of a waveform from dark to white, where high-frequency signals are concentrated, is made clearer than before. This results in sharper edges on vertical objects in the picture, improving the overall quality of the picture.

Detail enhancement

The detail enhancement circuit is a low-level, high-frequency, nonlinear emphasis circuit that reinforces the luminance signal. This allows recording of more detail of a scene, such as strands of hair or pebbles on a beach.

As noted, Fig. 8-12 shows a block diagram of video signal processor circuitry without detail enhancement. In this circuitry, the video signal simply passes through an equalizer. The circuitry with detail enhancement is shown in Fig. 8-13. The video signal passes through the equalizer plus a highpass filter and a limiter before being recombined in the adder. The output signal is a video signal that has enhanced high-frequency components. The circuit operates to increase the enhancement of objects in the picture in proportion of faintness. The fainter the object is in relation to its background, the more the detail enhancement increases. The chaff in Fig. 8-14 illustrates this relationship.

Circuitry without detail enhancement

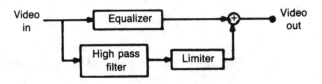

■ **8-13** *Circuitry with detail enhancement.*

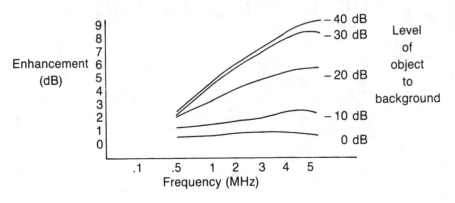

■ **8-14** *Detail enhancement (dB) curve chart.*

What does HQ do?

1. Increase white clip level (Fig. 8-15). HQ increases white clip level by 20 percent recording. This procedure results in sharper edges, particularly vertical edges and profiles.
2. Reduces color noise (Fig. 8-16). A noise reduction circuit doubles color output while increasing noise only 1.4 times. By improving color signal-to-noise ratio, color streaking and patches are reduced, even in minute details.

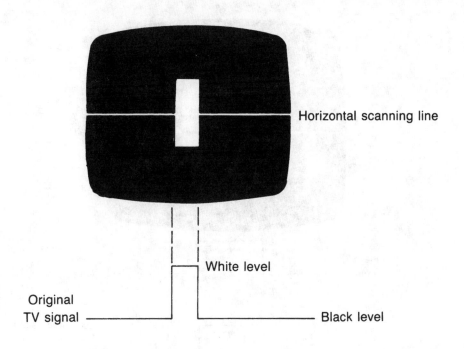

Horizontal scanning line

White level

Original
TV signal

Black level

New white clip level

Old white clip level

Pre-emphasis
of
signal recorded

X +
20%

X

Soft edge

Old
playback

Sharp
edge

New
playback

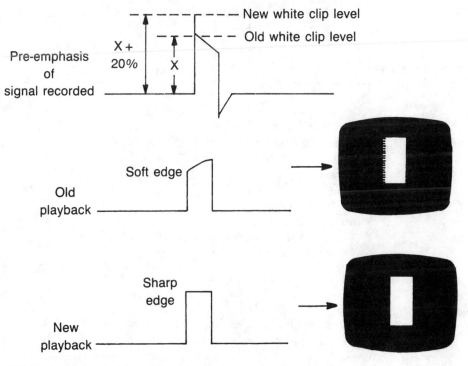

■ 8-15 *Increased white clip level waveforms.*

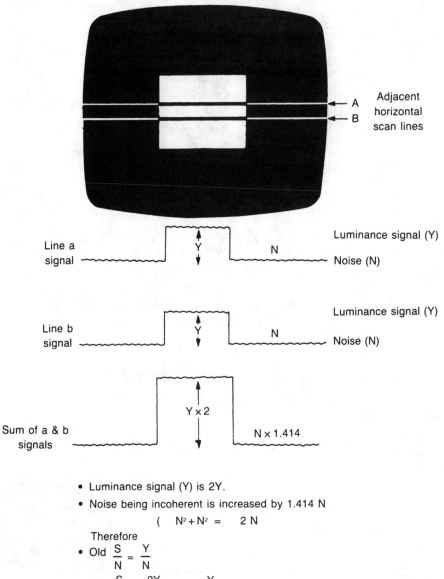

Line a signal ──── Luminance signal (Y) / Noise (N)

Line b signal ──── Luminance signal (Y) / Noise (N)

Sum of a & b signals ──── $Y \times 2$ / $N \times 1.414$

Adjacent horizontal scan lines

- Luminance signal (Y) is 2Y.
- Noise being incoherent is increased by 1.414 N

$$(\quad N^2 + N^2 = \quad 2\,N$$

Therefore

- Old $\dfrac{S}{N} = \dfrac{Y}{N}$
- New $\dfrac{S}{N} = \dfrac{2Y}{2\,N} = \quad 2\,\dfrac{Y}{N}$

Which is a 3 dB improvement

■ **8-16** *Luminance noise reduction waveform.*

3. Reduces luminance noise (Fig. 8-17). A noise reduction circuit doubles the luminance output while increasing noise only 1.4 times. The circuitry improves the signal-to-noise ratio by 3 dB, markedly reducing noise throughout the picture.

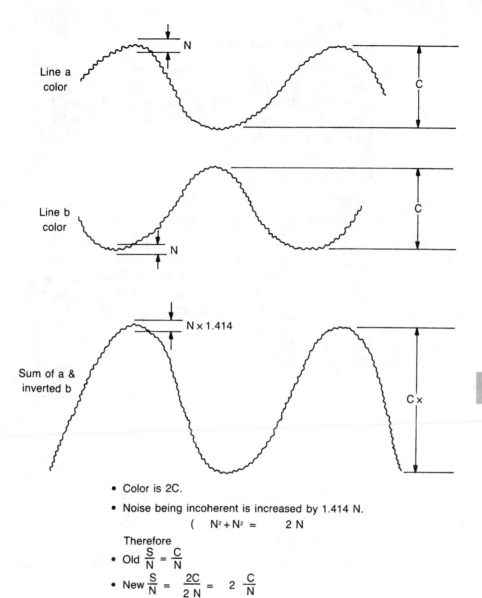

- Color is 2C.
- Noise being incoherent is increased by 1.414 N.

$$(\quad N^2 + N^2 = \quad 2\,N$$

Therefore
- Old $\dfrac{S}{N} = \dfrac{C}{N}$
- New $\dfrac{S}{N} = \dfrac{2C}{2\,N} = 2\,\dfrac{C}{N}$

Which is a 3 dB improvement

■ **8-17** *Color noise reduction illustration.*

Super-VHS video recorders

The Super-VHS VCRs contain new systems and circuitry that, when connected to a digital TV receiver, provide in excess of 400 lines of horizontal resolution in the RBG mode. When connected in the normal manner to a TV receiver, a significant improvement in the quality (resolution) of the picture is noticeable.

These new recorders look just like any other VHS recorder. The mechanical part of the deck is almost identical to those used in "standard" VCRs. Inside, most of the circuitry is the same for the mechacon, audio, servo, and tuner. The Super-VHS feature is accomplished by the following:

☐ New narrower gap heads.

☐ New oxide formula cassette.

☐ Expanded 1.6 MHz Y FM deviation.

☐ New nonlinear emphasis signal-processing circuitry.

The recorder has been designed for what might be considered a dual system. That is, it will record and play standard VHS tapes in addition to Super-VHS tapes. This is accomplished by a special video cassette with a new oxide formula. The cassette cartridge has an identification hole at the bottom (see Fig. 8-18) that functions to place the recorder in the Super-VHS mode. In this manner, it will record and play Super-VHS tapes and will also accept standard VHS tapes for record or playback.

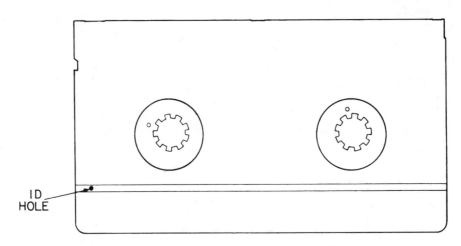

SUPER VHS CASSETTE

■ **8-18** *Drawing of Super-VHS cassette.*

In the video signal processing circuitry, changes have been made to increase the deviation of the luminance signal. In Fig. 8-19, the frequency range of the Y FM signal in a normal VHS recorder is from 3.4 MHz to 4.4 MHz, a range of 1 MHz. In the Super-VHS

VHS

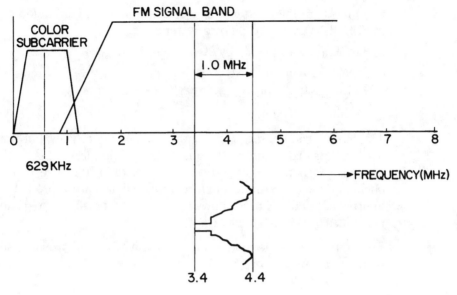

■ **8-19** *VHS FM signal band waveform.*

mode, the Y FM signal is from 5.4 MHz to 7 MHz, a range of 1.6 MHz as shown in Fig. 8-20. These are the major differences between standard and Super-VHS.

SUPER VHS

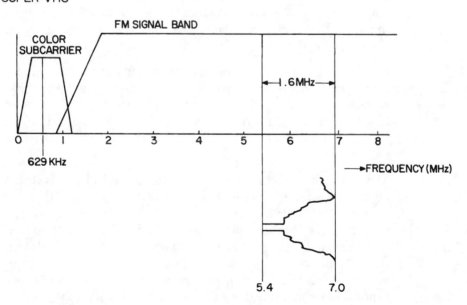

■ **8-20** *Super-VSH FM signal band waveform.*

Detailed HQ system analysis

Such newly developed technologies as the following are adopted in VHS VCRs of the HQ picture system:

☐ Level up for white clip by 20 percent.

☐ Y noise reduction (Y/NR).

☐ Chroma noise reduction (C/NR).

☐ Detailed enhancer.

TV pictures can be analyzed from various points of view, but the main factors that influence the quality of VCR pictures in playback are pulse response characteristic and S/N ratio. Pulse response characteristic shows how much recorded signals are reproduced in playback, and S/N ratio shows the amount of noise on the plane and edges of playback pictures.

Among the above mentioned four items, A and D improve pulse response characteristics, while B and C improve S/N ratio.

Luminance noise reduction (Y/NR)

VHS VCRs are equipped with an emphasis circuit as in the past. The emphasis circuit is composed of a preemphasis circuit and de-emphasis circuit.

Incoming video signals first go to the preemphasis circuit, which emphasizes high-frequency components of the signal. It then goes to the clipping circuit for frequency modulation as the input signal is recorded in FM waveform.

In playback, the demodulated video signals contain high-frequency noise components that are mainly generated in FM recording and playback. The de-emphasis circuit suppresses the high frequency to reduce noise, and the video signal returns to its original level because its high frequency was emphasized in recording.

The emphasis circuit improves S/N ratio in the manner as stated above. Conventional emphasis circuits are composed of C and R components that can be replaced with transversal filters of delay elements. The drawings in Fig. 8-21 and Fig. 8-22 illustrate pre-emphasis and deemphasis circuits and relationships between them. As shown by the drawings of transversal filter circuits, previously output horizontal signals are composed of outputs of conventional circuits. If a delay line of transversal filters is used to delay the horizontal scanning time at a unit, vertical signal compo-

(A) Pre-emphasis circuit of CR element (B) Frequency characteristic (C) Step response (D) Pre-emphasis circuit of transversal filters

■ **8-21** *Preemphasis circuits and waveforms.*

(A) De-emphasis circuit of CR element (B) Frequency characteristic (C) Step response (D) De-emphasis circuit of transversal filters

■ **8-22** *Deemphasis circuits and waveforms.*

sition is affected. According to the above principle then, a Y/NR circuit has been developed.

S/N ratio improvement

The next explanation is the idea of S/N ratio improvement. A TV picture is comprised of 525 scanning lines. In Fig. 8-23, for example, every scanning line is a signal waveform, shown by the corresponding arrows. In the figure, neighboring lines show little difference between correlation waveforms, except in the case of n-4, which is a bordering portion of the picture. If a noise is mixed with the signals in recording and playback (as shown later in Fig. 8-24), and the signals of line n and line n-1 are added, the *signal* becomes double, but the *noise* will be 2 times (it will not double, because noise occurs at random). Therefore, when the signal is re-

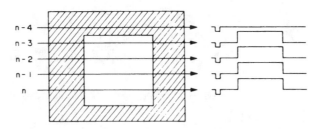

■ **8-23** *A drawing of how S/N ratio improvement is obtained.*

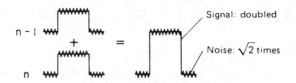

■ 8-24 *How S/N ratio can be set uniformly.*

duced to one-half of the original waveform, the noise becomes 2/2, and the S/N ratio is improved to 1/2 (3 dB in amount).

As the correlation between lines exists not only between two neighboring lines but also between the other two lines next to the neighboring ones, S/N ratio can be improved by adding these lines using a circuit like that in Fig. 8-25.

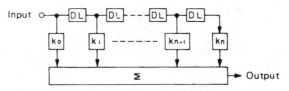

Note: DL is a 1H delay circuit

■ 8-25 *S/N ratio is improved by adding lines.*

In the above case, the adding is not done on the same rule, but the rate increases proportionally for two closer lines. The rate can be set uniformly such as shown in the example of Fig. 8-24. In the case of the addition on the same rule, this idea is realized by using a 1H delay circuit. Figure 8-26 shows an example.

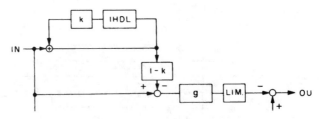

■ 8-26 *Block diagram of S/N limiter circuit.*

If a circuit such as in Fig. 8-27 is applied, vertical signals are added on the picture as shown in the waveforms of Fig. 8-28.

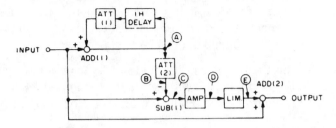

■ **8-27** *Y/NR circuit in recording system.*

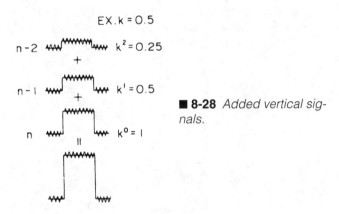

■ **8-28** *Added vertical signals.*

However, if an edge portion of a picture is vertically displayed, the waveform is moderate at its rise portion as in Fig. 8-29a, because the neighboring scanning line is added. This causes deterioration in its vertical frequency characteristic.

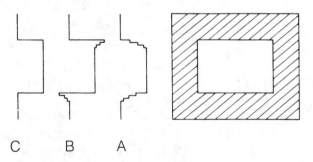

C   B   A

■ **8-29** *Treatment of signal waveform.*

To solve this problem, apply the principles of preemphasis and deemphasis explained previously for the vertical lines. Treat the signal as the waveform shown in Fig. 8-29b to correct the rise portion and increase it in vertical frequency characteristic. By this treatment, the waveform similar to the original, shown in Fig. 8-29c, is obtained in playback.

If this treatment is performed slightly just for low-level signals, S/N ratio is improved without deterioration in changeability. This treatment is done by the limiter (see block diagram of Fig. 8-26).

Principles of Y/NR

Refer to Fig. 8-27 for a circuit of the recording system. In recording, this circuit functions for precompensation of decrease in vertical resolution in playback. Waveforms observed at vertical rate are shown in Figs. 8-30a through e.

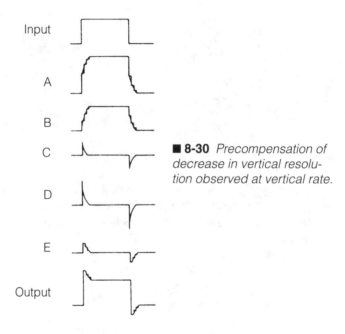

■ **8-30** *Precompensation of decrease in vertical resolution observed at vertical rate.*

Refer to Fig. 8-31. When a square waveform is input to a signal source, the output from the ADD (1) is moderate in its rise portion due to the cyclic lowpass filter that consists of ADD (1), 1H DELAY, and ATT (1). As this output is larger in amplitude than the original one, ATT (2) corrects it to have the same level as that of the original signal. The output of SUB (1) is a highpass signal that is the difference component between the original and lowpass signals. This highpass signal is amplified to have a level that compensates the playback signal, and after its amplitude is controlled by the limiter to secure the changeability, it is mixed with the original signal by ADD (2), to be sent as the REC signal.

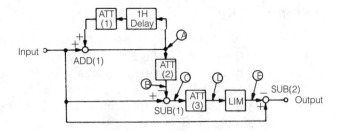

■ **8-31** *Block diagram of NR playback circuit.*

Figure 8-31, the playback circuit, functions to remove noise generated in recording and playback. When P.B. signals containing noise are fed to the input terminal, the cyclic lowpass filter composed of ADD (1), 1H DELAY, and ATT (1) removes the noise, and ADD (1) outputs a noiseless signal. If this signal is sent to ADD (2) in the same manner as in recording, the output of SUB (1) is a highpass signal containing noise.

Therefore, the circuit is designed so that the signal passes ATT (3) (which decreases noise effectively), the limiter (to secure changeability), and SUB (2) (which adds noise in reversed polarity). Through the above process, the same output signal as the original can be obtained, except it's without noise. The waveforms found throughout playback circuits are illustrated in Fig. 8-32a through e.

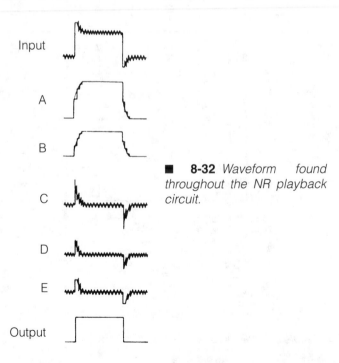

■ **8-32** *Waveform found throughout the NR playback circuit.*

Principles of color NR circuit

The principle of the C/NR circuit is the same as that of the Y/NR circuit. As the color signal is a 3.58-MHz signal whose phase is inverted every 1 H, the addition of lines is performed by inverting the phase every 1 H. For the edging portions, it is treated by applying the correlation with the luminance signal. The C/NR circuit is provided only for the playback system.

NR circuit chip

Refer to chip diagram shown in Fig. 8-33. The IC407 chip contains a dropout compensation circuit and a nonlinear de-emphasis circuit for playback in EL/LP mode, besides the Y NR circuit.

The Y signal is input through pin 22 of IC407 and is fed ADD (2) via the nonlinear de-emphasis circuit. The ADD (2) is a cyclic

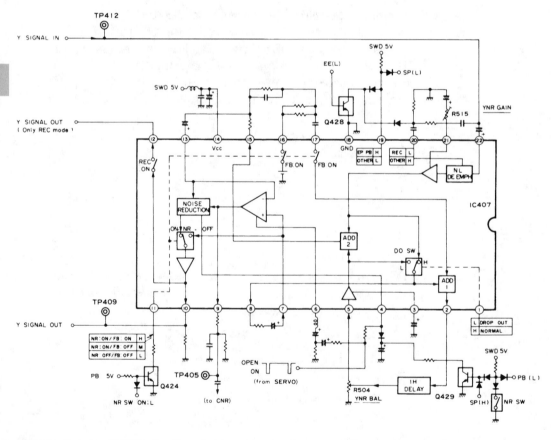

■ **8-33** *Details of Y/NR circuit and chip.*

comb filter composed of ADD (1) and 1H delay circuit, which is used for dropout compensation also. Signal output from ADD (2) is supplied to pin 13 through pin 15. This signal is sent to the noise reduction block and then output from pin 10. The signal output from pin 12 in REC mode is used for regular preemphasis.

MTS television sound system

The Multi-channel Television Sound (MTS) broadcast system features the transmission of stereo, bilingual, and voice/data signals. The spectral diagram of the MTS signal is shown in Fig. 8-34.

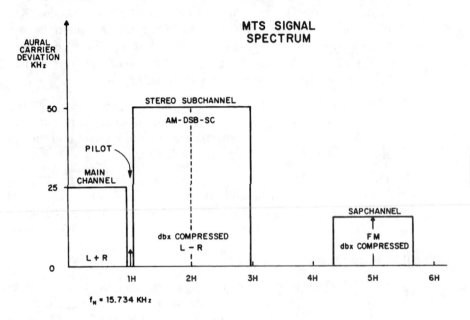

■ 8-34 *MTS signal spectrum.*

The MTS system is compatible with current transmissions and mono receivers. The L + R signal has 75 ms preemphasis and deviates the aural carrier 25 +/− kHz on 100 percent peaks. The next component of the MTS signal is the 15.734 kHz stereo pilot. The pilot deviates the aural carrier 5 kHz and is used in the receiver to detect the L − R stereo subchannel. The L − R channel is an AM double sideband suppressed carrier signal and deviates the aural carrier 50 kHz. The SAP subcarrier is an FM, 10 kHz deviated signal centered at 78.670 kHz (5FH). The SAP subcarrier deviates the aural carrier 15 kHz. Both the L − R and SAP audio signals are dbx encoded to reduce buzz and noise.

Circuit composition

The MTS demodulator is shown in Fig. 8-35. The audio demodulator is comprised of eight ICs. IC301 demodulates the IF signal to yield the baseband signal. Sub audio SAP demodulators and the L/R matrix are contained in IC302, IC303 and IC304 are for dbx noise reduction. The peripheral circuits required by IC303 and IC304 are contained in IC305, IC306, IC307, and IC308.

IF to MTS signal flow

The IF signal from the tuner/IF circuit is amplified by Q301 and Q302. The SAW (surface acoustic wave) filter (SAW 301) removes adjacent channel signal components, after which the signal goes to pins 8 and 9 of IC301. In IC301, the IF signal is amplified and applied to the video demodulator (AM detector). The phaselocked loop (PLL) circuit provides fully synchronous detection and features less audio noise than ordinary detector systems.

The signal from IC301 pin 28 goes to a bandpass filter (CF301), yielding the 4.5 MHz +/− 250 kHz FM audio signal. This is applied to IC301 pin 13. A limiter removes the AM component and the FM signal is demodulated to form the MTS signal.

The MTS signal from IC301 pin 19 is sent to IC302 pin 18, where it is amplified and distributed in two lines from pin 19. One of these becomes the main and subsignals, and the other becomes the SAP signal.

Main (L + R) and sub (L − R) signals

The MTS signal from IC302 pin 19 goes through LPF 301 (50 kHz) to remove the SAP component and then to pin 20 of IC302 to remove the pilot signal. It is then distributed in two lines. One goes as the main (L + R) signal to the pin 16 output and the other as the sub (L − R) signal to the stereo demodulator circuit. The main signal is sent via LPF 302 (15 kHz) to pin 12 of IC301, a voltage follower amplifier. The signal exits pin 15 and is routed to pin 12 of IC302, where it is applied to the matrix circuit. At the stereo demodulator in IC302, the subsignal is applied to the AM detector and mode switch and exits at pin 8. It passes through lowpass filter LPF303 (15 kHz) and is routed to pin 3 of IC303. The signal is inverted and distributed in three lines from pin 4. One of these goes to IC305 pin 4, the highpass and lowpass filter, then from IC305 pin 7 to RMS DET-1, to IC303 pin 7. RMS DET-1 controls the spectral expander VCA at pin 24.

The second route is via the wideband filter between IC305 pins 4 and 9 to pin 9 of IC303, then to RMS DET-2 at IC303, which controls the wideband expander VCA at pin 18.

In the third line, the signal from pin 4 of IC303 is sent to pin 26 and through a buffer stage, exiting pin 25 of IC303. IC306 contains a fixed deemphasis network between pins 10 and 5. A variable deemphasis circuit, which is composed of a frequency divider, includes VCA at pin 24 of IC303. Frequency response is precisely opposite to the variable preemphasis of the compressor circuit at the broadcast station. The signal from IC303 pin 19 goes through wideband expander VCA at pin 18 and fixed deemphasis, then exits pin 16 to be routed to the matrix at IC302 pin 11.

Matrix circuit and signal outputs

The matrix circuit of IC302 processes the main (L + R) signal applied to pin 12 and the sub (L − R) signal at pin 11. Outputs are obtained from pins 13 and 14. In this model, pin 9 is open and pin 17 is at ground potential. The pin 10 input selects between mono and stereo.

The audio signal outputs from pins 13 and 14 are supplied to the FM audio amplifier circuit. The circuit of Q310 and Q311 functions to mute noise during channel selection. This is controlled by the Tuner/Timer data control microprocessor.

The SAP signal

The bandpass filter BPF301 (5fh) extracts the SAP component from the MTS signal output from IC302 pin 2. A limiter removes the AM component, and after FM demodulation, the SAP signal appears at pin 7.

The SAP signal goes via the buffer amplifier of Q307 and Q308 to the dbx demodulator of IC304, IC307, and IC308. Within this circuit, the signal flow overlaps that of the subsignal. The FM carrier component is removed by the LPF (19.5 kHz) of R386 to C345, contained in Q307 and Q308. The dbx demodulator output from IC304 pin 16 goes through the voltage-follower amplifier between pins 13 and 15 to the normal audio circuit. Q312 is the SAP signal muting transistor. Muting is performed in three situations:

☐ Channel selection.

☐ SAP program absence from broadcast.

☐ Weak signal strength.

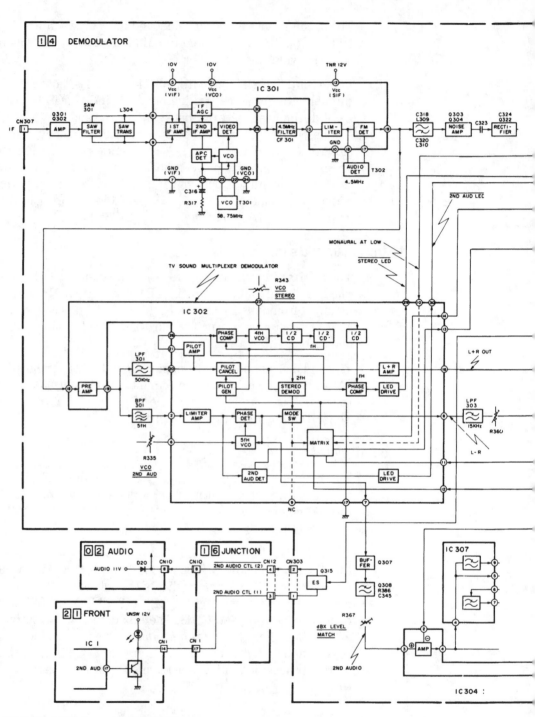

■ **8-35** *MTS decoder block diagram.*

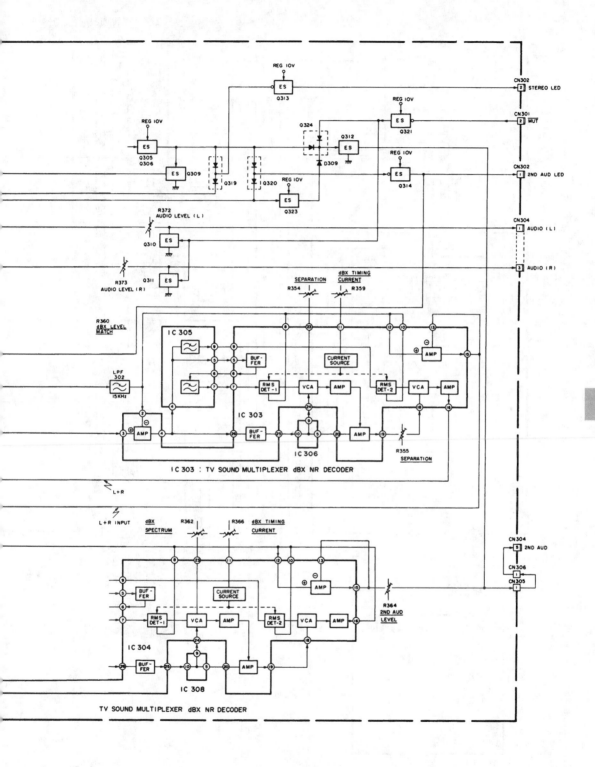

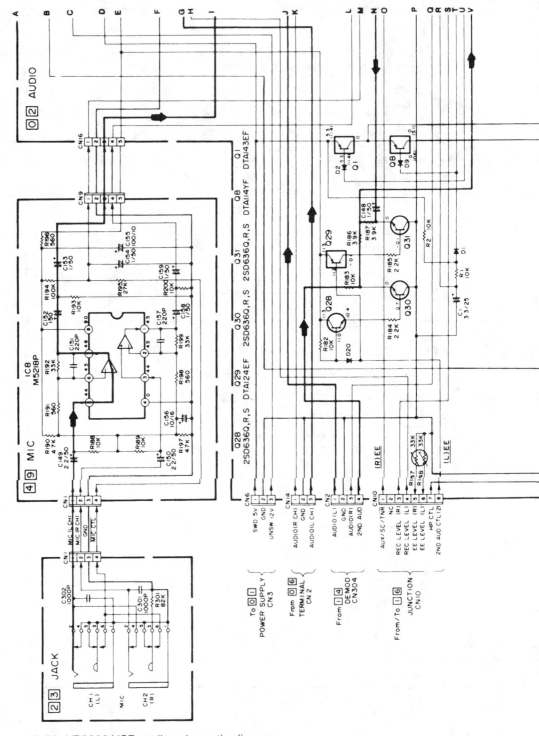

■ 8-36 *VR3300 VCR audio schematic diagram.*

Stereo and digital audio circuits

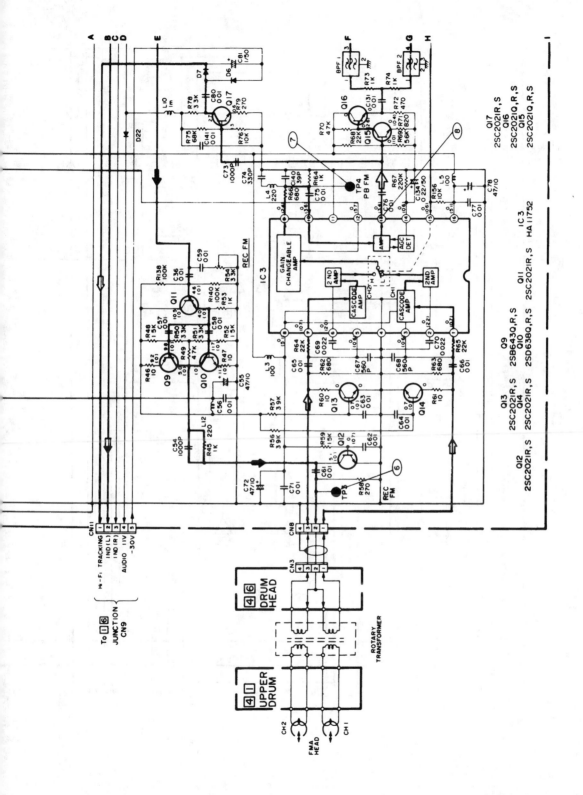

295

MTS television sound system

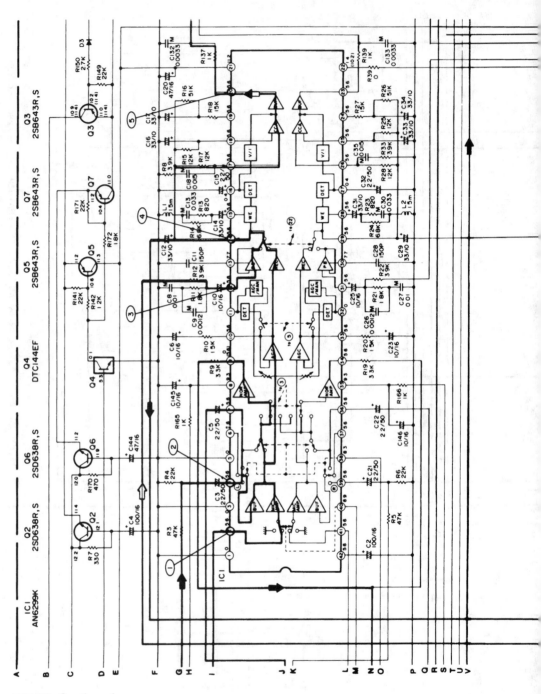

■ 8-36 *Continued*

Stereo and digital audio circuits

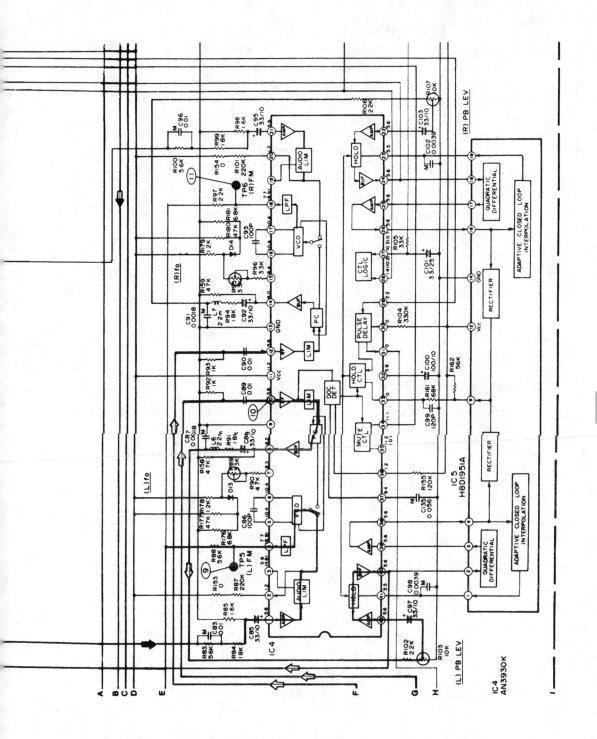

297

MTS television sound system

The weak signal detection circuit detects weak field strength to avoid erroneous STEREO/SAP LED lighting, and stereo/SAP signals with impaired signal-to-noise. The detector functions from the 100-150 kHz noise component of the MTS signal. At about 15 dBu, the stereo/SAP LED is extinguished, the SAP signal muted, and the monaural mode selected. The MTS signal from IC301 pin 19 is supplied to a bandpass filter composed of C318, C320, L309, and L310. The noise component output is amplified by Q303 and Q304, rectified by Q322 and smoothed by C324. The hysteresis amplifier of Q305 and Q306 shapes the dc signal waveform. Under weak signal conditions, the Q306 output is HIGH. This is distributed in three lines:

☐ To the stereo/SAP LED drive circuit for extinguishing the indication.

☐ To switch on Q312, to mute the SAP signal.

☐ To switch on Q309, to engage the monaural mode.

The complete stereo audio SAP signal circuit is shown in the schematic of Fig. 8-36.

MTS TV sound system reviewed

As previously noted, the MTS system features the transmission of stereo, bilingual, and voice/data signals. The waveform diagram in Fig. 8-37 illustrates the TV signal frequency spectrum.

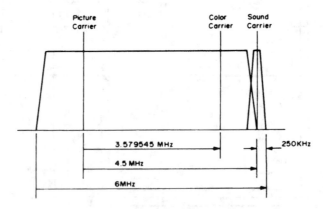

■ 8-37 *TV signal frequency spectrum.*

The MTS system is compatible with current transmissions and mono receivers. Figure 8-38 shows the spectral diagram of the MTS signal.

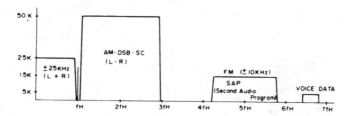

8-38 *Stereo sound baseband spectrum (United States).*

The L + R portion or main channel of the MTS signal is identical to the current mono signal. The L + R signal has 75 ms preemphasis and deviates the aural carrier 25 kHz.

The next component of the MTS signal is the 15.734 kHz stereo pilot carrier. The pilot carrier deviates the aural carrier 5 kHz and is used in the receiver to detect the L − R stereo subchannel. The L − R subchannel is an AM modulated double sideband suppressed carrier signal and deviates the aural carrier 50 kHz.

The SAP subcarrier is an FM modulated, 10 kHz deviated signal centered at 78.670 kHz. The SAP subcarrier deviates the aural carrier 15 kHz. Both the L − R and SAP audio signals are dbx encoded to reduce buzz and noise.

The waveforms for the MTS Baseband signals are illustrated in the drawings of Fig. 8-39.

Digital audio systems

With electronic systems, digital is a better approach than decimal. This is because two states can be manipulated and memorized easier than ten states represented by the ten different numbers we commonly use. In the world of electronics this would require ten different voltage levels, or analogs. The binary system requires only two numbers. These two numbers can be easily expressed electronically by opening or closing a switch. In terms of a voltage, this could be ON/OFF or HIGH/LOW.

In the decimal system (Fig. 8-40), each column of numbers represents a multiple of 10. The first column would be 1s, the second column 10s, the third column 100s, etc. In the binary system, each column represents 2s. The first column would be 1s again, the second column 2s, the third column 4s, etc. One binary digit, or bit, can represent two states. Two bits can represent four states, which is 2 to the second power. Three bits can represent eight

300

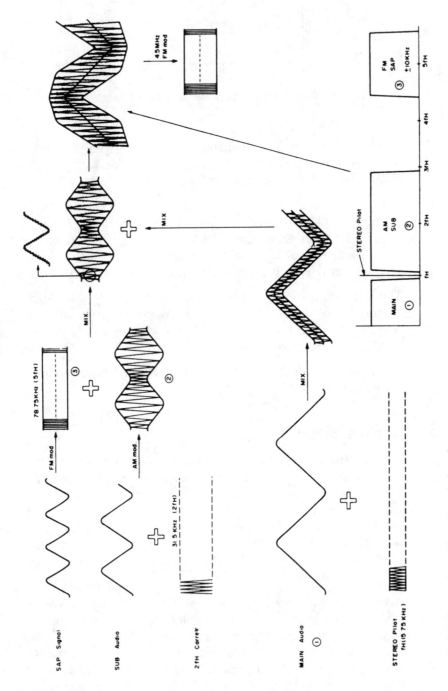

■ 8-39 *Illustration of baseband signals.*

DECIMAL SYSTEM

10^2 (100's)	10^1 (10's)	10^0 (1's)
0	0	0
0	0	1
0	0	2
0	0	3
0	0	4
0	0	5
0	0	6
0	0	7
0	0	8
0	0	9
0	1	0

■ **8-40** *Decimal number system.*

which is 2 to the second power. Three bits can represent eight states, which is 2 to the third power, etc.

Binary systems allow us, to electronically manipulate values while reducing the possibility of error because only two levels are used (see Fig. 8-41). Large numbers can be formed easily by using multiple bits. Thus, big numbers still require only two different levels in order to be expressed. The fact that only two levels are required, no matter how large the binary number is, makes digital storage much simpler than decimal.

BINARY SYSTEM

2^2 (4's)	2^1 (2's)	2^0 (1's)	DECIMAL NUMBER
0	0	0	0
0	0	1	1
0	1	0	2
0	1	1	3
1	0	0	4
1	0	1	5
1	1	0	6
1	1	1	7

■ **8-41** *Binary number system.*

Analog to digital process

Figure 8-42 illustrates how a digital PCM signal is produced from an analog waveform. The audio waveform, shown in the upper-left corner, can assume infinite values. In the analog-to-digital (A/D) process, these infinite values are converted to binary numbers consisting of only two levels, 1 and 0, formed into words capable of defining minute changes in amplitude of the analog signal.

In step 1, the signal is sent through a switch. The actuating voltage on the switch is shown as a series of short pulses called *sampling pulses*. In the instant the sampling pulse goes high, the switch is closed. The output of the switch is a series of pulses that have the same frequency as the sample pulse and whose amplitude is equal to the analog wave at the time it was sampled. This signal is still analog and is called *pulse amplitude modulation* (PAM).

In step 2, the PAM signal is converted to binary. This is called *analog-to-digital conversion*, or *quantizing*. The binary number system that we use is 2s complement.

In step 3, the 2s complement is modulated using nonreturn to zero (NRZ) modulation. For example, at time $t4$, the digital value is all 0s and the output signal is a LOW. At time $t5$, the digital word is all 1s and the output goes HIGH. Likewise, at time $t6$, when the one 0 occurs, the output goes LOW. In other words, the output voltage level stays HIGH as long as 1s are present and only goes LOW when 0s occur. This output is a pulse code-modulated (PCM) signal.

Negative digital numbers

Because binary numbers consist of either 1s or 0s, the sign of the number must also be a 1 or a 0. To perform binary addition and subtraction using negative numbers, the method of forming a negative number must be chosen correctly.

There are four different ways of expressing a negative number. One is sign plus magnitude. The 2s complement is the second line of the chart shown in Fig. 8-42. Offset binary, which is shown in the third line of the chart, is formed by assigning some arbitrary number a 0 value. One method of performing subtraction is with the use of complement numbers. Subtraction of a number B from a number A can be accomplished by the addition of the complement of B to A. In the last line of the chart, we see the 1 complement of a number is formed by taking the opposite of all the bits. It can be seen that a -1 is formed by taking a $+1$, which is a binary 0001,

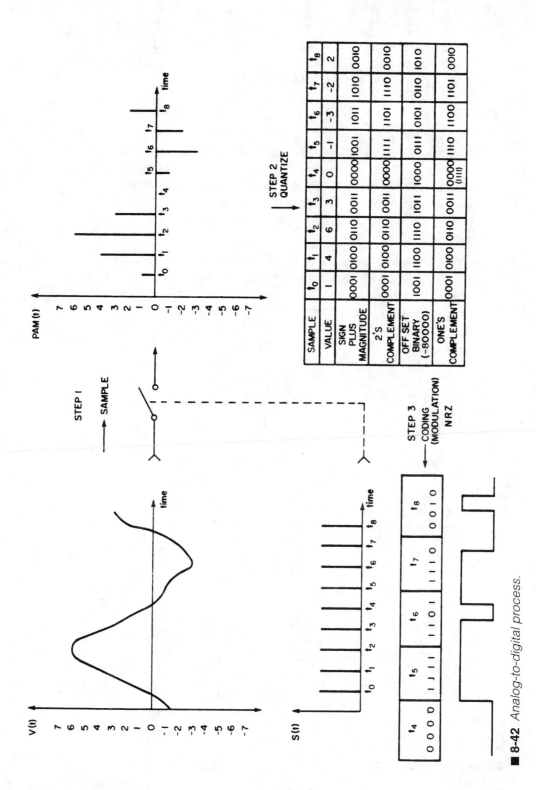

■ 8-42 Analog-to-digital process.

and reversing all of the bits until it becomes a 1110. A − 3 is formed by taking + 3, which is expressed in binary as 0011, reversing all of the bits, and making a 1100.

PCM block

A PCM block diagram in Fig. 8-43 shows the essential blocks for the record and playback of a PCM system. The analog audio is shown going into a lowpass filter (LPF), which keeps any high frequencies from entering the system. The next block shows a sample and hold circuit. The actuating signal on the sample and hold is the sampling pulse. The output from the sample and hold circuit is a PAM signal.

The next block shows an A/D converter. Here, the PAM signal is quantized into a binary code. Modulation follows. Because we are using a video recorder, it is necessary to make the PCM signal compatible with the frequencies used in this unit. Thus, we frequency-modulate the digital data using approximately the same frequencies that we used to frequency-modulate the luminance signal.

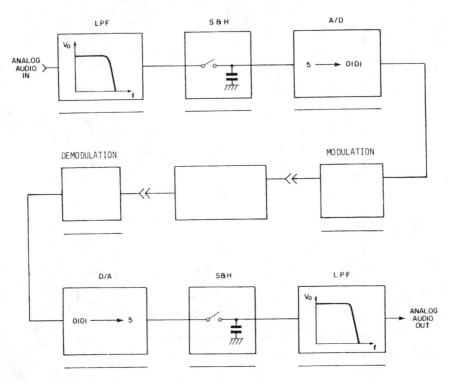

■ **8-43** *PCM block diagram.*

The middle block is our storage medium, in this case an 8-mm magnetic tape. In the playback mode, the FM signal comes off the magnetic tape and enters the demodulation block. Here it is demodulated into HIGHs and LOWs, so our digital ICs can process it.

The HIGHs and LOWs enter the digital-to-analog converter to be converted into a PAM analog signal. The PAM signal then enters another sample and hold circuit, where it is reconstructed into an analog signal. Finally, the analog signal goes through another LPF. This prevents any frequencies above 20 kHz that were created in the playback process from creating distortion in an audio amplifier.

Why use a lowpass filter?

The first block in the PCM block diagram (Fig 8-43), shows an LPF. This filter is called an *anti-aliasing filter*. Its purpose is to eliminate any high frequencies from entering the PCM system and creating extraneous and unwanted frequencies. This would cause severe distortion as we play back a PCM tape.

If we observed the analog input signal with an oscilloscope, we would see a series of sine waves at all different frequencies. They are shown as a simple sine wave in the diagram. However, if we connected this signal to a spectrum analyzer, which does measurements in the frequency domain, we would see that the output would contain frequencies from approximately 20 Hz to 20 kHz.

If we observed the sampling voltage going to the sample and hold switch, we would see a series of very quick positive going pulses. The period of these would be $t2$ minus $t1$. If we looked at this in the frequency domain with the spectrum analyzer, we would see one spike centered at the sampling frequency.

If we observed the PAM signal on the oscilloscope, we would see a series of discrete amplitude voltages that vary with the amplitude of the analog input signal at the sampling frequency.

This PAM signal is similar to an amplitude-modulated signal. In observing the PAM signal with the spectrum analyzer, you would see the original audio spectrum from 20 Hz to 20 kHz. You would also see an output located at the sampling frequency. However, just like the amplitude-modulated signal, you would see two sidebands centered around the sampling frequency. The width of the sidebands would be equal to the width of the audio spectrum. In the case shown, the sampling frequency is greater than twice the maximum audio frequency, which in the real world would be greater than 40 kHz.

The next waveform shows the frequency spectrum of the PAM signal when the sampling frequency is exactly twice the maximum audio frequency, or 40 kHz. Here the lower sideband is located just above the upper edge of the maximum audio frequency. These are all illustrated in the waveform drawings shown in Fig. 8-44.

In the next case, the sampling frequency is less than twice the maximum audio frequency, or less than 40 kHz. Now the lower sideband overlaps the upper edge of the audio spectrum, and an interference region is formed. This is shown as a shaded area. Inside this shaded area, beat frequencies will develop that would then become impossible to distinguish from original frequencies and could not be removed from the system. It is these frequencies that cause distortion during playback.

In the example, a 20-kHz input signal is sampled with a frequency of 31.5 kHz. Hence, the sampling frequency is less than twice the input frequency. When this happens, interference overlap would become an alias component. Here the alias component is shown as an 11.5-kHz signal. This is the difference between 31.5 kHz and 20

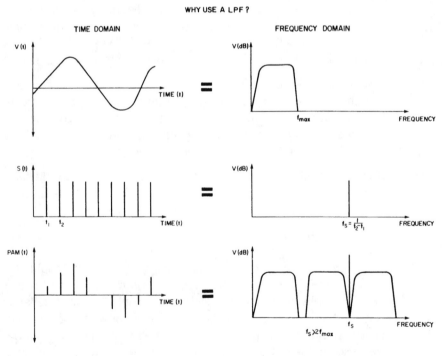

■ **8-44** *Illustration of the time domain and frequency domain.*

kHz. It is an unwanted signal. Once it is in the system it cannot be distinguished from any other audio signal that might be there. That is why it is called an alias component.

PCM ID

In addition to the optional PCM audio, the 8-mm standard format requires that all audio also be recorded as audio frequency modulation (AFM). The 8-mm AFM is identical to Beta hi-fi except that it is a single channel (mono). These two separate audio systems allow the machine to play back bilingual tapes, e.g., English recorded using AFM audio and Japanese recorded using PCM audio. The customer might then select which language to listen to by using the audio monitor switch on the front panel.

When a bilingual tape is played back, the STEREO indicator in the fluorescent display goes out and the BILINGUAL indicator comes on. The way the machine recognizes a bilingual tape is accomplished as follows: In the record process, two bits of every field are reserved for an ID code. If a normal PCM stereo recording is being made, these two bits are recorded as a 0 and a 1. If a bilingual recording is being made, the two bits change to a 1 and a 0. During playback, this ID code is read and tells the machine which indicator to light.

The ID code also distinguishes between stereo and nonstereo TV recordings. For example, if a nonstereo TV program was recorded and played back, the STEREO indicator would go out. The two-bit ID code, indicating a mono recording, is identified as two 0s.

IC154, shown in Fig. 8-45, acts as a data buffer. During playback it receives the ID data from IC102, alters its timing, and then sends the ID data to the system Control IC001. In record, the reverse takes place. The reason the timing must be altered is because system control IC001 is not locked to the PCM process timing.

During playback, the ID data block inside of IC102 strips off the two bits of ID data code that occur during every field. It then sends this information to IC154/pin 2. IC154/pin 14, XCE, is a chip enable signal that goes LOW during the time pin 2 is receiving data. IC154/pin 13, SCK, is a 30-kHz serial clock output used to clock the data into pin 2. SCK occurs during the time that pin 14 is LOW. The timing of this data transfer between IC154 and IC102 is controlled by the 30-Hz RCHG input at pin 4.

307

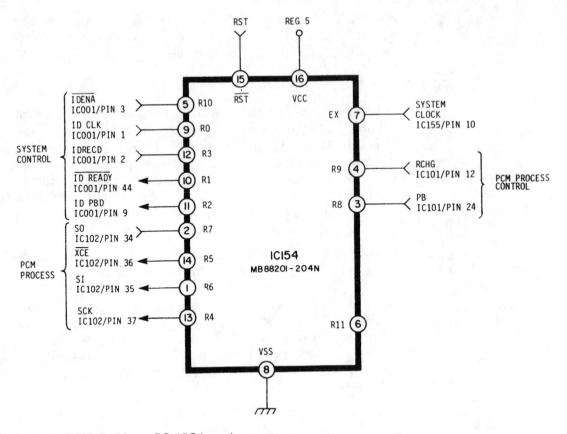

■ **8-45** *PCM ID chip on PC-15C board.*

IC001 receives the ID data from IC154/pin 11. During the time that IC154 sends the ID data, IC154/pin 10 goes LOW. IC154/pin 9 is used to clock the data into IC001. IC154/pin 5 is an input that goes LOW during playback and playback pause. This tells IC154 which playback modes are acceptable for sending ID data. During record, the same process takes place except the data flow is reversed.

9

VCR troubleshooting and performance testing

IN THIS CHAPTER YOU WILL LEARN HOW TO USE THE Sencore[1] VC93 all-format VCR analyzer. This chapter begins with using the VC93 for a fast, reliable method for performance testing your VCR or camcorder. Other sections cover overall VCR symptoms, servo testing and analysis, capstan speed problems, servo lock test, drum speed error test, and chroma troubleshooting.

VCR performance testing

For this performance test you connect the VC93 test lead to the VIDEO OUT and AUDIO OUT jacks of the VCR, insert the Servo Performance Test Tape of the format you are servicing, and you are ready to begin the tests.

Servos locked

This test checks if the capstan and drum are locked to each other. The playback audio is compared with the playback video for constant phase. A "BAD" servos locked reading is most likely caused by a problem in a phase loop.

Capstan speed error

This test determines if the capstan motor is running at the correct speed by measuring the playback audio from the Servo Performance Test Tape. The Servo Analyzer reading gives you a "GOOD/BAD" reading and also displays if the capstan speed is too fast ("+") or too slow ("−") and the percentage of speed error.

Capstan jitter

This servo test measures capstan speed along with measuring small speed variations that help separate speed loop problems from phase loop problems.

[1]Information in this chapter courtesy of Sencore, Inc.

Drum speed error

The drum speed error test checks for proper drum (head cylinder) rotation speed by measuring the playback vertical sync pulses. The digital display measures small variations in drum speed that helps separate speed loop problems from phase loop problems. The digital display indicates a "GOOD/BAD" reading with a percentage of error. This test is measuring the playback vertical sync or the SW30.

Servo problems

There are several symptoms that might or might not be servo related, but appear as though they are servo problems. They include:

☐ Jitter, or noise near picture top or bottom. Screen is blank, or audio is muted. Tape loads, but does not play.

☐ Machine begins to move tape, but then shuts down.

☐ VCR plays tapes it recorded okay, but will not play tapes recorded on another machine.

Symptoms that are probably not related to the servo include:

☐ Picture is always snowy or constantly noisy, but tracking control changes results (i.e. bad head signal).

☐ Picture has a noise bar that remains in the same position on the screen (i.e., tape path).

☐ Picture changes in brightness as tape plays; especially on copies of tapes.

Starting the check out

The Sencore VC93 shown in Fig. 9-1, improves troubleshooting effectiveness through a technique called "Functional Analyzing." In this technique you inject known good signals, supplied by the VC93, into the functional blocks. If the output returns to normal, you are injecting after the defective stage; if the output remains bad, your injection is before the defective stage.

Begin your troubleshooting by checking the machine for obvious operator problems. Check the position of the controls for a clue to the problem. A "Tracking" control that is all the way to one end of its range, for example, is a clue of a possible tracking problem.

310

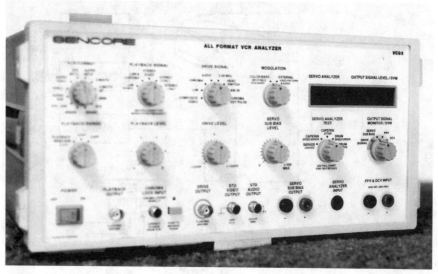

■ **9-1** *Photo of Sencore VC93 all-format VCR analyzer.*

After checking for obvious operator problems, place a work tape into the machine and observe the playback audio and video on a monitor connected to the machine's output.

Always troubleshoot a VCR or camcorder in the playback mode first. Many circuits are common to both playback and record. Ensuring that the playback circuits work eliminates most of the potential circuits that could be at fault. Additionally, to check the record circuits you must record a signal and play it back.

The overall VCR system

A VCR contains five major sections: luminance, chroma, audio, servo, and system control. The following is a brief description of each. Figure 9-2 illustrates these sections.

The luminance section receives the recorded information from the tape and processes it into a form that can be used by a TV or monitor. Inject the VC93 playback signals in the circuits between the video heads and the FM detector. Use the drive signals between the FM detector and the video output.

The chroma section processes the down-converted chroma signal so it can be combined with the luminance signal to create a color playback signal. The VC93 playback signal contains a down-con-

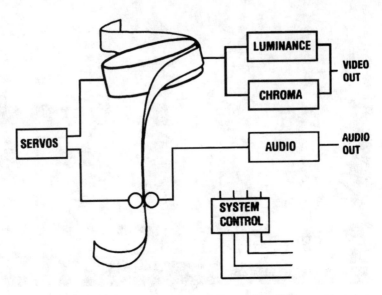

■ 9-2 *The five overall functional sections of a VCR.*

verted chroma signal and other key signals to troubleshoot the chroma conversion circuits.

The audio section converts the linearly recorded audio and FM modulated Stereo Hi-Fi signals to baseband audio. Use the VC93 playback "Stereo" signals to troubleshoot the Hi-Fi stereo section. Use the audio drive signals to troubleshoot the linear audio section.

The servo section controls the tape movement and adjusts the speed and phase relationship of the heads with respect to information recorded on the tape. The VC93 servo analyzer tests evaluate the servo operation and servo sub bias supply is used to further troubleshoot the servos.

The last section is the system control. This section has one or more microprocessors that monitor the entire operation of the machine and tell the various circuits what to do. The microprocessor program is unique to a particular model. This, combined with the digital signals involved, means signal substitution is not the best method for troubleshooting the system control section. Instead, troubleshoot this section with a voltmeter.

Identifying defective video heads

There are two symptoms of a defective video head, depending if one or both heads are at fault. The first symptom, caused by two faulty heads, is complete loss of video. In older VCRs the entire screen is snow. Newer machines often mute the video to produce a blank or solid color raster.

The second symptom appears when one head fails. In this case the machine produces a noisy playback picture. An important part of interpreting this symptom is that the noise must cover every part of the picture. If any section of the picture is clear (even if it's only a few inches somewhere on the screen) you do not have a defective head. The problem is likely servo or tape path related.

Symptoms alone do not prove when video heads are bad. Defects in other luminance circuits can produce identical symptoms. If the symptoms suggest bad video heads, use the VC93 playback signal to substitute for the head signal. Because the heads spin you can not inject the signal directly into them. Instead, inject at the output of the rotary transformer. This will determine if the problem is in the video heads/rotary transformer or if the problem is after the injection point. To determine if the circuits after the rotary transformer are working:

1. Insert a blank tape into the machine and press "PLAY."
2. Set the VC93 as follows:
    • VCR FORMAT to match format being serviced.
    • MODULATION to "Color Bars" or "External."
    • PLAYBACK RANGE to "Playback Head Sub."
    • PLAYBACK SIGNAL to "lum."
3. Connect the HEAD SUBSTITUTION TEST LEAD to the PLAYBACK OUTPUT jack.
4. Connect the HEAD SUBSTITUTION TEST LEAD to the ch. A and ch. B head pre-amps input of the VCR.
5. Observe the playback monitor for an improved picture.
6. Adjust the PLAYBACK LEVEL control for best picture.

A good picture shows that all the circuits after the injection point are good. This leaves either the rotary transformer or the video heads to be defective. Test the rotary transformer. If the rotary transformer is good, do the servo analyzer tests. If they do not indicate a problem, then replace the video heads.

If you do not obtain a good picture, a stage after the injection point is bad. The rotary transformers fail when either the rotating or stationary windings develop an open or short. Test the transformer by injecting the "lum" playback signal. If it passes the signal, the rotary transformer is good. See Fig. 9-3.

Servo functional analyzing

The following explains the VC93 servo analyzer tests, how they work, and how to localize a problem to the defective drum or capstan servo section.

The main difficulty in troubleshooting servo problems is determining which servo loop is at fault. Defects in one servo loop can produce symptoms that look like a problem elsewhere. In addition, nonservo-related problems can sometimes appear as a servo problem.

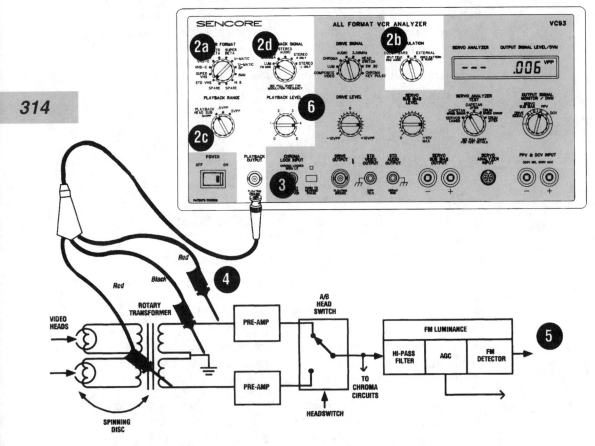

■ 9-3 *Set-up for injecting the PLAYBACK signal into the luminance sections*

The three step process

1. Using the VC93 servo analyzer tests to determine if a problem is servo related.
2. Using the same VC93 servo analyzer tests to localize the problem to the defective drum or capstan servo section.
3. Using the SC3100 waveform analyzer to check key signals to isolate the defective component or circuit within the bad servo section.

Understanding the servo analyzer tests

The VC93 uses five servo analyzer tests to determine if the servos are at fault or if the problem exists elsewhere. These tests are as follows:

☐ Servos locked tests.

☐ Capstan speed error test.

☐ Capstan jitter test.

☐ Drum speed error test.

☐ Drum jitter test.

All applicable tests should be done before troubleshooting further. The servo analyzer tests prove if the problem is servo related, and localizes the problem to the capstan or drum servo loop. The test results are displayed with a "Good/Bad" indication and a percent-of-error reading.

Servo analyzer test leads

The five servo analyzer tests are performed using either the servo performance test lead or the servo troubleshooting test lead. The servo performance test lead connects to the VCR's audio and video output jacks for a fast, easy, overall check of the servos. This test lead must be used with the Sencore servo performance test tape. This tape is recorded with a 479.520-Hz audio tone (vertical frequency × 8) and a 10 Bar Staircase video pattern. The audio signal on this tape is locked to the vertical sync pulse of the video. The VC93 monitors the change in frequency and phase of the audio and video signals in the servo analyzer tests.

The VCR must produce an audio and video signal in order for the VC93 servo performance test lead to work. A signal that has insufficient amplitude (or is missing) will give inconclusive results on the servo tests. If this happens, use the servo troubleshooting test lead.

The servo troubleshooting test lead connects to the key servo reference signals (CTL and SW30) to analyze the condition of the servos even if the audio and video signals are missing. SW30 and CTL are universal test points that are often marked in the VCR. You can usually find these signals without a schematic.

All servo tests should be done starting with the servos locked test. The servos locked test is the first test done to check the overall operation of the servos. Turn the servo analyzer switch to the next test and complete all servo tests for a complete check. Refer to block diagram in Fig. 9-4.

Note: Because 8 MM does not use a CTL pulse or linear audio, the capstan tests do not apply, immediately go to the drum servo test.

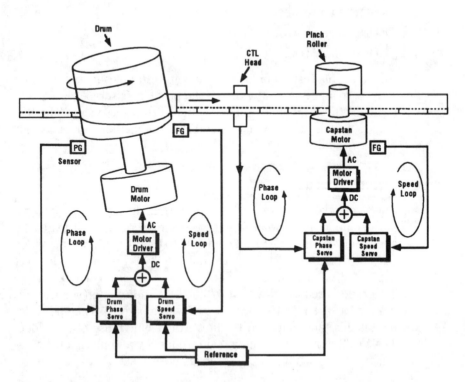

■ **9-4** *The capstan phase loop-and-drum phase loop must be locked to REF 30 for proper operation.*

Servos locked test

The servos locked test compares the change in phase of the CTL pulse to the SW30 pulse when using the servo troubleshooting test lead, or the phase of the audio signal to the video vertical sync pulses—when using the servo performance test lead and Sencore servo performance test tape.

VCRs lock the drum and capstan phase circuits to a common REF-30 source. The REF-30 signal is usually internal to the servo IC and can not be viewed or measured. Because the SW30 pulse is derived from the PG pulse and is more universal to VCRs, it is used to check servo locking. If both servos are operating properly, the capstan reference signal (CTL pulse) and drum signal (SW30 pulse) will be locked together. The VC93 compares these reference signals to each other to determine if the servos are locked, then they must be locked to the internal reference signal. If either servo is not locked to the internal reference signal, either the capstan or the drum loop is defective.

The result of the servos locked test are displayed as a percentage reading indicating how well the capstan and drum are locked. A reading greater than 1.5% will give you a "Bad" indication. A "Bad" servos locked reading indicates that either the capstan or the drum servo phase circuit is defective. A "Good" indication means that the servo phase circuits are operating properly.

The remaining servo analyzer tests further test the servo circuits and help localize problems to the capstan or drum servo loop. Note: The capstan tests should be done before the drum tests.

Capstan speed error test

The capstan error test checks how fast the tape is being pulled through the machine. The VC93 analyzes the frequency of the CTL pulse when using the servo troubleshooting test lead or the playback audio signal when using the servo performance test lead. This test checks the tape speed, not just the capstan motor speed. Mechanical problems such as tape drag or slippage, as well as capstan circuit problems, will cause this test to show a "Bad" indication.

The capstan speed error reading shows how far off the actual tape speed is from the desired speed. A percentage error of more than +/− 0.5% will give you a "Bad" indication. This indicates a problem in either the capstan phase or speed circuit. If a large percentage (more than 10%) error is displayed, this most likely indicates a phase loop or speed loop problem.

If this test produces a "Bad" indication, or no indication at all, the VCR has a problem in the capstan phase, speed loop, or it might be caused by a mechanical tape path problem. A capstan running at the wrong speed is often the result of a missing CTL or FG pulse, a bad motor driver, or a bad control circuit.

Capstan jitter test

The capstan jitter test analyzes for small speed variations. This test analyzes the short term variations in the capstan reference signal to determine how constant the capstan speed is. These short-term variations are called *jitter*.

The capstan jitter readings indicate the amount of tape speed variation. All VCRs have some speed variation due to tape stretch, the tightness of the capstan phase loop, and other mechanical variations. A "Bad" indication shows that the capstan speed variations are greater than 0.5%. This will produce unacceptable audio and might also affect video performance. A "Good" indication means that the capstan circuits are working.

If the test produces a "Bad" indication, it means that there is a problem with the electrical servo circuits or there is a mechanical problem. Excessive capstan jitter is often caused by a missing CTL pulse, excessive oxide buildup on the capstan or pinch roller, or a bad capstan servo control circuit.

Drum speed error test

The drum speed error test analyzes the frequency of the SW30 pulse when using the servo troubleshooting test lead, or the playback vertical sync pulses when using the servo performance test lead and Sencore servo performance test tape. It compares this signal to an internal 29.97-Hz reference signal. The SW30 signal is universal between VCRs and is derived from the drum PG signal that is used by the drum servo circuits.

The drum speed error reading displays how far the drum speed is off from the desired speed. Incorrect drum speed will cause the horizontal sync pulses in the playback video to occur at the wrong time. If the drum speed is slightly off, the symptom appears as a misadjusted horizontal hold control on the playback monitor. Modern VCRs sometimes mute the playback video when the drum speed is off.

318

The "Good/Bad" indication is based on the amount of frequency offset that can be tolerated by most television receivers. Speed errors greater than +/−0.10% produce a "Bad" indication. A "Good" indication means that the drum speed circuits are working properly. A "Bad" indication is most likely caused by a missing FG pulse from the drum motor.

Drum jitter test

The drum jitter test analyzes the frequency variation of the SW30 pulse when using the servo troubleshooting test lead, or the playback video vertical sync pulses when using the servo performance test lead. This test analyzes the drum servo reference signal for short term variations in the speed of the drum. These readings are shown in Fig. 9-5.

TEST	MAXIMUM ALLOWABLE PERCENTAGE READING
Servos Locked	1.5%
Capstan Speed Error	+/− 0.5%
Capstan Jitter	0.5%
Drum Speed Error	+/− 0.1%
Drum Jitter	0.1%

■ **9-5** *"Good/Bad" limits allowable for stereo analyzer tests.*

The percentage reading indicates how much speed variation there is in the revolving drum. A "Good" indication shows that the drum servo is working properly. A "Bad" indication shows that drum speed variations are greater than 0.10%, which will give unacceptable picture quality. Excessive drum jitter is often caused by problems such as bad bearings, excessive oxide on the drum, a missing drum PG signal, or a defective drum servo control loop.

Servo troubleshooting

Servo problems can be more difficult to identify and troubleshoot than other types of problems. Many servo problems produce symptoms of poor audio and/or poor video that can mislead a servicer into troubleshooting circuits that have no defect.

The VC93 provides five servo analyzer tests. Do the tests in sequence, starting with the SERVOS LOCKED test. Use the results of each test to zero in on the defect. Let's now review these tests.

Servos locked test

Determines if the servo capstan phase loop and drum phase loop are locked to the reference signal.

Capstan speed test

Determines if the capstan servo is operating at the correct speed. It identifies speed select circuit problems and other speed related problems.

Capstan jitter test

Measures how constant the capstan movement is. It helps identify capstan phase problems and mechanical problems including motor bearings, bad idlers etc.

Drum speed test

Determines if the drum is operating at the correct speed. It helps identify a drum that is operating too fast or too slow.

Drum jitter test

Measures how constant the drum rotation is. It helps identify drum related problems including bad motor bearings, bad drum phase loop, etc.

A summary of the servo tests and possible causes of "Bad" test results is provided in Fig. 9-6.

Understanding the servos locked test

Let's now go into more detail of the servos locked test. This test compares the change in phase of the CTL pulse to the SW30 pulse when using the "Troubleshooting Test Lead", or the phase of the audio and vertical sync pulse when using the "Performance Test Lead" and "Performance Test Tape." VCRs lock the capstan and drum phase circuits to a common REF 30 source. Usually the REF 30 source is internal to the servo IC and cannot be viewed or measured. Because the capstan and drum phase loops are locked to the same REF 30, they also must be locked to each other. The VC93 uses this fact to check servo lock.

SERVOS LOCKED	CAPSTAN SPEED ERROR	CAPSTAN JITTER	DRUM SPEED ERROR	DRUM JITTER	MOST LIKELY DEFECT
GOOD	GOOD	GOOD	GOOD	GOOD	NO SERVO DEFECTS*
GOOD	GOOD	GOOD	GOOD	BAD	DRUM MECHANICAL
GOOD	GOOD	GOOD	BAD	N/A	REFERENCE FREQUENCY
GOOD	GOOD	BAD	GOOD	GOOD	CAPSTAN MECHANICAL
GOOD	BAD	N/A	GOOD	GOOD	REFERENCE FREQUENCY
GOOD	BAD	N/A	BAD	N/A	REFERENCE FREQUENCY
GOOD	BAD	N/A	GOOD	BAD	REFERENCE FREQUENCY
BAD	GOOD	GOOD	GOOD	GOOD	CAPSTAN PHASE LOOP or DRUM PHASE LOOP
BAD	BAD	N/A	GOOD	GOOD	CAPSTAN SPEED LOOP or CAPSTAN MECHANICAL
BAD	GOOD	BAD	GOOD	GOOD	CAPSTAN PHASE LOOP or CAPSTAN MECHANICAL
BAD	GOOD	GOOD	BAD	N/A	DRUM SPEED LOOP or DRUM MECHANICAL
BAD	GOOD	GOOD	GOOD	BAD	DRUM PHASE LOOP or DRUM MECHANICAL
BAD	BAD	N/A	BAD	N/A	REFERENCE FREQUENCY
BAD	BAD	N/A	GOOD	BAD	REFERENCE FREQUENCY

*NOTE: A noise bar that occurs periodically at a rate of one minute or greater could be a capstan or drum phase problem.

■ **9-6** *Use the results of the five servo tests to determine the most likely cause of a servo problem.*

Note: The SW30 pulse is used for this test because it is derived from the PG pulse and is more universal to all VCRs.

Results of the servos locked test are displayed as a percentage reading that indicates how close the two servos are locked. This reading varies between brands and models. After testing several machines, however, you will develop a feel for what reading is typical for the machine you are testing. Generally older VCRs, that use analog servo circuits, show a higher percentage reading indicating a looser lock. Newer, digital machines, typically have a much smaller percentage reading indicating a more precise lock.

The "Good" and "Bad" servos locked readings are based on extensive research that shows that bad capstan or phase servo circuits produce errors greater than 1.5%. No "Good/Bad" reading is displayed if the servos are varying widely. If you see a "Bad" or no indication, suspect either a bad capstan or drum phase circuit. A "Good" indication means that the phase circuits are operating properly.

Regardless of the servos locked test results, do all the servo analyzer tests before you begin to troubleshoot the problem. The results of the other tests might further isolate the problem to the capstan or drum phase loop.

Special notes on the servos locked test

Note 1: Some newer machines have very tight capstan and drum speed loops that operate marginally even with a bad phase loop. The servos locked test has a minimum resolution and will not identify a problem where the reference signals change phase slower than one cycle per minute. If you observe a noise bar that comes and goes slower than once a minute, suspect a servo phase loop problem.

Note 2: Some machines use a combined PG/FG drum signal as illustrated in Fig. 9-7. The servos in these machines can lock incorrectly if the drum PG signal is missing, locked instead to the FG signal. Because the FG signal is a higher frequency, the servos can lock to one of several transitions. Depending on which transition it locks to, the drum will lock, but noise or a noise bar occurs in the picture. The VC93 servos locked test will indicate that the servos are locked. This is a correct diagnosis as the servo phase loops are locked.

The best way to prove this problem is to stop and start the VCR several times while observing the position of the noise bar. Because the servos can lock on any FG transition, the noise bar will change position each time the machine is placed into "PLAY." If you observe this symptom, check the combined FG/PG pulse for a missing PG pulse.

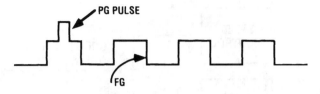

a. *NORMAL PG/FG PULSE*

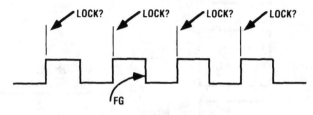

b. *PG PULSE MISSING*

■ **9-7** *Comparison of normal and abnormal PG pulses.*

Understanding the capstan speed error test

The capstan speed error percentage readings tell how far off the actual tape speed is from the desired speed. The percentage reading is very helpful in identifying the source of a speed error problem.

Readings that are off more than 10% suggest a problem in the capstan speed loop. Readings that are within 10% are caused by either a defective capstan speed loop or a defective capstan phase loop. (A defective phase loop will produce a wrong correction voltage. This voltage adds to the speed loop correction voltage to control the capstan motor. Because the capstan motor control voltage is slightly off, the motor speed is also off.)

To determine which loop is defective, play back a tape that contains segments of different tape speeds. If the speed circuits are working properly, you will obtain low capstan speed error percentages at all speeds. But if the speed circuits are not working properly, you will obtain large percentage errors at some tape speeds. Notice the block drawing in Fig. 9-8.

Note: The slowest tape speed will likely have a larger error than faster tape speeds, even in a correctly functioning machine.

If the capstan speed error percentages are only slightly greater than two percent for all tape speeds, the problem is likely to be a

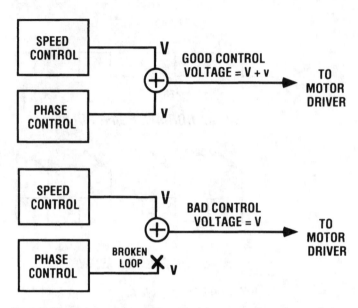

■ **9-8** *The phase-control loop adds a small amount of correction voltage to the speed control voltage.*

mechanical problem, such as excessive oxide buildup on the capstan, a hard capstan idler, or a loose capstan motor belt. If these check out OK, the problem is likely a defective capstan phase loop.

The "Good/Bad" readings are based on extensive research that shows that capstan speed errors greater than $+/- 0.5\%$ cannot be corrected by the phase loop circuitry. No reading will be displayed if the tapes speed is varying widely. If you see a "Bad" indicator, or no indication at all, use the percent readings to help determine what the capstan problem is.

Understanding the capstan jitter test

The capstan jitter analyzer tests the CTL pulse when using the "Troubleshooting Test Lead," or the playback audio when using the "Performance Test Tape," for minute speed variations. This test actually checks tape speed variations and not just capstan motor speed variations. Thus, it checks for mechanical problems such as tape drag and slippage, as well as for failures in the capstan circuits.

The capstan jitter percentage readings indicate the amount of tape speed variation. All machines have some speed variation due to tape stretch, the tightness of the capstan phase loops, and other mechanical problems. Excessive speed variations causes the linear audio to vary excessively in pitch.

The "Good/Bad" indications are based on research that shows that capstan speed variations greater than 0.5% give unacceptable audio performance and can affect the video picture playback.

Likely causes of "Bad" capstan jitter are as follows: a defective capstan phase loop, a slipping capstan belt or idler, or a bent or dirty capstan shaft. Also suspect a possible bad capstan motor. A bad motor bearing or a motor winding can cause the capstan to "catch" as it turns. The lack of a "Good/Bad" reading indicates that one of the above potential defects is causing severe capstan jitter.

Understanding the drum speed error test

The drum speed error test analyzes the frequency of the SW30 pulse when using the "Troubleshooting Test Leads," or the playback video vertical sync pulse when using the SERVO PERFORMANCE TEST LEAD and "Performance Test Tape." It compares the signal to an internal 29.97-Hz reference. The SW30 signal is universal for most VCRs and comes from the drum PG signal that is used by the drum servo circuits.

The drum speed error percentage reading displays how far the drum speed is off from the desired speed. Wrong drum speed will cause the horizontal sync pulses in the playback video to occur at the wrong time. If the drum speed is close to correct, the symptom appears as a misadjusted horizontal hold control on the playback monitor. A large drum speed error will cause the playback video to be so bad that no conclusion can be made by observing the playback monitor. Newer VCRs will mute the playback video when the drum speed is far off.

The good/bad readings are based on the amount of frequency offset that can be tolerated by most television receivers. A speed error more than 0.10% will produce a "Bad" indication. If the drum speed is varying widely, no "Good/Bad" indication is displayed. The most likely reason for a "Bad" drum speed error reading is a missing FG pulse from the drum motor.

Understanding the drum jitter test

The drum jitter test analyzes the SW30 signal, or the playback video vertical sync pulse when you use the SERVO PERFORMANCE TEST LEAD, for minute speed variations. The percentage reading indicates how much the speed of the revolving drum varies. Excessive drum speed variations cause the picture to appear to "breath" in and out.

The "Good/Bad" indication is based on research showing that drum speed variations greater than 0.10% give unacceptable picture quality. No "Good/Bad" reading is displayed if the speed is varying widely. This indicates a problem in the drum servo. A "Bad" indication also means the drum servo is bad. The most likely reason is a problem in the drum phase circuit or a mechanical defect such as a bad motor bearing, a drum that is out of balance, or a spot of dirt build-up or oxide on the drum surface.

Troubleshooting the VHS-C format

The only main difference between VHS and VHS-C machines is the configuration of the video heads. The VHS-C format was developed to make the mechanical section physically smaller for use in camcorders. To get this size reduction, the video head cylinder was reduced in size. To make the information recorded on the tape compatible with standard VHS machines, VHS-C uses four video heads instead of two.

Use the same procedures to troubleshoot and signal inject into VHS-C machines as used for standard VHS machines. The only procedure that is different is injecting a signal between the rotary transformer and the headswitcher. Because VHS-C machines have four outputs instead of two, you will need to jumper the corresponding video heads into pairs so you can inject the signal into all four heads simultaneously. Figure 9-9 shows the connections for injecting at the output of rotary transformers in VHS-C machines.

Overall chroma checks

If you see color, all playback color circuits are working. If you cannot obtain color when playing a test tape that contains a color pattern, the chroma signal is getting lost before the VC93 injection point.

If the "Checking Overall Chroma Operation" test does not produce color, either the color circuits are defective or one of the chroma signals is missing or incorrect.

Note: Some VCRs separate the chroma and luminance signals inside the enclosure containing the head pre-amps. These VCRs have separate luminance and chroma signal output pins. Connect

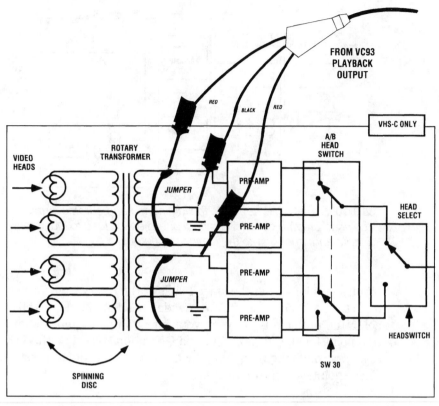

FROM VC93 PLAYBACK OUTPUT

RED BLACK RED

VHS-C ONLY

VIDEO HEADS

ROTARY TRANSFORMER

A/B HEAD SWITCH

JUMPER

PRE-AMP

PRE-AMP

HEAD SELECT

PRE-AMP

JUMPER

PRE-AMP

HEADSWITCH

SPINNING DISC

SW 30

■ **9-9** *VHS-C VCRs use four video heads. Jumper the four heads into pairs for signal injection.*

one of the "red" HEAD SUBSTITUTION TEST LEADs to the luminance pin and the other "red" lead to the chroma pin. Refer to Fig. 9-10 for these connections.

Identifying head related chroma problems

A wrong or missing color problem can be caused by a frequency response problem in a stage before the chroma and luminance processing circuits. To test the stages before the chroma conversion circuits follow these procedures:

1. Set up the VC93 for checking overall chroma operation.
2. Set the PLAYBACK RANGE control to "Playback Head Sub."
3. Move the HEAD SUBSTITUTION TEST LEAD to the rotary transformer output.
4. Observe the playback monitor for color.
5. Adjust the PLAYBACK LEVEL control for best color.

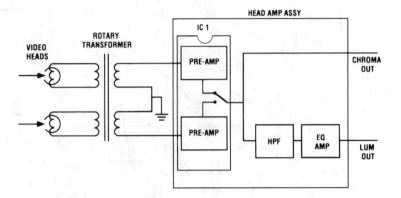

■ 9-10 *Some VCRs separate the luminance and chroma signals in the head preamp circuits. Connect one injection lead to the chroma output and other one to the luminance output.*

If you see color, either the video heads or the rotary transformer is defective. A likely cause is a worn video head. If you do not obtain color, the problem is between your injection point and the injection point for the "Overall Chroma Operation" test. Use the VC93 to step through the remaining stages to isolate the defective stage. These points are shown in Fig. 9-11.

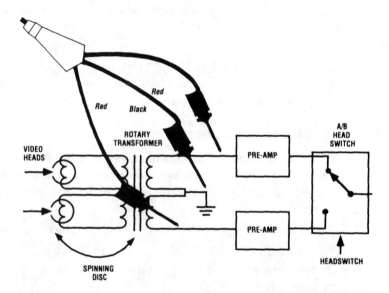

■ 9-11 *If you cannot obtain color from a VCR tape, but the VC93 proves that the color circuits are working, move to the test leads to head pre-amp input to determine if the preamps are functioning.*

Troubleshooting chroma conversion problems

If the "Overall Chroma Operation" test does not produce color, you need to troubleshoot the chroma conversion circuits. Before substituting for these signals, check these three additional things:

☐ Power supply voltage to the conversion IC.

☐ Grounds to the color conversion IC.

☐ The record/playback selector signal.

If these signals are correct, you need to troubleshoot the color conversion section. First, supply a down-converted chroma signal from the VC93 to the VCR chroma conversion circuits. Now substitute the 3.58-MHz color oscillator, substitute the chroma key pulse, and then sub in the SW30-Hz pulse.

Substituting the chroma key pulse signal

VHS, Beta, and 8-MM VCRs use a horizontal sync pulse to operate the chroma phase shift circuits. No or poor playback color occurs if the phase shift circuits are not functioning. Some VCRs feed the

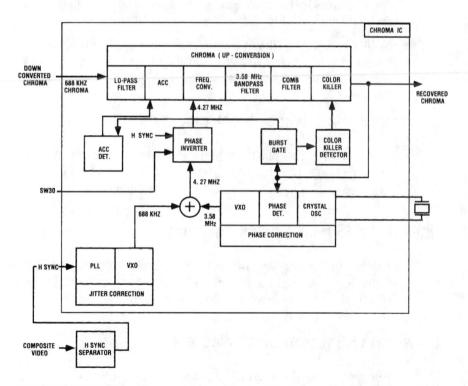

■ 9-12 *Some VCRs use a sync separator to obtain the horizontal sync pulse for the chroma circuits.*

composite video from the luminance circuits to the color IC and extract the horizontal sync pulses inside the IC. Other VCRs have the sync separator outside the chroma IC. The block diagram of a VCR chroma system is shown in Fig. 9-12.

Zenith VCR model VRM4150 self diagnostic

An internal program in these Zenith VCRs microcomputer chips will display error codes that indicate an area in the VCR where a problem might exist. This new feature will provide technicians additional information that will help to locate a defect more rapidly. To see this error code, use the remote hand control and press MENU. When the word "MENU" comes up on the Fluorescent display, press 4,3,2,1 and then "Enter". The self diagnostic procedure will sequence through nine checks. During this time, the fluorescent display will display "Diagnosing". If there is no faults in these checks, the display will indicate "NO DATA". If an error is detected, a code number will flash for approximately five seconds and then the VCR will return to its initial mode of operation. The self diagnosis feature can be activated with the VCR on or off. Refer to the chart in Fig. 9-13 for a description of the error codes.

Note: Error codes will not be stored in memory. The program must be re-initialized each time for self-diagnosis. All pin numbers in the chart refer to IC501, the mechacon IC.

Universal 8-MM playback block diagrams

Figure 9-14 is the playback luminance/chroma circuit block diagram. Figure 9-15 is the playback audio circuit block diagram. Figure 9-16 is the playback servo circuit block diagram.

Universal VHS record block diagrams

Figure 9-17 is the record luminance/chroma circuits block diagram. Figure 9-18 is the record audio circuit block diagram. Figure 9-19 is the record servo circuit block diagram.

Universal VHS record block diagrams

Figure 9-20 is the record luminance/chroma circuits block diagram. Figure 9-21 is the record audio circuit block diagram. Figure 9-22 is the record servo circuit block diagram.

DISPLAY	DESCRIPTION	DETECTION SIGNAL	CIRCUIT STATUS	SERVICE POINTS
"*D 1*"	• TAPE LOADING ERROR	• Mode SW Pos. A, B, C Pins 20, 21, 22 • Load Motor Pins 87 (+), 88 (−)	• Mode SW Position not changed within 6 seconds after cassette loading attempt.	• Loading motor, mech. problems gears, timing. • Mode SW Position Circuitry
"*D 2*"	• TAPE LOADING ERROR	• Mode SW Pos. A, B, C Pins 20, 21, 22 • Load Motor Pins 87 (+), 88 (−)	• Mode SW Position not changed within 6 seconds after cassette loading attempt.	• Mode SW contacts
"*D 3*"	• CST LOADING ERROR	• Mode SW Pos. A, B, C Pins 20, 21, 22 • Load Motor Pins 87 (+), 88 (−) • CST SW PIN 29	• Mode SW Position not changed within 6 seconds after cassette loading attempt.	• Mode SW contacts
"*D 4*"	• CST LOADING ERROR	• Mode SW Pos. A, B, C Pins 20, 21, 22 • Load Motor Pins 87 (+), 88 (−) • CST SW PIN 29	• CST SW must be activated within 3 seconds, otherwise unit shut down will occur.	• CST SW or Connector Contacts
"*D 5*"	• DRUM MOTOR ERROR	• Head SW (30Hz) Pulses PIN 19	• Drum Motor (slow start). • Motor must be up to SPEED within 3 seconds of operation.	• Drum Motor and Control Circuits
"*D 6*"	• REEL ROTATION ERROR	• Take-up Reel Pulses PIN 25	• CFG Signal present but take-up pulses are missing (capstan motor running).	• Capstan belt • Idler and reel gears damaged • T/R Sensor
"*D 7*"	• CAPSTAN MOTOR ERROR	• CAPSTAN FG Pulses PIN 32	• NO CFG SIGNAL	• Capstan control ckt. • Motor assm. • IC201 ckt.
"*D 8*"	• HEAD/PRE-AMP ERROR	• PB-FM Waveform PIN 100	• No PB-FM One head missing Dirt on head.	• Clean head • Replace head • Pre-Amp
"*F 1*"	• TUNER/VIDEO ERROR	• Composite Sync PIN 23	• Composite sync not detected, RF or VIDEO.	• Loss of C-Sync, Tuner Line Video Path

VCR should be in PB mode for D5, D6 and D7. Stop mode F1, Cassette loading D3, D4 and tape loading D1, D2.

■ **9-13** *Zenith VCR self diagnosis chart.*

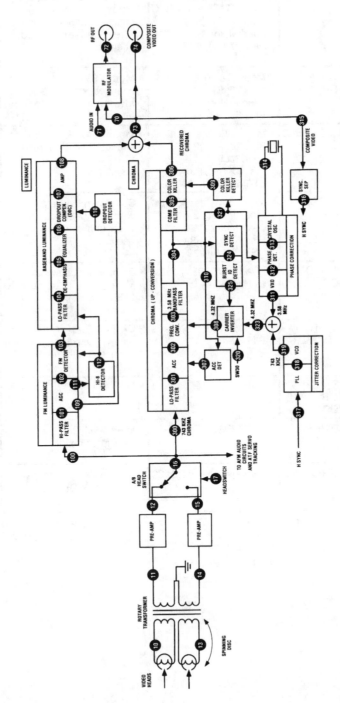

■ 9-14 Playback luminance/chroma circuit block diagram.

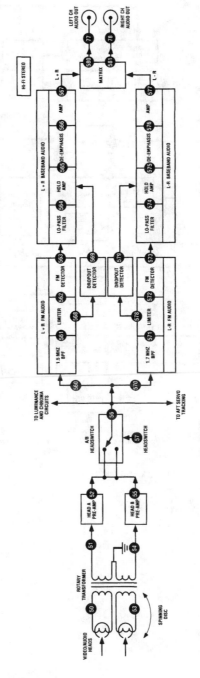

■ **9-15** Playback audio circuit block diagram.

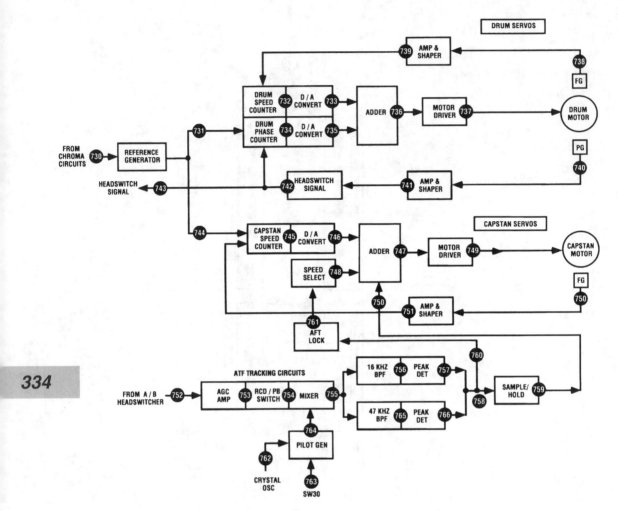

■ **9-16** *Playback servo circuit block diagram.*

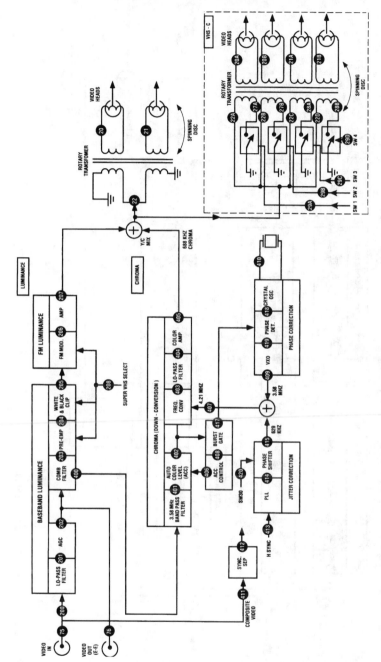

9-17 *Record luminance/chroma circuits block diagram.*

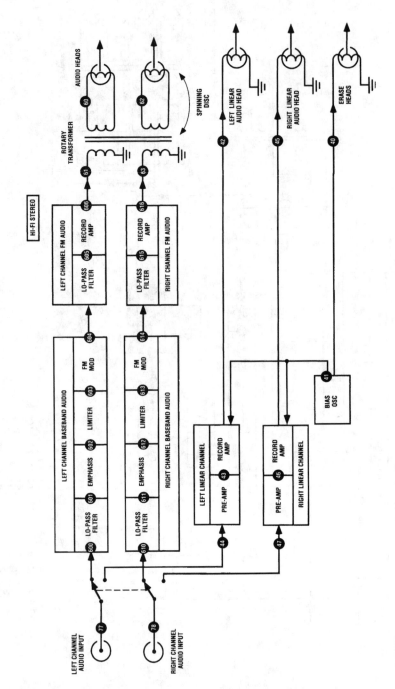

336

■ 9-18 Record audio circuit block diagram.

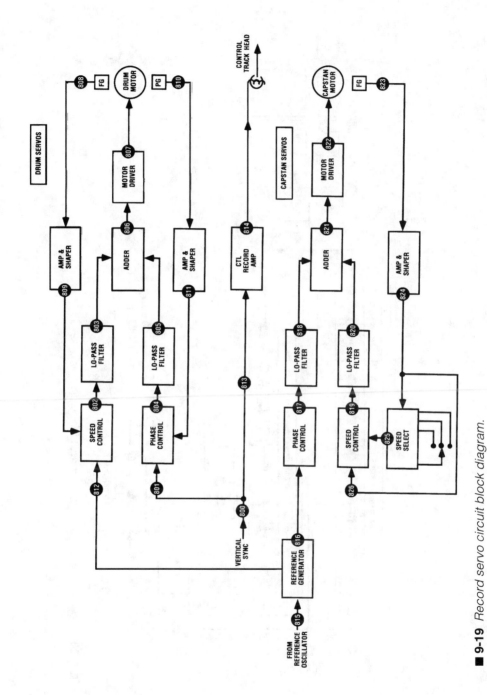

■ 9-19 Record servo circuit block diagram.

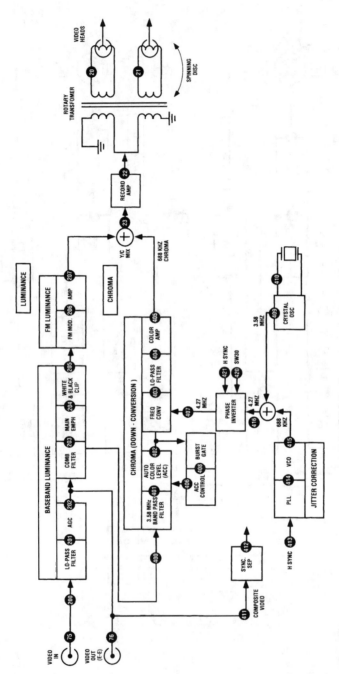

■ 9-20 *Record luminance/chroma circuits block diagram.*

338

VCR troubleshooting and performance testing

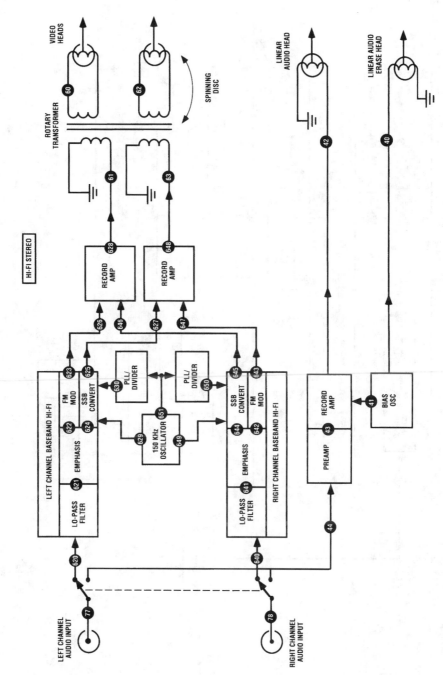

339

■ **9-21** *Record audio circuit block diagram.*

Universal VHS record block diagrams

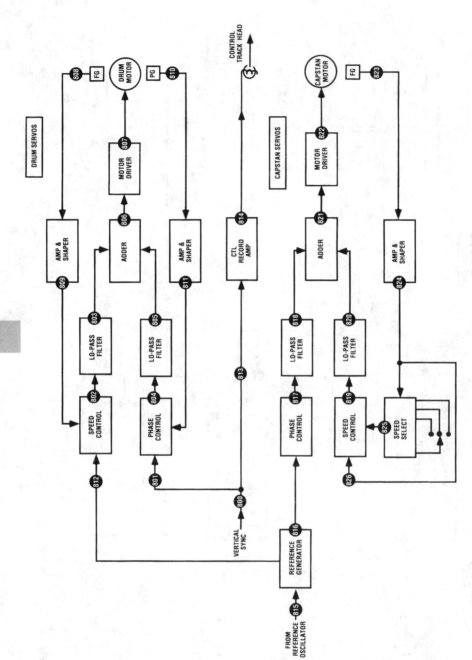

■ **9-22** Record servo circuit block diagram.

10

Video camera and camcorder analyzer techniques

THIS CHAPTER BEGINS WITH VIDEO CAMERA BASICS AND A brief discussion of how they operate. The chapter continues with help in understanding the various types of video cameras and it includes functional signal block diagrams.

Also featured is the Sencore[1] CVA94 "Video tracker" camera analyzer that is used for performing digital waveform measurements, vectorscope measurements for error-free color camera checks, "hum" test for power supply checks, "chroma noise" tests, and "burst frequency and "frequency error" to quickly identify reference oscillator problems. Plus information on the Sencore VR940 video reference color bar chart.

The last portion of this chapter covers other video camera adjustments. This includes sync generator frequency, sync level settings, burst levels, black level setup, and adjustments made with a scope or the Sencore SC3100 waveform analyzer.

Video cameras are used everywhere in greater and greater numbers. Consumers are hooked on capturing all of their unforgettable moments with portable, convenient camcorders. Business and industry are increasing their use of video cameras for security, process control, and remote monitoring.

Considering that service needs for electronic equipment generally lag sales by about three years, the growth opportunity for video camera servicing is just hitting its stride. There are excellent opportunities for both consumer and business. Although prices have decreased, cameras and camcorders are still not inexpensive, and owners are willing to pay to have them serviced rather than re-

[1]Information in this chapter courtesy of Sencore, Inc.

place them. Any servicer with basic video experience, and a minimal amount of training and test instrument investment can easily capture a profitable segment of this growing service market.

Camera basics

There are two basic types of video cameras, classified according to the type of pickup device they use to convert light to electrical signals. The two camera types are those with tube-type pickup devices and those with solid-state pickups. There are two types of solid-state pickups, CCD (Charge Coupled Devices) and MOS (Metal Oxide Semiconductor), although CCD is now more popular. One type of CCD camcorder is the model CVR315 EASYCAM shown in Fig. 10-1.

■ **10-1** *Photo of model CVR315 EasyCam video camcorder.*

Until the mid eighties, all video cameras used a tube-type device for image pickup. As solid state imaging ICs became available, they were quickly incorporated into portable consumer cameras for their small size and light-weight advantages. Industrial cameras in this time period still primarily used tube-type pickups for their better sensitivity in low light, their better resolution, and their lower cost.

However, as the cost of solid-state imaging ICs decreased and their resolution and light sensitivity increased, consumer, industrial, and even broadcast users have converted almost entirely to solid-state CCD pickup cameras. Other advantages of CCD pickups are their increased ruggedness, decreased image lag, lower power drive circuits, higher level output signal, and simplified support circuits.

How video cameras work

Regardless of the type of pickup device, the operation of camera circuits is very similar from one video camera to another. The simplified block diagram shown in Fig. 10-2 is the interconnection of major circuits and signal flow for all camera types. Besides the lens and pickup device, the signal processing and control circuits are similar to circuits found in other consumer video products.

Camera types

You are likely to encounter video cameras in one of two basic configurations. Either the camera will be a stand-alone device or it will be part of a camcorder.

Stand-alone video cameras are used by industry for monitoring, surveillance, or for developing video signals to be taped onto commercial video recorders. Video broadcast studios also use stand-alone video cameras in their broadcast studios. Early consumer

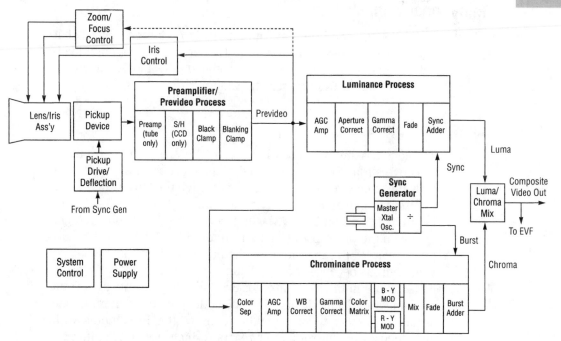

■ **10-2** *Simplified video camera block diagram.*

cameras were of the stand-alone type, but they have been almost entirely replaced by camcorders.

Camcorders are in heavy use by consumers as well as by broadcasters for field use. Camcorders combine a camera, a VCR record/playback section, and a small viewfinder to form a convenient unit for recording and playing back video images as shown in Fig. 10-3. The major difference with a camera found in a camcorder is that it shares some of its support and control circuits (power supply, system control, Y/C mix) with the VCR section.

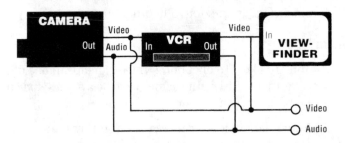

■ **10-3** *A camcorder is a combination of a camera, VCR, and viewfinder.*

344

Functional signal blocks

The following is a brief description of the operational block that make up a typical video camera. Note that, depending on individual camera design, the order of some of the blocks might be slightly different from camera to camera.

Lens/iris/motors

The lens assembly focuses light from the scene onto the light-sensitive surface of the pick-up device. The auto-iris circuit controls the amount of light that passes through the lens by operating a motor to open and close the iris diagram. Notice Fig. 10-4. Under bright lighting conditions the iris controls the amount of light falling on the pick-up and thus the amplitude of the prevideo output signal. Proper operation of the auto-iris circuit is crucial for video output because the iris diaphragm is spring-loaded closed, and a failure in the iris control or drive circuit prevents light from reaching the pickup device. The focus drive circuit generates the signals necessary to operate the focus motor. In cameras with auto-focus, the control circuit reacts to high frequency information in the prevideo signal, or to an infrared or LED sensor. The

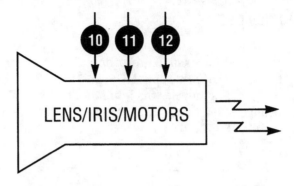

■ **10-4** *The lens, iris, and control motors control the light passed through to the pick-up device.*

zoom drive circuit generates the signals necessary to operate the zoom motor by reacting to input from the camera zoom switches.

Sync generator

The sync generator provides synchronization for all the other camera circuits. The output signals are developed by dividing down the signal from a master crystal oscillator. The master oscillator typically runs at two, four, or eight times the 3.58-MHz chroma burst frequency. The sync generator provides horizontal and vertical drive signals to the pickup device, composite sync and burst for the video output, and 3.58-MHz subcarrier reference signals for the R-Y and B-Y modulators. Refer to the block diagram in Fig. 10-5.

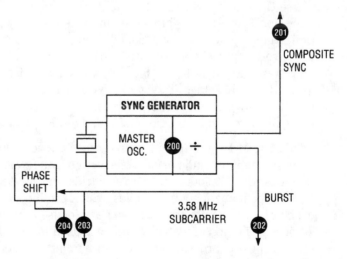

■ **10-5** *The sync generator provides timing signals for the remaining camera stages.*

Pick-up devices

There are three major types of image pick-up devices that have been used in consumer, broadcast, and industrial video cameras. These are vacuum tube, MOS, and CCD. MOS and CCD are solid state pick-up devices made up of a large number of photodiodes arranged horizontally and vertically in rows and columns as illustrated in Fig. 10-6.

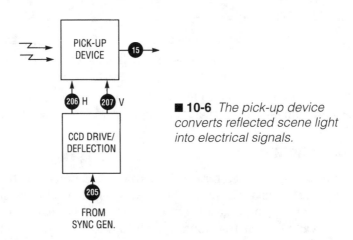

■ **10-6** *The pick-up device converts reflected scene light into electrical signals.*

Tube pick-up devices (Vidicon, Saticon, and Newvicon are common types) use magnetic yoke deflection and a high voltage supply to scan an electron beam across a light-sensitive surface. These tube pick-ups suffer the same scanning irregularities that television CRT's have, plus more, and require many scan correction circuits to produce an acceptable output signal. Also, the very low level output signal from tube pick-ups (200 uV or less) requires an extremely high-gain, extremely low-noise preamplifier as the first signal stage. Tube pick-ups have been entirely replaced by solid-state CCD and MOS devices in consumer cameras and are being phased out of most broadcast and industrial cameras.

Solid-state MOS and CCD pick-up devices are very similar to each other in operation and performance with only a few significant differences. Conversion of light to electrical energy takes place at each of the individual photodiodes, which produce a small electrical charge when light from the scene is focused on their exposed surface. A method of matrix scanning is used to repeatedly collect each of these charges and assemble them into a video signal. The scanning method used to collect these charges is one of the major differences between MOS and CCD pick-ups.

MOS devices use a scanning method that results in three or four signal output lines. These lines carry white, yellow, cyan, and green color output signals (no green for older three-line devices). One disadvantage of MOS devices is that the output signals are at a fairly low level (40-50 mV) and require low-noise preamps to bring the signals up to levels usable by standard signal processing circuits.

CCD devices use a scanning method that results in a single video output line. This signal contains all the necessary luminance and chrominance information needed to generate NTSC composite video. Also, the level of the output signal is high enough that no preamp is required. An advantage of CCD devices is that they have been more reliable than MOS devices.

With all types of pick-up devices, when color is desired, a multi-colored filter is placed in front of the pick-up device's light-sensitive surface. This, along with the scanning of the device, results in the production of an extra high frequency signal that carries information about color in the scene.

Prevideo process

Preamplifiers(s) immediately follow tube and MOS pick-up devices to increase the signals to levels usable by the following signal circuits. CCD devices produce a relatively high output and do not require a special preamplifier. CCD and MOS pick-up devices both develop their output signals(s) by collecting the charge on individual photodiodes. This causes the output signal to occur as isolated voltage charges corresponding to each photodiode that was scanned, with no signal between the individual charges. CCD and MOS cameras both use sample and hold circuits to detect the level of each charge and hold that level until the next one arrives. This results in a continuous output signal. Also, the sequence of colors obtained from the color filter is rearranged to produce a continuous luminance output signal. Note this process in Fig. 10-7.

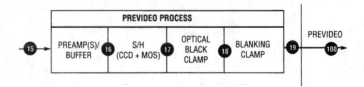

■ **10-7** *The prevideo process section converts the signal into a form useable by the luminance and Chrominance Process sections.*

An optical black clamp circuit clamps the signal to a black reference level at the end of each horizontal scan, while the pick-up device is scanning an optically black area of the pick-up surface. A blanking clamp circuit clamps the entire blanking period to a fixed level with respect to the black level. The difference between this blanking level and the black reference level is called black setup.

Luminance process

Gamma correction in the luminance process section compresses the white portion of the video signal to correct for the black compression that occurs in the CRT of all television sets and monitors. Note these luminance process blocks in Fig. 10-8. This causes the intensity of the image picked up by the camera to directly correspond to the brightness of the image on the TV screen. At low light levels when the iris is fully open, the AGC amplifier reacts to a control voltage from the AGC detector to maintain a constant signal level at the amplifier output. This circuit normally amplifies only during low light levels when the iris has reached the end of its range and cannot open any further to hold the prevideo signal level constant. The aperture correction circuit (edge correction) senses transitions from black to white and vice versa in both the horizontal and vertical directions. Fast spikes are added to these transition points in order to increase the sharpness of edges in the picture. The fade circuit causes the video signal to slowly decrease to black when the camera fade button is released. At the end of the luminance process section, horizontal and vertical sync are added to the video signal to form a composite sync and luminance (Y) signal, which is sent to the Y/C mixer stage.

348

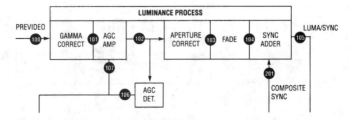

■ **10-8** *The Luminance Process Section converts the prevideo signal into video with sync.*

Chrominance process

The chrominance signal is separated from the rest of the video signal at the start of the chrominance process section. This is done with a filter in tube type cameras, and with sample and hold circuits or signal delay/summing circuits in MOS and CCD cameras. The signal at this point usually consists of R-Y and B-Y signals during alternating horizontal scan periods, or of individual red, green, and blue signals. White balance correction controls the level balance between the red and blue signals to ensure reproduced white picture areas of the scene are being scanned. Gamma correction corrects the amplitude of the chroma signal the same as it does the luminance signal. The chroma AGC amplifier receives a control signal from the AGC detector in the luminance section and amplifies the signal during low light conditions when the iris is unable to open further to hold the signal level constant. Note block circuit diagram in Fig. 10-9.

The R-Y/B-Y separation circuit produces separate R-Y and B-Y signals from the color input signal(s). These two signals are then sent to modulators in which they are modulated into separate 3.58-MHz subcarrier signals. The R-Y subcarrier is shifted 90 degrees from the B-Y subcarrier. The two modulated subcarriers are then added to produce a composite chrominance signal. After going through a fade circuit similar to that in the luminance section, burst from the sync generator is added to the chroma signal and is sent to the Y/C mixer stage.

The auto white balance circuit analyzes the color composition of light arriving at the camera, by examining the output of either the

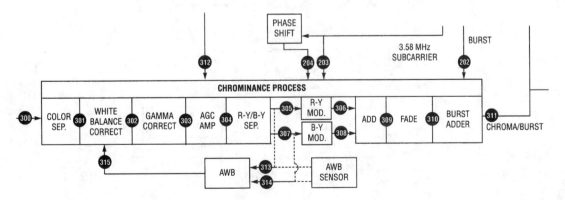

■ **10-9** *The Chrominance Process Section separates the color portion of the prevideo signal and converts it into 3.58-MHz subcarrier chroma with burst.*

R-Y/B-Y separate circuit of the white balance sensor. After averaging the color out over a period of time, it applies a correction signal to the red or blue channel in the white balance correction stage.

Video camera troubleshooting

You can start troubleshooting video cameras with the CVA94 Sencore camera video analyzer by framing the camera on reference test charts and viewing the video output signal on the CVA94s waveform and vector displays, a video monitor, or by viewing internal camera signals on an oscilloscope. You will use the CVA94 and the video monitor to identify degraded signals at the camera output and help localize the section of the camera that is causing the signal defect. You will then need an oscilloscope to signal trace the reference signal through the suspected sections of the camera to isolate the defective stage.

The video camera functional analyzing troubleshooting guide (troubles tree shown in Fig. 10-10), will help you identify which tests and measurements to perform with the CVA94 at the cam-

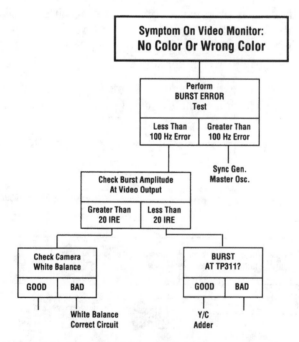

■ **10-10** *The troubleshooting guide will assist you in servicing video cameras with the CVA94.*

era's video output, which mechanical parts to check, and which sections of the camera to signal trace with the waveform analyzer to localize and isolate camera problems. Begin your troubleshooting by selecting the "trouble tree" branch that best fits the camera symptoms you observe. A photo of the Sencore CVA94 camera video analyzer is shown in Fig. 10-11.

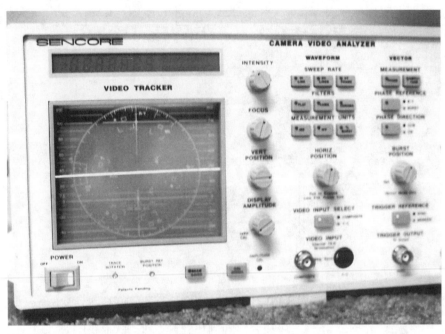

■ 10-11 *The Sencore CVA94 camera video analyzer.*

Camera sync levels

A video camera adds horizontal and vertical sync pulses to the luma and chroma signals it generates to produce composite video. Video monitors and other devices that receive signals from the camera's video output expect the sync signals to be at the industry standard amplitude of 40 IRE, or 286 mV in a 1 VPP video signal. Note video waveforms shown in Fig. 10-12.

If the sync signals are not at the proper level, loss of sync or signal distortion might occur in the devices to which the camera video signal is applied.

If the camera's sync level is not correct, suspect the sync generator or the sync adder stage. The sync level adjustment, if there is

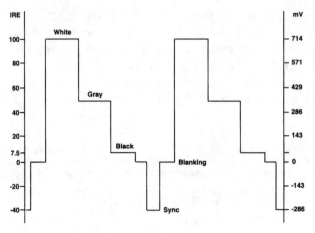

Sample video signal at 2H horizontal sweep rate.

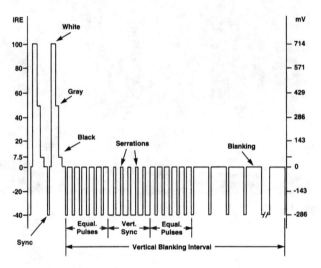

■ **10-12** *Sample video 2H horizontal sweep and 2V vertical sweep rate waveforms.*

one, should be checked first. Checking camera sync level can be done by following these procedures:

1. Cap the camera's lens.
2. Press the CVA94 1H SWEEP RATE button.
3. Press the LUMA FILTERS button.
4. Press the IRE or mV MEASUREMENT UNITS button.
5. Adjust the DELTA BAR POSITION buttons to place the left edge of the CRT DELTA BAR in the middle of the horizontal sync pulse.

6. Adjust the DELTA BAR WIDTH buttons to highlight the rising right edge of the sync pulse including a portion of the back porch just to the left of the burst (do not include any of the burst signal).

7. Check to see that the sync level displayed on the CVA94 LCD DISPLAY is within 5% of 40 IRE or 286 mV.

Checking luminance level

The gain of the camera's luminance stages are normally adjusted to set the amplitude of maximum white level signals so that the total amplitude between the maximum white level and the sync tip level is 1 VPP. Note the waveform graticule shown in Fig. 10-13.

If the camera's luminance level is misadjusted, the picture will be too light or too dark.

If the camera's luminance level is wrong, but the chroma levels are correct, suspect one of the circuits in the LUMINANCE PROCESS

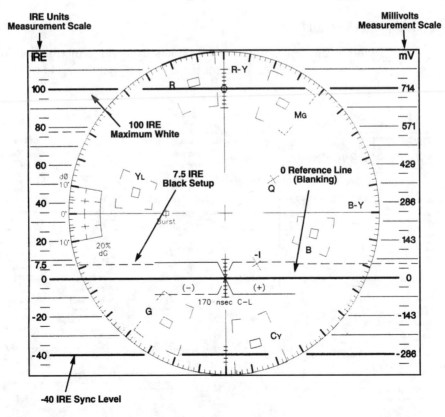

■ 10-13 CRT screen marking of Sencore CVA94.

section. The luminance level adjustment should be checked first. If both luminance and chroma levels are wrong, suspect the circuits in the PREVIDEO PROCESS section. The front panel controls of the CVA94 is shown in Fig. 10-14. You can check the camera's luminance level by following these steps:

1. Frame the camera on the GRAY SCALE CHART.

2. Press the CVA94 1H or 2H SWEEP RATE button.

3. Press the LUMA FILTERS button.

4. Press the IRE or mV MEASUREMENT UNITS button.

5. Adjust the DELTA BAR POSITION buttons to place the left edge of the CRT DELTA BAR just to the right of the chroma burst.

6. Adjust the DELTA BAR WIDTH buttons to highlight approximately half of the white bar, which is just to the right of the burst signal.

7. Check to see that the luminance level displayed on the CVA94 LCD DISPLAY is within 5% of 100 IRE or 714 mV.

Checking black setup level

To aid in the separation of sync signals from video modulation in video-receiving devices, the maximum black signal level in a cam-

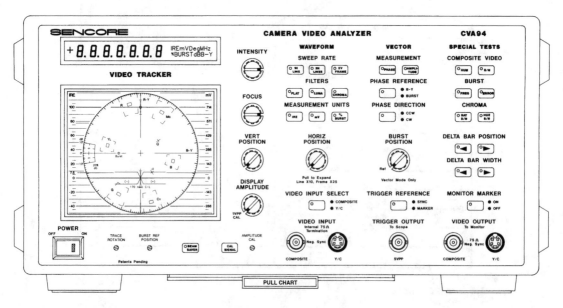

■ **10-14** *Front panel controls of Sencore CVA94 analyzer.*

Video camera and camcorder analyzer techniques

era's composite video signal is held slightly above the blanking level. The difference in level between blanking and maximum black is referred to as black setup (sometimes also called *pedestal* level).

If this proper level of setup is not maintained, sync instability might result in some receiving devices, or the picture will be too light or too dark.

If the camera's setup level is wrong, suspect the blanking control stage in the PREVIDEO PROCESS or LUMINANCE PROCESS sections. The setup level adjustment, if there is one, should be checked first. The Sencore VR940 video reference level light box is shown in Fig. 10-15. To check the camera's black setup level:

1. Cap the camera's lens.
2. Press the CvA94 1H SWEEP RATE button.
3. Press the LUMA FILTERS button.
4. Press the mV MEASUREMENT UNITS button.
5. Adjust the DELTA BAR POSITION buttons to place the left edge of the CRT Delta Bar just to the right of the chroma burst.

355

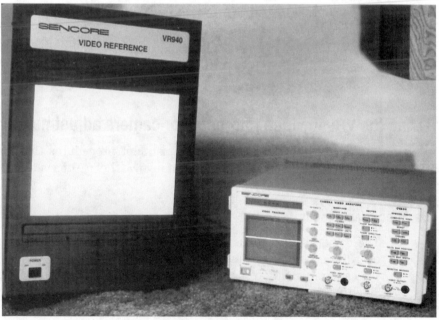

■ **10-15** *Photo of CVA94 and VR940 light box.*

6. Adjust the DELTA BAR WIDTH buttons to highlight the black signal level.

7. Check to see that the black setup level displayed on the CVA94 LCD DISPLAY is within 5% of 53 mV.

Checking burst level

A camera adds chroma burst to its composite video output signal. This is used in video-receiving devices as a reference to which their chroma demodulator circuits are locked. The NTSC standard amplitude that video receiving circuits are designed for is 40 IRE or 286 mV in a 1-VPP video signal.

If the camera's burst signal is not at the proper level, loss of color sync or no color might occur in the devices to which the camera video signal is applied. If the camera's burst level is wrong, suspect the burst adder stage. The burst level adjustment, if there is one, should be checked first. To check the burst level adjustment:

1. Cap the camera's lens.
2. Press the CVA94 1H or 2H SWEEP button.
3. Press the FLAT or CHROMA FILTERS button.
4. Press the mV MEASUREMENT UNITS button.
5. Adjust the DELTA BAR POSITION buttons to place the left edge of the CRT DELTA BAR at the left end of the chroma burst signal.
6. Adjust the DELTA BAR WIDTH buttons to place the right edge of the CRT DELTA BAR at the right end of the chroma burst signal.

Using the VR940 video reference for camera adjustments

Lets now look at some general adjustment procedures that are most common for video cameras that involve the use of a video reference test chart.

Back focus/auto focus

The camera's back focus adjustment controls the focus tracking over the entire zoom range. The auto focus adjustment controls the camera's ability to properly focus on objects at any distance. To check the camera's back focus:

1. Install the VR940 NEUTRAL DENSITY FILTER and FOCUS CHART.

2. Frame the camera on the FOCUS CHART at a distance of 6 feet.

3. Adjust the zoom control to wide angle.

4. With the focus control set to manual, adjust for best focus.

5. Adjust the zoom control to telephoto.

6. Check to see that best focus is still obtained at the same point on the focus control.

7. If necessary, adjust the back focus adjustment following the camera manuals procedure.

To check the camera's auto focus

1. Install the VR940 NEUTRAL DENSITY FILTER and FOCUS CHART.

2. Frame the camera on the FOCUS CHART at a distance of 6 feet.

3. Set the camera's focus control to "Auto".

4. Reframe the camera on the PAPER FOCUS CHART at a distance of 15 to 20 feet.

5. Check to see that optimum focus is obtained at both distances.

6. If necessary, adjust the auto focus adjustment following this procedure.

Blooming

The camera's blooming adjustment minimizes the amount of bright smearing that occurs in the picture. Overcorrection of the adjustment can cause loss of color, however, so proper color operation needs to be checked after adjusting the blooming control.

Luminance level

The gain of the camera's luminance stages is normally adjusted to set the amplitude of maximum white level signals so that the total amplitude between the maximum white level and the sync tip level is 1 VPP.

There are luminance level adjustments at one to three points along a camera's luminance signal path. The first (and sometimes only) level adjustment sets the automatic iris operation. If there are additional luminance level adjustments after the iris adjustment, all but the final adjustments are made while monitoring an internal test point with an oscilloscope. The final (or only) adjustment in the signal path (and in the adjustment procedure) sets the luminance level at the video output connector. This adjustment can be set by monitoring the video output signal with the CVA94.

White balance

When a video camera is scanning an uncolored portion of a scene, it should produce an uncolored white, gray, or black output image. It should do this whether outdoor light (bluish) or indoor incandescent light (reddish orange) is lighting the scene. To accomplish this, video cameras have internal controls that set the balance of the red and blue chroma process circuits to correct for different colored light sources. Often, cameras have two fixed white balance setups, one for indoor and one for outdoor light.

Because both indoor and outdoor light sources tend to vary in color, video cameras usually also have circuits (auto white balance) that automatically adjust the red and blue chroma process circuits to compensate for changing or nonstandard light sources.

Without these correction circuits, objects might appear the correct color when viewed under one type of light source but would appear a wrong color when viewed under another type of light source.

Camera white balance is always adjusted while monitoring the camera's video output. Improper white balance can be observed on a waveform display as chroma subcarrier riding on parts of the signal that should be a shade of gray. Or, it can be observed on a vector display as the center dot of the vector trace offset away from the center cross of the vector CRT display. For adjustment, the vector display is usually more convenient to use.

Chroma phase

If the camera's chroma process section does not produce the correct phase chroma signals, the hue of some or all colors produced by the camera will be incorrect. This is checked by viewing the camera's chroma output on a vector display while the COLOR BAR CHART, with known chroma phases, is viewed with the camera.

The color bar phases specified for a consumer camera usually do not agree with the vector target boxes for a standard broadcast signal. Consult the camera manufacturer's service information for the expected color output of any particular color camera.

Chroma amplitude

The amplitude of the modulated R-Y and B-Y chroma signals determines the amplitude of chroma signals produced when colors of different saturation or intensity are viewed by the video camera. If the chroma output signals are not at the proper amplitude, colored objects will appear washed out or too intense. This is checked by viewing the chroma output on a vector or waveform display when a color bar pattern with known chroma amplitudes is viewed with the camera.

Service information usually specifies this adjustment to be made with a vectorscope, and often specifies the chroma amplitude as a percentage of burst. Because standard vectorscopes have no amplitude graticule calibration, this has been a difficult measurement for servicers to verify in the past.

Frequency response

A camera's frequency response or resolution relates to its ability to sharply display small objects in a scene. The frequency response of a particular camera is limited by the design of the pick-up device and the signal amplifiers. The pick-up device usually does not change in a way that affects the frequency response, but defective amplifier circuits or aperture correct circuits might affect it.

Typical frequency response of consumer cameras is in the range of 4 MHz or 320 lines of resolution. Poor frequency response results in a picture that appears fuzzy or out of focus.

Video noise

A video camera's pick-up device (CCD) and signal process circuits always generate a small amount of noise. This small amount of noise mixes with the desired signal to produce a video output signal with a signal-to-noise (S/N) ratio of 40 dB or greater in typical new camera's. Excessive video noise in the signal (low S/N rating) results in grainy or noisy video displays. Cameras producing a video SIN ratio or less than 35 dB will require correction for satisfactory performance in consumer applications.

If the camera's video noise and chroma noise are both excessive, suspect the pick-up device, prevideo process section, or iris control. In cameras with digital signal processing, this can be caused by loss of one of the data lines carrying the digital signal. If the video noise is excessive, but the chroma noise is acceptable, suspect the luminance process section, especially the AGC amp.

Chroma noise

Small amounts of noise normally produced in the pick-up device, prevideo process, and chroma process sections combine to produce noise in the chroma output signal. Because both the amplitude and the phase of the chroma signal carry information about colors in the scene, both amplitude and phase noise have an effect on the quality of the chroma signal. Amplitude noise affects color saturation and produces a grainy effect in highly saturated colors, similar to video noise. Phase noise affects color hue and produces a smeary effect in highly saturated colors.

Typical new cameras produce chroma saturation and hue signal-to-noise (S/N) ratios of 40 dB or greater. Cameras producing chroma S/N ratios of less than 35 dB will require correction for satisfactory performance in consumer applications.

If the camera's chroma noise and video noise are both excessive, suspect the pick-up device, prevideo process section, or iris control. In cameras with digital signal processing, this can be caused by loss of one of the data lines carrying the digital signal. If both chroma saturation and hue noise is excessive, suspect the color matrix circuits. The color balance adjustment, if there is one, should be checked first.

Adjusting sync generator frequency

The sync generator adjustment method specified in service data usually involves connecting a frequency counter to the master oscillator test point, which will be running at some multiple of the burst frequency. The CVA94, however, allows you to make this adjustment without connecting to the internal test point. Instead, the CVA94 samples the camera's burst signal at the video output connection. Because no probe is connected to the oscillator test point, there is no chance of loading down the oscillator frequency with a probe's capacitance, plus you do not have to hunt for the test point. You only need check for proper frequency on the CVA94 burst frequency display, or adjust for zero error on the burst error display. To adjust the camera's sync generator frequency:

1. Press the CVA94 BURST ERROR button in the SPECIAL TESTS section.
2. Adjust the camera's sync generator frequency control to set the frequency error reading to zero.

Adjusting sync level

Camera information might specify to connect an oscilloscope to an internal camera test point while making this adjustment. The adjustment can be made just as correctly and more conveniently, however, by simply connecting the CVA94 to the camera's video output connection.

Checking composite video hum

Low frequency variations in the dc level of a camera's video output signal can cause undesired brightness variations on a video monitor. The brightness variations might occur in the form of dark bars across the screen, often known as *hum bars*. These low frequency variations are often caused by unfiltered ripple coming from the ac power line. Cameras with hum levels over 5% definitely should be corrected, because they will cause objectionable pictures on some monitors. Hum levels over 3% probably indicate developing problems and should be investigated.

If the camera's hum is excessive when operating from the power adapter, but is acceptable when operating from the battery, suspect the power adapter or input power filters. If the camera's hum is excessive, regardless of the power source, suspect the camera's dc regulator/converter or the sync generator.

Adjusting burst level

The camera service data might specify to connect an oscilloscope to an internal camera test point while making this adjustment. The adjustment can be made just as correctly and more conveniently, however, by connecting the CVA94 to the camera's video output connection. To adjust the camera's burst level:

1. Cap the camera's lens.
2. Press the CVA94 1H or 2H SWEEP RATE button.
3. Press the FLAT or CHROMA FILTERS button.
4. Press the IRE or mV MEASUREMENT UNITS button.
5. Adjust the DELTA BAR WIDTH buttons to place the left edge of the CRT DELTA BAR at the left end of the chroma burst signal.
6. Adjust the DELTA BAR WIDTH buttons to place the right edge of the CRT DELTA BAR at the right end of the chroma burst signal.
7. Adjust the camera's burst level control to obtain a CVA94 LCD DISPLAY reading of 40 IRE or 286 mV.

Note: Burst level also can be measured visually on the CVA94 CRT in the waveform mode by setting the back porch of the sync pulses to the 0 IRE reference line with the VERT POSITION control and adjusting the camera's burst level control to cause the burst to fall between the 20 and -20 IRE graticule lines. Notice CRT graticule shown in Fig. 10-13.

Adjusting black setup level

In most cases the camera's service literature will tell you to connect an oscilloscope to an internal camera test point while making this adjustment. The adjustment can be made just as correctly and more conveniently, however, by simply connecting the CVA94 to the camera's video output connection.

Adjusting indoor/outdoor white balance

Camera white balance is always adjusted while monitoring the camera's video output. Improper white balance can be observed on a waveform display as chroma subcarrier riding on parts of the signal that should be a shade of gray. Or, it can be observed on a vector display as the center dot of the vector offset away from the center cross of the vector CRT display. For adjustment, the vector display is usually more convenient to use.

Adjusting chroma phase

Because the color bar phases specified for a consumer camera usually do not agree with the vector target boxes for a standard broadcast signal, making this adjustment with a standard vectorscope can become a guessing game. The CVA94 simplifies this adjustment by providing a digital chroma phase measurement that is expressed with the same phase reference and direction as the measurements stated in the service data. This makes it easy to verify that the camera's chroma phases agree with those given in the service information. To adjust the camera's chroma phase:

1. Frame the camera on the COLOR BAR TEST CHART.

2. Press the CVA94 VECTOR MEASUREMENT PHASE button.

3. Set the PHASE REFERENCE button to B-Y if service notes phase measurements are made starting at the right edge (B-Y) of the vector display or to BURST if they are made starting at the left edge (BURST) of the vector display.

4. Set the PHASE DIRECTION button to CCW if service data phase measurements are made in a counter-clockwise direction or to CW if they are made in a CLOCKWISE direction.

5. Adjust the DELTA BAR POSITION buttons to place the CRT DELTA BAR on the color bar to be measured.

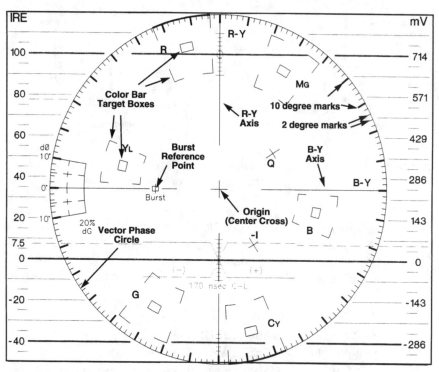

10-16 *CVA94 vector display.*

6. Adjust the camera's chroma phase control to obtain a CVA94 LCD DISPLAY reading which agrees with the phase specified in the service information.

Note: Chroma phases can also be measured visually on the CVA94 CRT VECTOR DISPLAY as shown in Fig. 10-16.

Adjustments made with a waveform analyzer (triggered)

Some camera adjustments (e.g. auto iris, AGC, shading, aperture) are made while viewing a signal at a test point within the camera circuits rather than at the camera video output. Most of these test points are before the sync adder stage, though, so triggering an oscilloscope on the test signal is difficult without an external trigger reference signal for the scope. Note drawing shown in Fig. 10-17. To simplify these adjustments, view the signal with the waveform analyzer, triggering from the CVA94 TRIGGER OUTPUT.

Because the signals within the camera processing stages will be different from camera to camera depending on circuit design, ex-

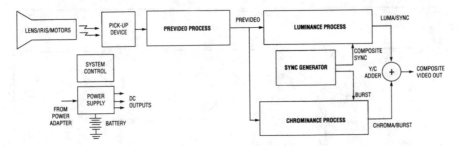

■ **10-17** *An external trigger is needed when scoping signals in the chrominance section or in the luminance section before the sync adder stage.*

act waveforms and voltage levels cannot be given. Instead, the
general procedure to be used with the waveform analyzer and
CVA94 camera video analyzer is outlined by the following. To
make camera adjustments while viewing internal camera signals:

1. Connect the camera video output signal to the CVA94
   COMPOSITE or Y/C VIDEO INPUT.
2. Connect the CVA94 TRIGGER OUTPUT to the external trigger
   input of the WAVEFORM ANALYZER.
3. Set the WAVEFORM ANALYZER'S trigger mode control to
   "AUTO", trigger source control to "EXT", and the trigger
   polarity control to "-".
4. Connect the WAVEFORM ANALYZER input probe to the
   camera. This test can be monitored while making the
   adjustment.
5. Make the camera adjustment while viewing the waveform on
   the WAVEFORM ANALYZER.

Adjustments made with a waveform analyzer (direct)

Some camera adjustments (e.g. power supply voltages, drive
pulse frequency) involve dc voltage and frequency measurements
within the camera circuits. These adjustments are best made directly with a Sencore SC61, SC3080 waveform analyzer, or the
SC3100 waveform and circuit analyzer as shown in Fig. 10-18.

Video monitor adjustments

Some camera adjustments (e.g. back focus/flange back, blooming)
are best made while viewing the camera output signal on a video

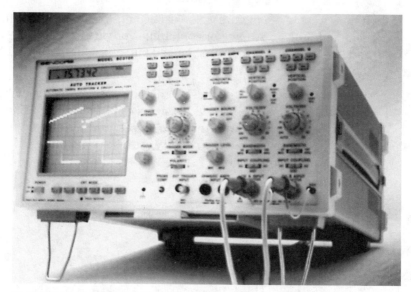

■ **10-18** *Sencore SC3100 waveform analyzer.*

monitor. The CVA94 provides COMPOSITE and Y/C VIDEO OUTPUTs to simplify connection of your video monitor to the camera video output.

The complex composite video signal

One of the most complex waveforms in electronics is the NTSC composite video waveform. This signal includes many different parts: video, horizontal and vertical sync pulses, horizontal and vertical blanking pulses, and special signals that all combine into one waveform.

The composite video signal

The composite video waveform contains all the information needed for the complete CRT picture, line by line and field by field. This signal is used in the picture tube to reproduce the picture on the scanning raster. Composite video consists of video information plus the blanking and sync pulses needed for synchronized picture reproduction.

In producing a picture on the CRT, the raster is scanned twice, 262-1/2 lines for each interlaced field, for a total 525 lines per frame (a complete picture). Not all 525 lines contain picture information, however. A few horizontal lines of video at both the top and bottom of the screen are blanked out, and some are "used up" in vertical retrace.

Two important characteristics of the composite video signal are polarity and amplitude. The video signal can have two polarities: A positive sync polarity, with the sync pulses "up" (as shown in Fig. 10-19A) and a negative sync polarity, with the sync pulses "down" (as shown in Fig. 10-19B).

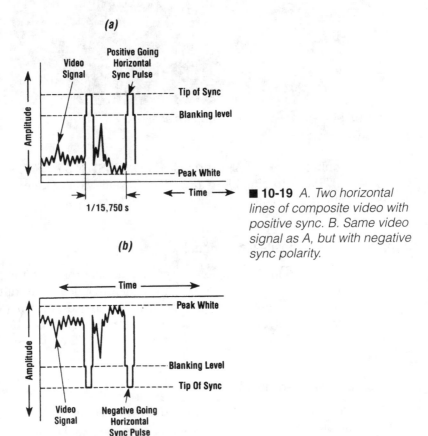

■ 10-19 A. Two horizontal lines of composite video with positive sync. B. Same video signal as A, but with negative sync polarity.

The signals both contain the same picture information, the only difference is the polarity. Negative sync polarity is standard for the input or output signals of video equipment such as TV cameras, video control equipment, and video ports on monitors and VCRs.

For either polarity, the white parts of the video signal are opposite to those of the sync pulses. The black parts of the video are closer to the blanking and sync tip levels, which are considered blacker than black.

The standard input/output amplitude for the video signal in the video equipment noted above is 1 VPP into 75 ohms. The requirements for the composite video signal at the inputs for different CRTs, however, varies from 30 to 150 VPP or more for larger tubes.

In relation to time, the composite video signal can be divided into two different parts: the horizontal interval and the vertical interval.

Horizontal interval

The composite video signal, shown at the horizontal rate in Fig. 10-20, consists of a complex series of waveforms. This series of waveforms shown represents one line of picture information that takes 63.5 microseconds to occur (15,750 Hz).

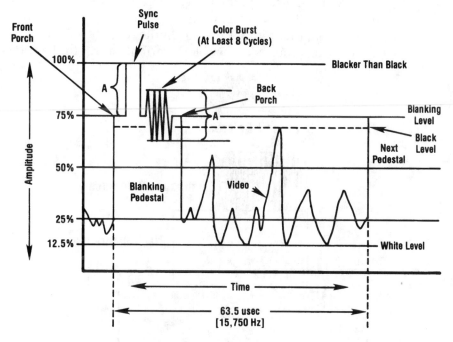

■ **10-20** *Detail of one horizontal line and sync (+sync).*

Vertical interval

After the CRT is filled with lines of video, the scanning beam must return to the top of the screen so it can start all over again. This repositioning time, called the *vertical blanking interval*, is composed of 21 horizontal lines, which are not displayed. This portion of the composite video waveform is extremely important because it contains timing pulses, FCC regulated test signals, source identification codes, reference signals, and information regarding caption availability for the deaf during a broadcast.

Test signals in the vertical blanking interval

Most TV stations transmit special reference signals during the vertical blanking interval. The two most common signals are the VITS (vertical interval test signal) and VIRS (vertical interval reference signal).

VITS

The VITS are used by the TV networks to assist in evaluation of various parameters of their broadcast system's performance. The VITS are transmitted during active operation to ensure continuous quality and accuracy in terms of color and distortion. In regards to transmitter performance, the quality of the VITS frequency determines if a problem warrants a trip to the transmitter site.

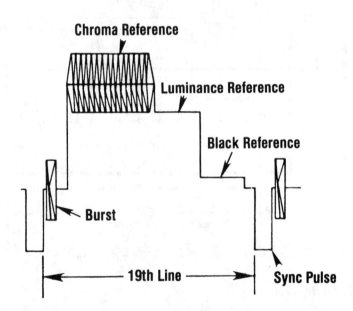

■ 10-21 *The VIRS consists of a chroma reference, a luminance reference and black reference.*

The VITS is typically located on line 17 and 18 during vertical blanking. The type of test signal is by individual networks or the FCC and depends on transmitter operation, area served, etc.

VIRS

The VIRS is used to establish the correct values of chroma amplitude and phase, luminance and black setup levels. The VIRS, transmitted on line 19 of the vertical blanking interval (see Fig. 10-21), includes special reference values for these parameters.

Because it is part of the transmitted signal, the VIRS is available at the TV receiver. Circuits have been developed to detect the VIRS reference, and automatically set the receiver levels to the correct values.

A summary of the National Television Systems Committee (NTSC) television standards is shown in Fig. 10-22.

Summary of NTSC Television Standards	
Horizontal scan frequency, Hz	15,734.26
Vertical scan frequency, Hz	59.94
Color Subcarrier, MHz	3.579545
Channel Band width, MHz	6
Lines per frame	525
Lines per field	262.5
Frames per second	30
Fields per second	60
Fields per frame	2
Video Band width, MHz	4.2
Video signal	AM
Audio signal	FM
Video Modulation	Negative
Aspect ratio	4:3

■ **10-22** *NTSC television standards.*

Camera troubleshooting tree guide

Figure 10-23 has symptoms on monitor of a blank screen or all snow. Figure 10-24 has monitor screen symptoms of no color or the wrong color.

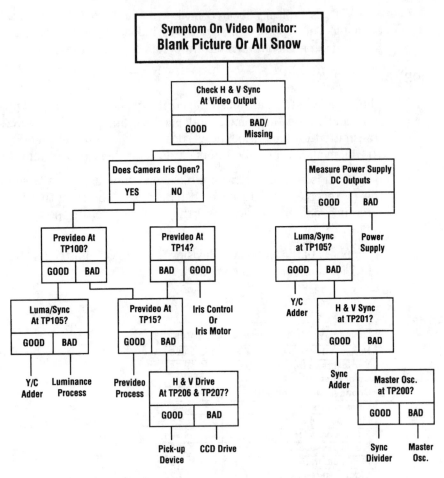

■ **10-23** *Trouble tree for camcorder with blank picture or all snow.*

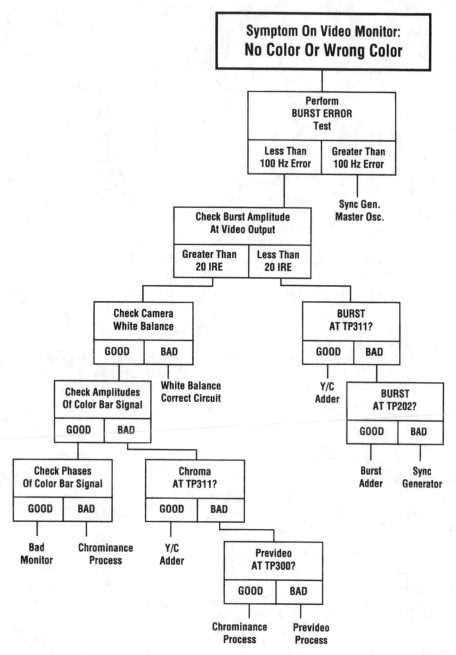

■ **10-24** *Camcorder trouble tree with monitor symptom of no color or wrong color.*

Camcorders

11

THIS CHAPTER COVERS THREE DIFFERENT BRANDS OF camcorders. For practicality, I only highlight some of the special circuits that are not found in basic VCR machines. The chapter begins with the Sony model CCD-V8 camcorder that contains special circuits such as the autofocus and power control. Then the Zenith model VM6200 camcorder system and troubleshooting chart are covered. The chapter then covers the RCA model CC011 camcorder circuit operations. This chapter concludes with details of Zenith's VM8300 camcorders and VHSC camcorders.

Sony camcorders

The following information and illustrations are courtesy of Sony Corp. of America. Charge-coupled device (CCD), solid-state pickup elements are now used in most camcorders. The CCD camera can be divided into eight main circuits as follows:

☐ The CCD.

☐ Timer.

☐ Sync generator.

☐ CCD driver circuits.

☐ Process.

☐ White balance and iris control.

☐ Matrix.

☐ Encoder.

All of these circuits are ICs with support components for adjustments, effectively simplifying the overall complex circuitry of the CCD camera.

Timing

The block diagram in Fig. 11-1 shows the timing generator (IC707) and its relationship to the other main circuits of the camera. Its function is to generate all the necessary timing pulses for the op-

373

eration of the camera circuits. The output signals on the left of the timing IC block (CX23047A) are used to develop all the pulses needed to operate the CCD imager. XSG1 and XSG2 are used to develop a 60-Hz pulse. The combination of this 60-Hz pulse and the 180 degree out-of-phase 15.75-kHz pulses from XV1 to XV3 produce the V1 and V3 pulses that shift charges out of the sensors to, and down, the vertical registers.

Also shown on this side of the block are the XV2 and XV4 outputs. These two signals are 180 degrees out of phase at a frequency of 15.75 kHz. They are processed to produce the V2 and V4 pulses, whose only purpose is to shift charges down the vertical registers.

The final two outputs shown on the left side of the block are XH1 and XH2. The pulses originating at these points are 180 degrees out of phase and have a frequency of 9.55 MHz. They are processed to develop the H1 and H2 pulses for shifting, or clocking, the charges from the H1 and H2 registers at the bottom of the CCD imager to the output.

374

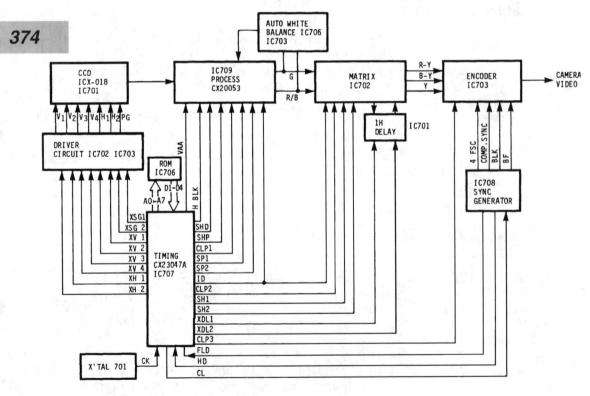

■ **11-1** *Overall block diagram of CCD camera.*

To operate the output circuit of the CCD, which functions similar to a sample-and-hold circuit, a switching pulse is needed. This switching pulse, called the *PG pulse*, is developed from the XH1 pulse.

On the right side of the timing IC block, the outputs to all the other operational blocks of the camera are shown. The first seven outputs are used primarily by the process IC. Their purposes are as follows:

- [ ] H.BLK (horizontal blanking) introduces the horizontal blanking pulse in the signal.
- [ ] SHD (sample/hold data) extracts the CCD imager pixel data from the output signal.
- [ ] SHP (sample/hold precharge) extracts the CCD imager precharge level from the output signal.
- [ ] CLP1 (clamp 1) is used to clamp the green, red, and blue signals prior to multiplexing and insertion of horizontal blanking.
- [ ] SP1 and SP2 (separation) separate green picture information from the red and blue information.
- [ ] ID separates the red and blue information. It is also used in the color multiplexing circuits of the matrix IC.
- [ ] VAA Pulse (vertical area available) is used together with CLP1 in the clamping process.

The last six outputs of the timing generator are used mainly in the matrix circuits:

- [ ] CLP2 and CLP3 (clamp) clamp G and Rg signals.
- [ ] SH1 and SH2 (sample and hold) operate the sample and hold circuits in the matrix IC that develops the Y signal.
- [ ] XDL1 and XDL2 (delay) drive a clock in the 1H delay IC that is instrumental in matrixing the G and RB signals.

At the bottom of the timing IC block are two input pulses: HD (horizontal drive) and FLD (field drive). These two pulses from the sync generator inform the timing generator of line and field timing.

The main support circuits of the timing generator are the 28.6-MHz clock and IC706. IC706 is a ROM, programmed to correct flaws created in the CCD during manufacture. Each CCD has its own ROM specially programmed, because the flaws are not at the same spot on every CCD.

Power control circuits

The power control circuits (see Fig. 11-2) are located on the DS-10 board, which uses the inputs from the CAMERA ON and VTR ON switches together with the output from the mode control CPU to switch power to the appropriate circuitry in response to the input commands.

The unregulated 6 Vdc is applied to CN12 (pin 8), providing B + to the DS-10 board power control circuit is present. When the CAMERA POWER ON switch is activated, the CAM POWER SW at CN012 (pin 1) goes low. This input is applied to Q602/B, turning Q602 on and producing a high at Q602/C. This high is applied to an inverter within IC601 (pins 10 and 11) and a second inverter, IC601 (pins 12 and 13), and the resultant high output at pin 12 is applied to the clock input IC602 (pin 11). Q602 and the two inverters within IC601 form a chattering prevention circuit that prevents chattering of the switch contacts from affecting power control and produces a fast rise time to clock the flip-flop in IC602. A similar chatter prevention circuit consisting of Q601 and two inverters in IC601 is used for the VTR POWER SWITCH input. The output of the second circuit is applied to the clock input of the second flip-flop at IC602 (pin 3).

When power is applied to the CCD-V8, the two D-type flip-flops in IC602 must be reset, so the Q outputs at pin 1 and pin 13 are both low. This is done by the reset circuit, consisting of Q604 and an inverter in IC601 (pins 8 and 9). When the UNREG voltage is applied to the DS-10 board, power is applied to the power control circuit, and the increasing UNREG voltage is coupled across C603 to the base of Q604, turning it on. This produces a low at Q603/C while C603 is charging. This low is inverted by IC601 (pins 8 and 9), and the high at IC601 (pin 8) is applied to the CLEAR inputs of the D-type flip-flop (pins 4 and 10). When the CLEAR line is high, the Q outputs (pins 1 and 13) are both made low. When the capacitor C603 fully charges, Q604 is no longer held on, and the collector goes high, removing the high reset from the D-type flip-flops.

When the CAMERA POWER switch is operated on the CCD-V8, the LOW CAM POWER SW is inverted by the chatter prevention circuitry and applied as a clock signal to pin 11 of the D-type flip-flop, LC602 (pins 8 - 13). The D-type flip-flop is connected with the Q output (pin 12) and D input (pin 9) connected. In this configuration, the flip-flop will act as a toggle circuit, and the output Q will change states at every positive-going clock transition. Therefore, the clock pulse produced when the CAMERA POWER switch

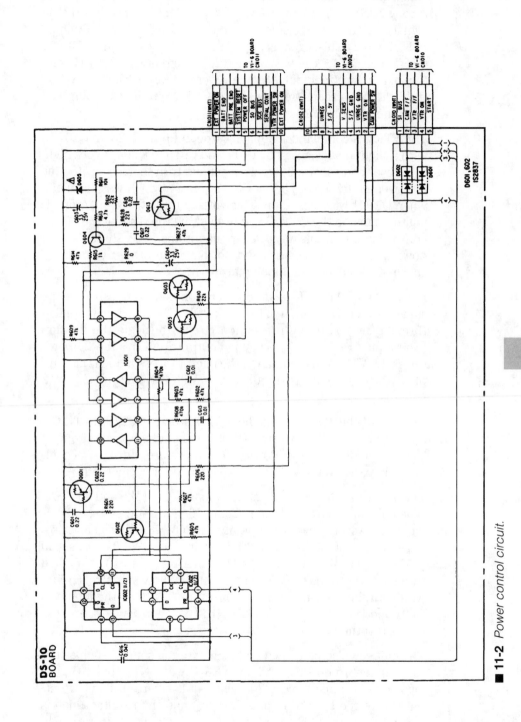

■ 11-2 *Power control circuit.*

is activated produces a high at IC602 (pin 13), which is applied through CNIO (pin 2) and CAM F/F, to the Mode Control CPU. The signal goes through the steering diode D602, making VTR ON high (CN12, pin 2).This signal is applied to the dc-dc converter DR901 to produce 5 Vdc, which is applied to the circuits in the VTR section of the unit (including the Mode Control CPU). The Mode Control CPU, upon receipt of the high CAM F/F, produces a high VTR ON command at CN10 (pin 4).

The VTR ON command is applied through R627, R628, to Q604/B, with a slight delay caused by C617 and C615. If Q604/B goes high, the power control circuit would reset and turn the power off. This is prevented by the S/S 5 V at CN 12 (pin 7), which was produced when DR901 was turned on. This voltage turns Q613 on, grounding the signal to the Q604/B and preventing the reset of the power control circuit. In this way, the circuit automatically shuts the unit down if the systems control did not have the proper voltage because of DR901 not functioning properly.

Power ON in the VTR mode of operation is accomplished in the same manner. The action of the VTR POWER ON switch is applied through the chattering prevention circuit to toggle the flip-flop IC602 (pins 1 through 7). The high output from this flip-flop produces the VTR F/F command to the microprocessor and is also applied through D601 to produce the VTR ON output at CN12 (pin 2).

Because of the action of the diodes D601 and D602, the VTR ON signal to the dc-dc converter, DR901, is held high, even if one of the inputs is made low. Power OFF, therefore, must be accomplished by the systems control through a different method.

The mode control CPU constantly monitors CAM F/F and VTR F/F. When one of these signals goes from a high to a low state, indicating that the switch has been operated, a POWER OFF signal is produced by the mode control CPU, as well as a LOW VTR ON signal at CN10 (pin 4). The POWER OFF at CN11 (pin 5) is applied through R612 to the Q604/B. This resets the power control circuit, with both flip-flops producing a low output. With all the inputs to D601 and D602 low, VTR ON at CN12 (pin 2) is also low, and the power is turned off.

The unit can also be turned on by an external source, such as the tuner timer. To do this, EXT POWER ON at CN11 (pin 10) goes high. This is applied to Q603/B, where it is inverted. The low output is inverted by IC601 (pins 5 and 6) and the resultant high is applied through D602 to produce a high VTR ON. This is applied to the dc-dc converter, DR901, and powers the unit. Q605 is turned on during reset to prevent the unit from being turned on by an external POWER ON command while in the reset mode.

If the UNREG voltage goes above 9 V, D605 will conduct, placing a high on the Q604/B. This puts the power control circuit in the reset mode and shuts the power off.

CPU power control

The mode control CPU, IC601 on the SK-3 board, is used to control power turn-on together with the power control circuit on the DS-10 board. Refer to Fig. 11-3.

Whenever power is applied to the CCD-V8 unit, even when the power is turned off, SYS 5 V at CN604 (pin 7) is present to maintain memory. This voltage is applied to IC601 (pin 26) and through L601 to IC605 (pin 14) and IC601 (pin 58). This provides power for IC605 and IC601 to maintain memory. However, the microprocessor IC601 does not operate until the power is turned on, as it receives no clock input to process data. The clock for IC601 is a 400-kHz crystal oscillator consisting of IC605 (pins 10 through 13) and associated components. This crystal oscillator applies a clock signal to IC601 (pin 57).

When the power is off, S/S 5 V at CN604 (pin 8) is low. This low voltage is applied to the cathodes of D601. D601 is forward biased, and the input to the crystal oscillator IC605 (pin 13) is grounded, preventing the oscillator from functioning. When power is turned on, the power control circuit in the DS-10 board produces a VTR ON command through the steering diodes and turns S/S 5 V on. When this occurs, the high at CN604 (pin 8) is applied to the cathodes of D601, reverse biasing them and allowing the oscillator to function.

When S/S 5 V goes high, IC601 must also be reset to ensure proper operation. This is accomplished by the reset circuit consisting of IC604 (pins 5 through 7) and positive reset is maintained for the charge time of C612.

The inputs from the power control, either VTR F/F or CAM F/F, are applied through the scan matrix to IC601, which produces a HIGH VTR ON or CAM ON at pins 18 or 19.

An additional circuit consisting of Q606 and Q607 is turned on when S/S 5 V goes high. This connects the SYS 5 V and S/S 5 V while the unit is operating. The other functions of IC601 are in reference to the mode control CPU.

379

380

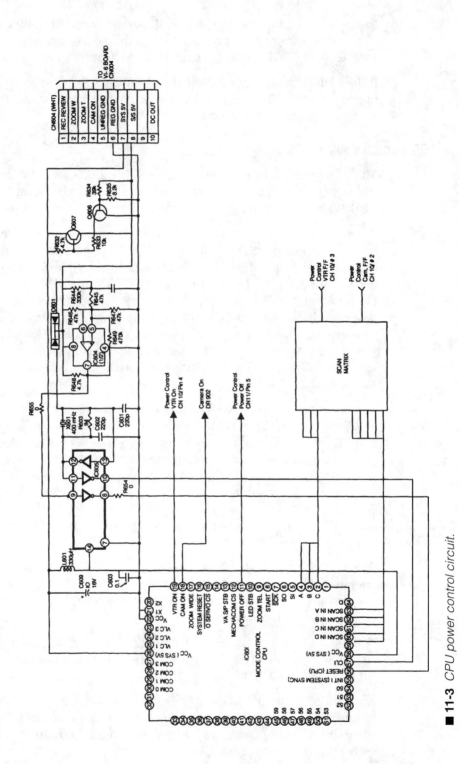

■ 11-3 *CPU power control circuit.*

Systems control CPUs

The systems control in the CCD-V8 contains three CPUs. These three CPUs are all tied together to form the systems control circuit as shown in the Fig. 11-4 block diagram. The first CPU, the mode control CPU, receives the inputs from the key matrix and controls the liquid crystal display (LCD). The second CPU is the mechanical control CPU, which controls the mechanisms and servos. It also monitors the mechanical mode switches, RF SW PULSE and CAPSTAN FG to ensure that the mechanism is functioning properly. Because of the interrelation between the commands and the mechanical controls, the three microprocessor's must constantly communicate with each other as well as with the individual circuits they monitor or control. This is taken care of by the communications CPU, which synchronizes the two other CPUs with each other and any external circuitry connected to the CCD-V8.

The main control in the CCD-V8 is the mode control CPU, a UPD7503G microprocessor. This CPU monitors a 4×8 input key matrix that detects when a command is received for the unit to execute an operation. This CPU also directly controls the LCD, and through the inner bus (a series of data communication lines between the microprocessors) controls the digital servo, the LED drivers, and the video/audio switching circuits.

The second CPU in the CCD-V8 is the mechanical control CPU, an MB88505 microprocessor. This microprocessor, as its name implies, controls the mechanism of the unit, including the capstan/drum servo, capstan driver, capstan FG, ATF servo, the drum driver, loading and control motors drivers, and the brake solenoid. In addition, it controls the tape top/end LED sensors and monitors the RF switching pulse and capstan FG signals to ensure that the motors are correctly operating. In the event of a malfunction of any of the motors, the malfunction is detected by the mechanical control CPU, using these inputs, and the unit will be shut down. The mechanical control CPU monitors the mode sensors with a 3×4 matrix.

The mechanical control CPU communicates to the digital servo and the mode control CPU through the inner bus. In this way, the mechanical control CPU not only receives commands from the mode control CPU but also informs it in the event of a servo malfunction to shut the unit down.

This bidirectional communication between the mode control and mechanical control CPUs is controlled by the third CPU. This is the communications CPU, an MB88201BF microprocessor. Using

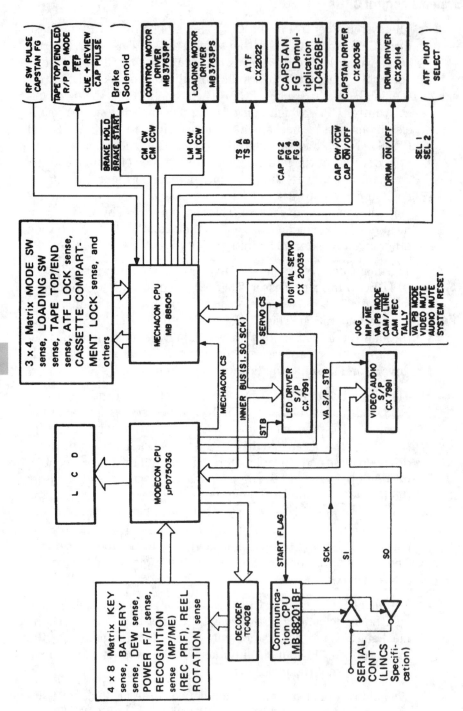

■ 11-4 *CPU input/output signals.*

the START FLAG input, the communications CPU generates a serial clock (SCK) that synchronizes the SERIAL IN (SI) and SERIAL OUT (SO) signals between the two microprocessors. The communications CPU also controls the direction of the SERIAL IN and SERIAL OUTPUT lines to an external data communications line. This data communication line can be used by such accessories as a tuner/timer or remote control units.

Input key matrix

The systems control begins its operation, when the commands are initially sensed by the systems control, it begins its operation in the input matrix of the mode control CPU. Refer to Fig. 11-3.

The mode control CPU uses four scan-in and four scan-out ports to sense the inputs from 28 different sources. In order to do this, an input matrix, which is an array of diodes, transistors, and an integrated circuit is used. These devices are arranged in such a way that a unique input is received for each sensor to be detected.

To achieve this input detection, four scan-out pulses are generated within the mode control CPU at the scan-out ports (pins 2, 3, 4, and 64). These scan-out pulses are in the form of a 4-bit binary word, which is sequenced by the microprocessor. This 4-bit word is applied to a TC4028BP IC. This IC is a binary-to-decimal decoder. It converts the four binary bits from the scan-output to ten outputs, one unique output for each binary word. Of the ten possible binary outputs, only eight are used in the CCD-V8 system.

In order to make the transistor switches in the key input matrix operate, the C scan-output at pin 2 is connected through pull-up resistors to the input key matrix. The sequence of operation that takes place in sensing a command can be considered, assuming the playback button is pressed. The SYSTEMS SYNC at pin 55 of the mode control CPU synchronizes the timing of the key matrix and other functions of the microprocessor. When the systems sync goes high, the binary code for 6 (A = 0, B = 1, C = 1, D = 0), is output at the scan-out to the input key matrix. The binary/decimal decoder converts this to a high at Q6, which is applied to the three reel-sensor transistors at the bottom of the key input matrix. During this time, the input from the reel sensors is detected by the Q6 output, turning on the three transistors. If a reel sensor is active, the transistor will conduct, bringing down the high voltage applied to the scan-inputs during this time. After the initial period, when the Q6 output is high, the mode control CPU changes the scan-output code to indicate a binary 1. This binary 1 is converted by the binary/decimal decoder to a high output for Q1. This Q1 output is applied to the first row of switches, which includes the playback switch. The closed playback switch routes the signal through the isolation diode to the C input at pin 61 of the mode control CPU. Whenever the C input of the mode control CPU goes high during Q1 time, this must be because of a closed playback switch.

Drum and capstan FG circuit

The signal from the FG and PG coils in the drum and FG device in the capstan are very small. Notice the waveforms shown in Fig. 11-5. To be usable in the digital servo circuit, they must be amplified and shaped into 5-V square waves. This is the function of the drum and capstan FG circuit. The drum FG output is a 50-mV PP sine wave. The drum PG output is a 1-mV PP pulse. This signal is so small that it should be considered unmeasurable. The capstan FG signal, being produced by a DME device, is approximately 50 MV PP.

The capstan FG signal consists of CAP FG (+) and CAP FG (−). These two signals are applied to the inputs of a comparator that produces a shaped output that exits pin 9 as a 960-Hz 5-V square wave. The ×2 logic circuit in IC501 is not used in this unit. The frequency of the output at pin 9 is the same as the capstan FG input.

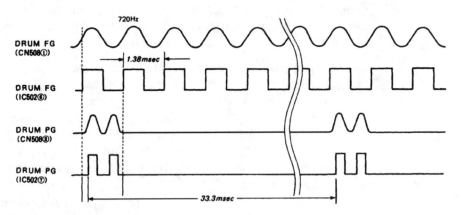

■ **11-5** *Drum PG and FG waveforms.*

The bias for the drum PG and FG coils is produced within IC501. Refer to Fig. 11-6 for IC501 circuit. The bias voltage exits IC501/pin 1 and is applied to the D COM connector of the FG and PG coils. The outputs from these coils are applied to independent preamplifiers and comparators at IC501 (pin 2) and IC501 (pin 4).

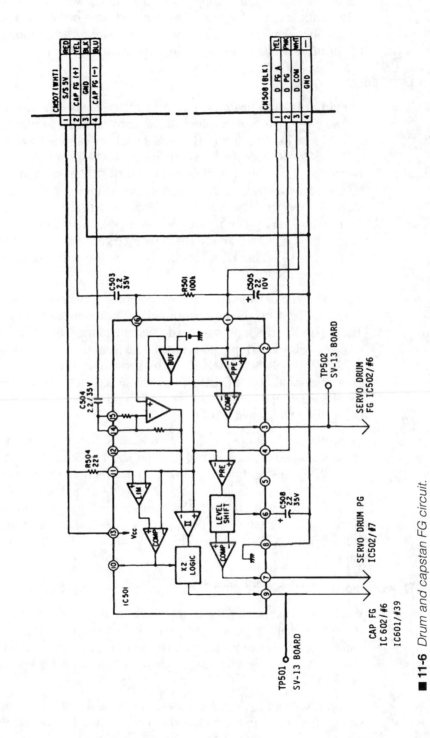

■ 11-6 *Drum and capstan FG circuit.*

Sony camcorders

These pulses are both amplified to 5-V square waves that are in phase with the input signals. The drum PG and drum FG outputs exit IC502 (pin 7) and IC502 (pin 3), and are directly applied to the drum servo IC (refer to drum servo IC502 in Fig. 11-7.

Drum servo

The drum PG and drum FG signals, which have been shaped into 5-V square waves, are applied to the capstan/drum servo. IC502 (pins 6 and 7). Within the IC, these signals are applied to a drum counter. The IC uses the 3.58-MHz input at IC501 (pin 2) as a time reference. This signal is obtained from the color section of the video processing circuit. It is applied to a drum reference generator that produces the reference signal for the drum speed and drum phase servo. During the record mode, the drum reference generator is synchronized to the reference vertical signal entering IC502 (pin 19). This signal is derived from the video sync of the signal to be recorded.

The switching between modes is accomplished within IC502. The mode control CPU directly controls IC502 via pins 24 through 27. These pins are the inputs from the system control. Pin 27 is a serial clock (SCK BUS), pin 26 is serial data from the serial input bus (SI BUS), pin 24 is connected to the serial output bus but is also data entering IC502 (SO BUS), and pin 25 is the chip enable input (D SERVO CS), which allows the IC to read the data entering the other three pins when it is low.

In normal mode of operation, data is always present on all four of these lines and repeats at a rate of 60 Hz. The data (SO) occurs when the chip select input is low. This data is read into the memory by five groups of clock pulses that occur when the chip select signal is low. Each group of clock pulses contains eight negative clock pulses. These clock pulses are always present and do not change from mode to mode. The SI signal at pin 26 also provides data to IC502.

Drum servo input troubleshooting

When troubleshooting these inputs, it is important to observe their presence. If pulses are present (note that in the record mode, no pulses are present on the SI line), the system control, which produces these pulses, can be assumed to be operating properly.

The output from the drum counter, as well as the output from the dc bias memory, are applied to the motor driver IC to control the

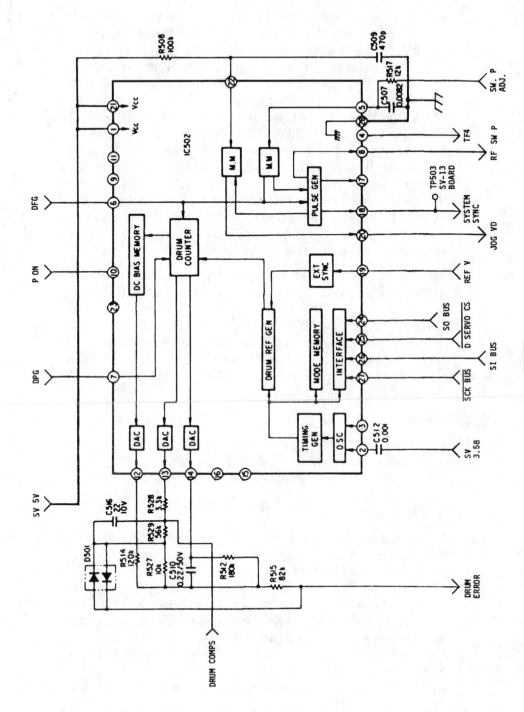

387

■ **11-7** *Drum servo circuit.*

Sony camcorders

speed. The dc bias memory produces a dc output at pin 12. This output voltage changes from mode to mode and is controlled by the serial data from the system control. Once the mode has been entered, no change should be observed in the dc voltage at pin 12. The voltages at pins 13 and 14 are dc voltages produced by digital-to-analog converters within IC502. These voltages are the speed and phase servo outputs at pins 13 and 14, respectively. They change the dc level according to the speed and phase of the motor. The three outputs from the IC are combined in a resistor and capacitor matrix to produce the drum error voltage that is applied to the drum and controls the speed of the motor through the drum driver circuit. A diode D501 is incorporated into the circuit so that if the speed servo voltage becomes very high, indicating a severe difference in speed, the speed servo will bypass the resistor capacitor network and directly control the drum error and the motor speed to quickly lock up the motor.

The drum FG input (DFG) is also used to produce the RF switching pulse. The DFG is applied to a multivibrator within IC502. The positive transition of this multivibrator is controlled by the capacitor at IC502 (pin 5). This is the RF switching pulse adjustment. By adjusting the resistance of the switching pulse adjustment, the time constant attached to pin 5 will change. This controls the slope of the signal at pin 5 and, therefore, the position of the RF switching pulse that exits at pin 8.

Capstan servo

IC502 contains both the drum and capstan servo circuitry. The capstan FG, from the FG amplifier, is applied to IC502 (pin 23) through IC603, a programmable divider. In the play mode, this programmable divider is set to divide capstan FG frequency by one. Refer to circuit in Fig. 11-8.

In the CCD-V8, the capstan drives the tape in the cue and review modes of operation. The capstan servo is also used during this mode, but to speed up the tape, a programmable divider, IC602, is incorporated in the capstan FG input circuit. In the cue mode, commands from the mechanism control CPU place a high FG8 input at IC602 (pin 2). This changes the programmable divider from a divide-by-1 to a divide-by-9 circuit. Because the capstan FG has been divided by 9, the servo within IC502 will automatically increase the output voltage to restore the capstan FG to the proper frequency. This results in speeding up the tape for the cue mode.

In the review mode, high FG2 and FG4 commands from the mechanism control CPU change the programmable divider to divide by

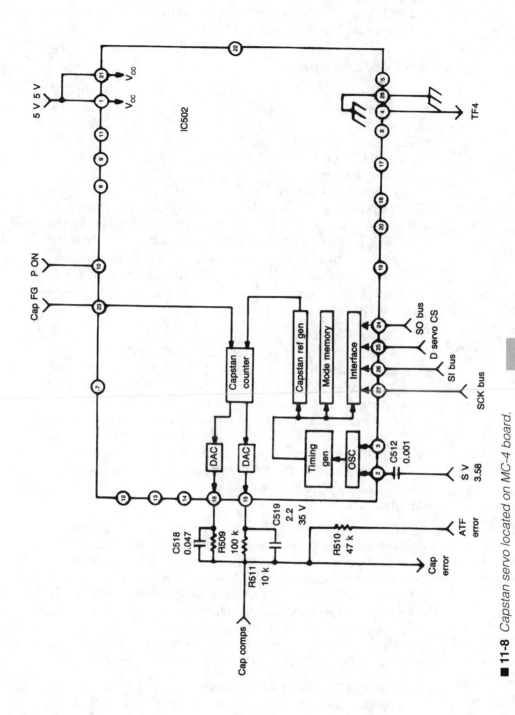

389

■ 11-8 Capstan servo located on MC-4 board.

Sony camcorders

7, and the speed of the capstan is changed accordingly by the capstan servo circuit. The capstan motor also drives the reel tables through a timing belt.

Capstan free-speed compensation

The capstan error voltage, which was produced in the capstan servo circuit by combining the capstan speed, phase, and AFT servo outputs, is amplified by IC504 (pins 1 through 3) and applied to the noninverting input of an amplifier within IC502 (the capstan motor driver IC). IC502 compares the capstan error voltage at pin 23 with the capstan bias at pin 24. The capstan bias is produced in a resistor divider network consisting of R575, R576, RV504, RV503, and RV574. This voltage divider produces a dc voltage that is buffered by IC504 (pins 5 through 7) and applied to IC502 (pin 24). The capstan bias voltage is adjusted so that the capstan motor turns at the correct speed if the ATF servo output is missing. This is essential to allow the ATF servo the ability to correct in either direction for a relatively large range to ensure proper servo lockup.

Autofocus

One way to have autofocus relies on the camera emitting a light beam to place a spot at the center of the scene to be picked up by the camera. This is accomplished by a sensor within the camera. The sensor consists of two photo sensors that are only sensitive to the light beam projected by the camera. If the image is in focus, the outputs of these two sensors is equal (because of the mechanical coupling between the light-emitting diode and the sensors). The focus servo tries to adjust the position of the lens to achieve this desired output.

In the real world, several problems arise that must be eliminated by the autofocus circuit for reliable focus operation. One of these is caused by random light sources interfering with the beam emitted by the camera. Another problem is the varying intensity of the light reflected back to the camera. Light intensity changes with distance from the camera to the object, as well as with differing colors and reflectivities of the image being picked up by the camera. Another problem is that of moving or vibrating objects. These three problems are eliminated by the autofocus circuitry.

The output of the IRED (infrared-emitting diode) is controlled by IC12 (pins 32 and 33), the autofocus controller. Refer to the block diagram in Fig. 11-9. The IRED is turned on and off at a rate of

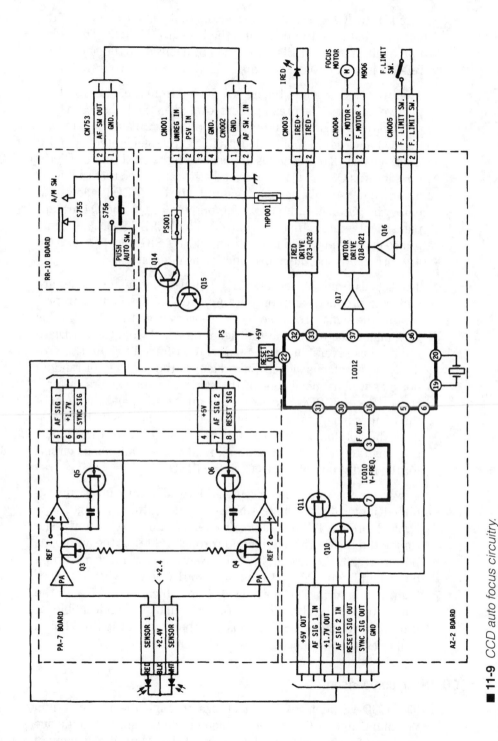

■ 11-9 *CCD auto focus circuitry.*

Sony camcorders

12.5 kHz by IC12 to enable the sensors to distinguish between the light coming from the IRED and the background light. The radiation emitted by the IRED is focussed by a lens onto the object that reflects it back into the sensors.

The two sensors detect the light reflected from the object and apply two signals to the preamplifiers located on the PA-7 board. The output from the preamplifiers are chopped by Q3 and Q4. Transistors Q3 and Q4 are controlled by SYNC SIG generated by IC12. This signal is 12.5 kHz and is in phase with the IRED drive. Therefore, it turns Q3 and Q4 on at exactly the same time that the IRED is turned on, so only the light generated by the IRED is allowed to enter the autofocus circuitry. The outputs of Q3 and Q4 are applied to integrators that are controlled by RESET SIG. This signal is also generated within IC12. When RESET SIG is low, Q5 and Q6 are turned off, and the amps function as integrators.

The signals from the sensors are integrated, i.e., many pulses are added to produce ramp outputs AF SIG 1 and AF SIG 2, representing the signals received by the sensors from the IRED. In this way, minor fluctuations in the signal returned from the IRED are removed as several hundred pulses are averaged together to produce the autofocus signal. This removes the problem of random noise as well as movement of the subject in the camera. After a short time, the IRED and SYNC SIG pulses stop, and the level of the ramp is maintained. It is this level that is used by IC12 to produce the autofocus control. After the signal is used, RESET SIG goes high to reset the integrators, readying the autofocus circuit for the next group of pulses from the IRED.

Power for the autofocus circuit is derived from camera unregulated input. This voltage UNREG is the battery voltage. It is applied through THP001 (a positive temperature-coefficient resistor) to the IRED drive circuit and through PS001 to the autofocus power circuit. Because UNREG is always high, it must be turned on and off by the P 5-Volt input at CN001 (pin 2). When this goes high, Q15 is turned on, turning Q14 on, which applies 5 V to the autofocus circuit. Because autofocus might be undesirable at times, an AUTOFOCUS switch is incorporated on the RR-8 board that allows the autofocus circuit to be defeated.

CCD/UNcon comparison

The CCD-V8 achieves its small size through the use of 8-mm videotape and a CCD camera, which employs a solid-state device in place of a conventional glass pickup tube. Thus, it has many of the advantages a transistor has over a vacuum tube, including long life, reduced sensitivity to vibrations and shocks, no warm-up time, small size and weight, invulnerability to magnetic fields, and low power consumption. In addition, because of the solid-state structure, there is no lag or blooming in the CCD device.

In the Trinicon, however, lag and blooming are characteristic of the photoconductor that senses the light. This is because very bright images produce leakage across the surface of the photoconductor, which is impossible in the CCD device. Notice the comparison chart in Fig. 11-10.

	CCD (Charge Coupled Device)	Pickup Tube
1. Life & Reliability	Long-life	Heaters progressively weaken from beam radiation.
2. After Image & Scorch	No after image. No scorch with lengthy photographing of the same object or with strong light.	Characteristics of photoelectric film make this unavoidable.
3. No Figure Distortion	As picture elements are arranged regularly and they also perform selfscanning, exact geometrical figure can be obtained.	With beam-scanning, it is difficult to inspect accurately both the center and perimeter by scanning.
4. Vibration Proof & Impulse Proof	Strong because it uses the semiconductor chip.	Weak because it uses glass tube, filament and socket.
5. Picturing Time	Fast because it uses no heater.	Needs time for heater warm up
6. Size & Weight	Small and light.	Needs length for emitting beams and also space of coil for focussing and deflection.
7. Use in an Electro-magnetic Field	No influence.	Easily influenced by electronic beams.
8. Electric Power Consumption	Low power consumption because it is composed of semiconductor.	Large amounts because the heater and coil, etc. are used with several hundreds of voltages.
	Structure of CCD Camera	Structure of Pickup Tube Camera

■ **11-10** *Comparison of CCD and pickup camera tube.*

An additional advantage of the CCD device is that it is not prone to damage by the customer. The CCD-V8 is much less sensitive to image burns. This is because the CCD element is a single piece of metallic circuit. If a bright light is focused on the surface of a photoelectric film in a Trinicon, the heat generated could damage it.

On the other hand, the CCD chip easily dissipates any heat because it is made of metallic silicon and is very difficult to damage in normal use.

CCD imager

The heart of the camera in the CCD-V8 is the pickup element itself. This is a single chip of silicon that is constructed to be light sensitive and to be able to sequentially draw information from this device to produce the video signal.

The CCD element is divided into discrete elements that can be individually accessed. This is similar in concept to the sampling of a digital audio signal where the information is sampled at specific points. In the CCD imager, the picture is sampled at individual points both horizontally and vertically. Refer to Fig. 11-11 and Fig. 11-12. In the CCD-V8, there are 532 columns of elements and 504 rows of elements, resulting in a total of over 250,000 elements. These elements are called *pixels* and are the smallest amount of detail that can be resolved by the CCD pickup. The CCD imager used in the CCD-V8 contains 60,000 pixels—more than the pickups used in previous CCD cameras-resulting in much higher resolution of detail. In the CCD-V8, several rows and columns of the picture elements are not used. These elements are black-masked and form the black border at the top, beginning, and end of each

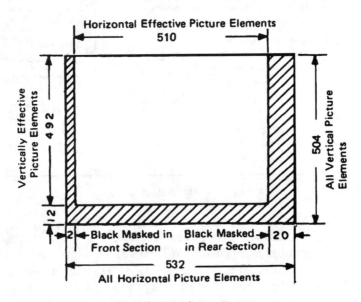

■ **11-11** *All horizontal picture elements.*

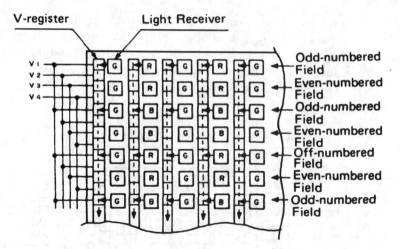

■ 11-12 *Each pixel is placed behind a colored filter.*

line. This leaves an array of 492 by 510 active picture elements in the CCD-V8.

In order to produce color, each pixel is placed behind a color filter. The color alternates as shown in Fig. 11-12. In the first two rows, the elements are colored green and red. In the third and fourth rows, the elements are green and blue; and in the fifth and sixth rows, the elements again go to green and red. This results in an optimal pattern for both high-resolution luminance and color signals. The two identical rows are alternated between fields in order to perform odd/even line interlacing. The individual light receptors are addressed using a series of pulses to move the information from the light receptor to the vertical register, down the vertical register, and ultimately out of the device.

The CCD imager must transfer the signal produced by the light across the device to the output lead. A principle of charge coupled devices is employed. These devices are similar to a bucket brigade. The charge from one device is passed to the adjacent device by varying the voltage on both devices. From the second device, the charge can be transferred to a third device, without loss, by again varying the voltage in the proper manner. The individual sensors are MOS transistors constructed in such a way that in the presence of light, electrons flow from the gate to the sensor, where they are accumulated. For this device to produce a video output signal, the charge from each sensor must be moved to the output port at the correct time.

Because of an inversion that takes place in the lens, the top of the image is seen at the bottom of the diagram. The charge from the sensors must be moved to the output from the bottom row first, followed by the next row above it and the row above that, and ultimately to the top row of the diagram shown in Fig. 11-13.

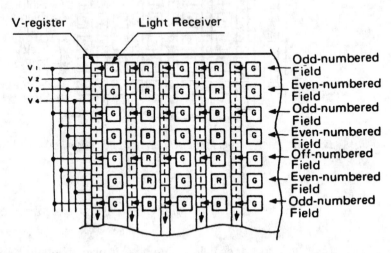

■ **11-13** *The two identical rows are altered between fields to perform odd/even line interlacing.*

The last step in the CCD process involves moving the charge from the last horizontal register to an output buffer and at the same time converting the charge to a voltage. In the diagram of Fig. 11-14 at time A3, H1 and PG are high. This causes the switch to close, which charges the capacitor on the gate of the output MOS to 18 Vdc. CCD output at this time is high. During time B3, H1 and PG go low. When this happens, the switch opens, and the charge in the H1 register starts discharging the capacitor. The amount that the capacitor discharges is exactly proportional to the amount of charge in the H1 register and will determine the final voltage on the capacitor. This voltage is then buffered by the output of the CCD. The output is proportional to the intensity of the light striking a particular sensor and represents the video level. At time C3, H1 and PG go high, and the process repeats. The resultant output is a series of amplitude-modulated pulses that are very similar to a PAM output seen in digital audio. The signal take-out circuit is shown in Fig. 11-15.

A photo of the new Canon 8-mm digital camcorder is shown in Fig. 11-16 and the remote control 8-mm Canon camcorder UC1 is

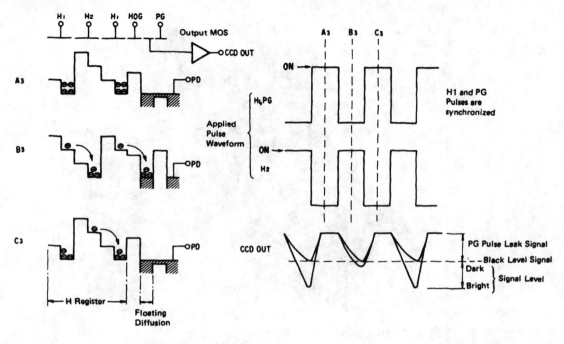

■ **11-14** *Waveform transfer mechanism from H register to output.*

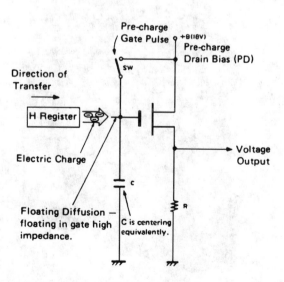

■ **11-15** *Signal takeout point.*

■ **11-16** *Canon 8-mm video camcorder.*

■ **11-17** *Canon 8-mm camcorder UCI.*

shown in Fig. 11-17, and the new Zenith VM8000 camcorder is shown in Fig. 11-18.

■ **11-18** *Zenith VM8000 camcorder.*

Zenith camcorders

The Zenith VM6200 is a high-quality, high-performance camcorder. It is one of the smallest units that features both record and playback modes. The following information and illustrations are courtesy of Zenith Electronics Corp.

This camcorder uses the newly developed 1/2-inch CCD pickup element and high-quality (HQ) VHS picture improvement technology that makes one-hour recordings possible. The unit has full automatic operation that includes auto focus, auto white balance, auto filter, and auto iris. For convenience, the auto focus and auto filter circuits are switchable from auto to manual.

The VM6200 has an SP-EP switch. This allows a full hour of recording in the EP mode using a TC-20 tape. The unit comes with the VAC415, a 1,000 mAh battery. The weight of the unit with the battery is approximately 3.9 pounds.

Recording color signal flow

The color circuits block diagram is shown in Fig. 11-19. IC4 selects the color signal input for either external or

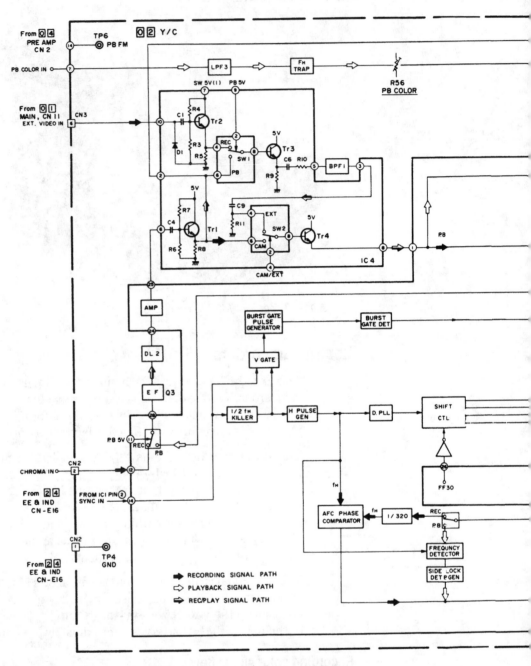

11-19 *Portion of color system in Zenith VM6200 camcorder.*

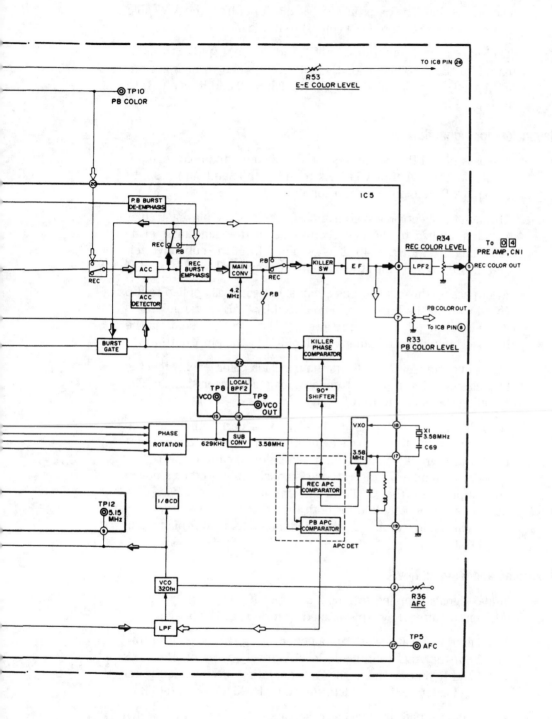

TO IC8 PIN (26)

R53
E-E COLOR LEVEL

TP10
PB COLOR

IC 5

R34
REC COLOR LEVEL

To 0 4
PRE AMP, CN1

REC COLOR OUT

PB COLOR OUT

To IC8 PIN (8)

R33
PB COLOR LEVEL

PB BURST
DE-EMPHASIS

REC   PB
ACC
ACC
DETECTOR

REC
BURST
EMPHASIS

MAIN
CONV

PB
REC

KILLER
SW

E F

LPF2

4.2
MHz

PB

BURST
GATE

KILLER
PHASE
COMPARATOR

LOCAL
BPF2

TP8
VCO

TP9
VCO
OUT

90°
SHIFTER

VXO

X1
3.58MHz

C69

PHASE
ROTATION

SUB
CONV

629kHz   3.58MHz

3.58
MHz

REC APC
COMPARATOR

TP12
5.15
MHz

1/8CD

PB APC
COMPARATOR

APC DET

VCO
320fH

R36
AFC

LPF

TP5
AFC

401

Zenith camcorders

camera section. The external video signal input from the A/V cable is applied via Y/C board CN3 (pin 6) to IC4 (pin 10).

The signal goes through switch SW1, and from IC4 (pin 5) to band-pass filter BPF1. This yields the color component, which goes to IC4 (pin 3). This signal is applied through SW2 and IC4 (pin 8) to IC5 (pin 1).

Playback color signal flow

Lowpass filter LPF3 separates the down-converted color component from the signal played back from the tape and supplies it to IC5 (pin 20). R56 adjusts for a suitable level.

Through the electronic switch and ACC of IC5, the playback color signal is applied to the main converter. In the same manner as recording, 3.58-MHz + 40 fH is obtained from the subconverter, and the output becomes the 3.58-MHz color signal.

After passing through various circuits, the signal is applied to IC5 (pin 1) where the burst level is reduced by 6 dB (reversing the 6 dB increase applied during recording). The result is sent via the playback amplifier and killer switch of IC5 to the pin 7 output.

Color level control R33 adjusts the playback color level, while the trap circuit of L8 and C35 attenuates main converter leakage. The resulting signal is sent to IC8 (pin 8).

Hybrid IC5 has been specially developed for performing the main color circuit functions of this camcorder. In addition to contributing toward circuit miniaturization and simplification, the reduction in the number of discrete parts considerably reduces power consumption. Further cost and space reductions are achieved by using the same delay line for both the camera and recorder sections. The audio circuits are shown in the system block diagram in Fig. 11-20.

Recording audio signal flow

Audio signals supplied to pins 2 and 4 of IC303 are input to the ALC circuit after they are switched by IC303 pin 1's voltage.

Level is adjusted at the line amplifier according to the control voltage from the ALC (automatic level control) detector. Resistors R338 and R336 determine the ALC operating level, while the response characteristics are determined by R339, C325, and C326.

LPF removes high-frequency noise (above 12 kHz), after which the signal goes into two lines. One of these is amplified by the monitor amplifier and supplied as the E-E signal to the earphone and 8-pin A/V connector (this is also sent to the ADC detector). The other signal line goes via R314 (REC LEVEL) to the recording amplifier. The REC amp serves to compensate for high frequency loss in the tape recording and playback processes. From this point, the signal is mixed with ac bias and is supplied to the audio head for recording.

Playback audio signal flow

The signal played back by the audio head is supplied to the playback equalizer amplifier, which functions for low frequency level compensation. This output goes via the PB LEVEL control to the line amplifier. Afterward, signal flow is the same as the E-E output.

Mode control

The four operating modes (REC, E-E, PB, and PB MUTE) of the audio circuit are switched by the mechacon circuit. The modes of IC301 are determined by the states at pins 15 and 16.

The mechacon section of the VM6200 has quite a few features, such as a feature that monitors the functioning of the mechanical operations and provides necessary signals and voltages to carry out these operations.

One feature of the mechacon circuit is an expanded emergency mode. This feature indicates where a failure has occurred by causing different LEDs to flash when the unit goes into auto-stop. In the VM6200, there are 27 possible emergency modes. The expansion of the emergency mode gives the service technician a more accurate answer to the cause of a certain failure. These 27 modes are indicated in the chart shown in Fig. 11-21.

The VM6200 also has expanded test mode capabilities. If a failure occurs during one of these test modes, the unit enters one of the emergency modes. A special switch activates these test modes.

The mechacon circuit in the VM6200 also has a feature that allows it to use less battery power in the OFF mode. The previous camcorders used 4 mA/h and 2 mA/h respectively when OFF. The VM6200 uses only 100 μA/h when OFF. This means that while the VM6200 is between recording segments, it uses less battery power than its predecessors. (However, it is still recommended that the battery be removed for extended storage.)

403

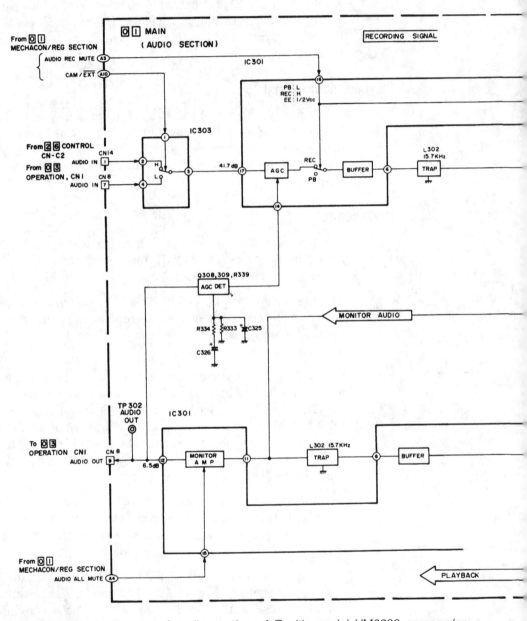

■ **11-20** *Block diagram of audio section of Zenith model VM6200 camcorder. Courtesy Zenith.*

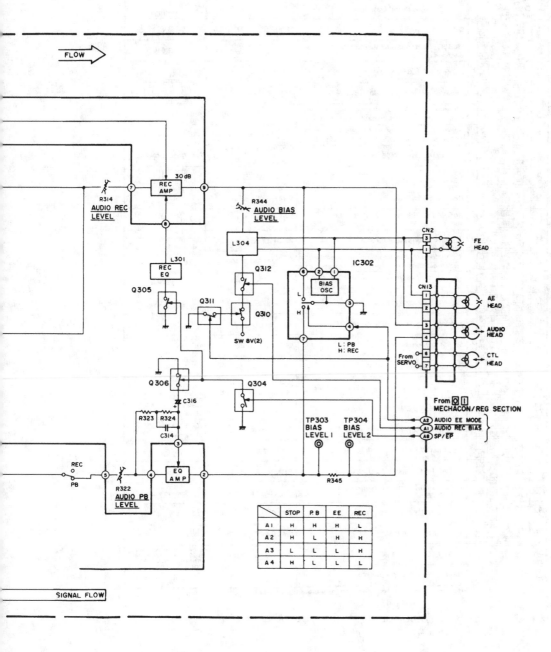

FLOW ➡

30 dB

REC
AMP

R314
AUDIO REC
LEVEL

R344
AUDIO BIAS
LEVEL

L304

CN2

FE
HEAD

L301

REC
EQ

Q305

Q312

IC302

BIAS
OSC

L

H

L: PB
H: REC

CN13

AE
HEAD

AUDIO
HEAD

CTL
HEAD

Q311

Q310

SW 8V(2)

From
SERVO

Q306

Q304

C316

R323 R324

C314

TP303
BIAS
LEVEL 1

TP304
BIAS
LEVEL 2

From ◉ ▯
MECHACON/REG SECTION

AUDIO EE MODE
AUDIO REC BIAS
SP/EP

A2
A1
A8

405

REC

PB

R322
AUDIO PB
LEVEL

EQ
AMP

R345

	STOP	P.B	EE	REC
A1	H	H	H	L
A2	H	L	H	H
A3	L	L	L	H
A4	H	L	L	L

SIGNAL FLOW

VM6200 TROUBLESHOOTING IN EMERGENCY CONDITIONS

No.	Emergency Indication	Cause and Description	Check Point	Normal Waveform & Voltage
1		Defect in 8 V SW	CP3(◙①), IC1, Q1(②⑧)	M3 DC 8 V *
2		SW REG's input voltage for the drum/capstan motor is abnormal.	CP2	M2 DC 9.6 V *
3		No input of P.B. CTL pulse	CTL signalling system	S11 ⊢33.3msec⊣ DC 5 V / DC 0 V
4		Unstable input voltage of IC406 pins 27 – 31	IC406 pins 27 – 33 (◙①) (in other modes than (TEST) No chattering must be confirmed.	DC 5 V or D.F.F. DC 5 V
5		Abnormal CAM SW's input in the condition that the loading motor is rotating in the direction of loading.	CAM SW, CN19(◙①)	CN19 H:5 V L:GND
6		Abnormal CAM SW's input in the condition that the loading motor is rotating in the direction of unloading.	Same as above	Same as above
7		Abnormal inputs of IC406 pins 2 – 9	IC403~405 DC 5 V / DC 0 V High level = 5 V approx.	Port A input check.
8		No CAP. FW FG input at back-spacing in QUICK REV. mode.	CN16, IC411(◙①)	S12 DC 5 V / DC 0 V *
9		Dew sensor is faulty.	CN21, IC407(◙①)	N1(CN21-③) NORMAL:L DEW :2.5 V
10		No CAP. FW FG input at REC start in the assembly mode.	CN16, IC411(◙①)	S12 DC 5 V / DC 0 V *
11		Abnormal clock pulse of IC401 pin 6	IC401(◙①)	CN9-⑦, R487 DC 5 V OSC RANG 0.1msec
12		The set does not enter the next mode 10 sec after the loading motor has started rotation in the loading direction.	CAM SW, IC403(◙①)	CN19-②,③,④ H:5 V L:GND
13		The set does not enter the next mode 10 sec after the loading motor has started rotation in the unloading direction.	Same as above	Same as above
14		Defect in the CASSETTE SW	CN4-①(◙①)	ON :L OFF:5 V
15		Abnormal input voltage of IC401 pins 27 – 32	IC401 pins 27 – 32	H:5 V
16		SW 5 V (01) level is abnormal.	CP1, SW REG BLOCK	IC405 pin 2 DC 5 V (at POWER ON)
17		Abnormal level of LOAD 9.6 V	CP5	IC405 pin 8 DC 9.6 V (at POWER ON)
18		KEY data inputs (IC401 pins 2 – 9) are unstable.	Pins 3, 5, 7, 9, 12, 14, 16, 18 of IC403, 404 or 405	DC 5 V in Stop mode
19		There is SERVO EMERGENCY input.	Servo circuit	IC405 pin 17 ≈ 5 V at normal
20		There is AUDIO EMERGENCY input.	Audio circuit	IC405 pin 17 ≈ 5 V at normal
21		No DRUM FF input	IC401-㊽(◙①)	DC 5 V
22		No CAP. FW FG input in PB quick review	Refer to Item No. 8.	Refer to Item No. 8.
23		No CAP. FW FG input at back-spacing in REC PAUSE mode.	Same as above	Same as above
24		Defect in REC SAFETY SW	CN3-①	ON :L OPEN:5 V
25		Abnormal H level at IC401 pin 60	IC411, 410, 408	H:5 V L:GND
26		No TU REEL SENSOR INPUT	CN20, IC409, IC403	Set the oscilloscope's range to 20 ms, and confirm that there are periodic SCAN inputs to IC403 pins 3 and 12. DC 5 V
27		No SUP. REEL SENSOR INPUT		

※ : "M" and "S" in the right column show checker lands on circuit boards. Check them referring to pattern figures.

■ **11-21** *Troubleshooting guide for Zenith VM6200 camcorder.*

The addition of the EP mode of operation, necessitates some changes in the servo circuit. The changes are mostly within the servo IC and are switched by the mechacon circuit.

RCA camcorders

A brief circuit description is given on some of the CC011 camcorder systems in the following section of this chapter. The operating controls for the CC011 are shown in Fig. 11-22. The following information and illustrations are courtesy of RCA Corp.

Color multiplexing system using a single Newvicon

The incoming light to the camera passes through the lens, the optical system (in which the color temperature conversion filter is installed), the infrared (IR) cut filter, and the crystal filter (where it is imaged on the surface of a striped filter). Note the block diagram in Fig. 11-23. The human eye is sensitive to electromagnetic waves from 380 run to 780 nm in wavelength (the visible region). In addition, the human eye also discerns the wavelength difference as a color difference. The human eye responds to the light between 400 nm and 500 nm wavelengths as predominantly blue information, the light between 500 nm to 600 nm as green information, and the light between 600 nm and 700 nm as red information.

Horizontal sawtooth/parabola generator circuit

This circuit (see Fig. 11-24) generates the horizontal sawtooth/parabola waveforms required in the bias light shading correction circuit and the dynamic focus correction circuit. The horizontal, sawtooth/parabola signals are also supplied to the process circuit board for signal shading correction.

Bias light shading correction circuit

The light intensity of the bias light LEDs is not uniform along the entire target surface but has a shading. The red and blue signals are obtained by detecting the modulated signal from the preamplifier output signal with a bandpass filter; therefore, the red and blue signals are free from bias light shading. However, the luminance signal is obtained by removing the modulated signal with a trap circuit. Thus, the luminance signal receives the adverse effect of the bias light shading. The bias light shading correction circuit supplies horizontal sawtooth and parabola signals B to the process circuit board to correct color shading at low illumination levels. Refer to circuit diagram in Fig. 11-25.

Dynamic focus correction circuit

The level of focus voltages that produces the best focus of the beam in the center part of the pickup tube is different than that

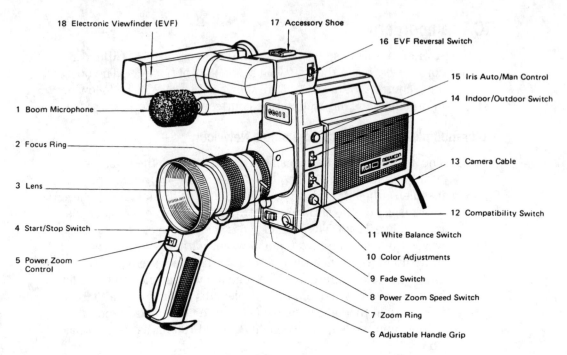

18 Electronic Viewfinder (EVF)

17 Accessory Shoe

16 EVF Reversal Switch

15 Iris Auto/Man Control

14 Indoor/Outdoor Switch

13 Camera Cable

12 Compatibility Switch

11 White Balance Switch

10 Color Adjustments

9 Fade Switch

8 Power Zoom Speed Switch

7 Zoom Ring

6 Adjustable Handle Grip

1 Boom Microphone

2 Focus Ring

3 Lens

4 Start/Stop Switch

5 Power Zoom Control

■ **11-22** *Operating control location for RCA CC011 camcorder.*

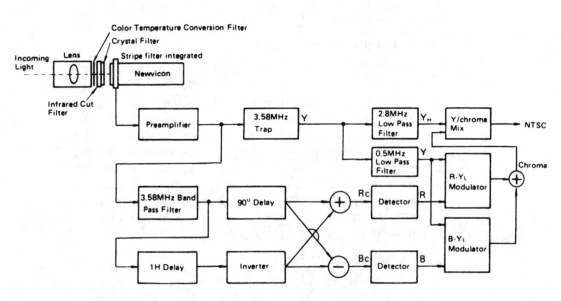

■ **11-23** *Block diagram of multiplexing system using a single tube.*

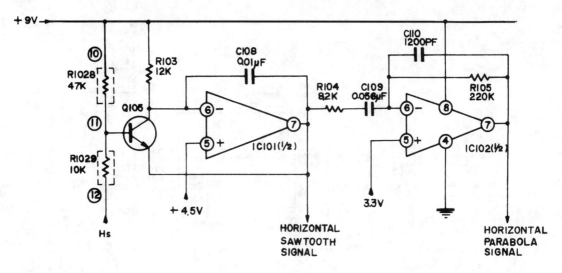

■ 11-24 *Horizontal sawtooth/parabola generator circuit.*

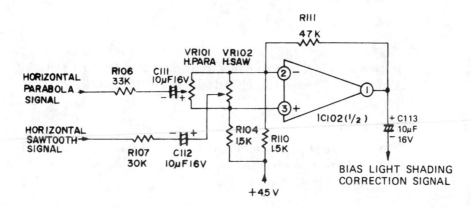

■ 11-25 *Bias light shading correction circuit.*

which produces best focus at the edges of the pickup tube. Therefore, the modulation depth for the center and the edges of the pickup tube differ. When the camera is directed at an evenly illuminated white object, the red and blue signals modulated at 3.58 MHz do not have a uniform level, and as a result, color shading appears. Note Fig. 11-26.

The dynamic focus correction circuit supplies horizontal and vertical sawtooth and parabola signals to the focus grid (grid-4). These signals, together with the dc focus voltage from the high-voltage circuit, focus the electron beam along the entire scanning area to correct the unevenness of modulation.

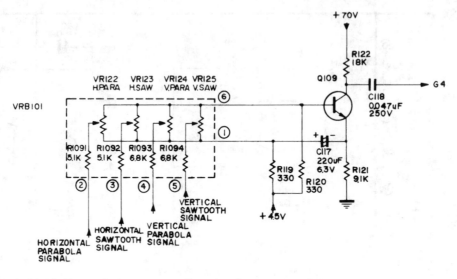

■ **11-26** *Dynamic focus correction circuit.*

Note: Dynamic focus should be adjusted when color nonuniformity is seen in the picture of a white card after bias light shading is properly adjusted and electrical focus and beam alignment are properly set.

Fade circuit

The fade circuit generates a triangle pulse that is applied to IC207 (pin 4). Here it controls the video signal, creating the "fade in" or "fade out" effect. Refer to Fig. 11-27 for the fade timing chart waveforms.

Shading correction circuit

Even when dynamic focus is applied to grid-4 (G4), the nonuniformity of modulation (due to uneven focus on the target) cannot be completely eliminated. The shading correction circuit compensates for residual picture shading using horizontal (H) and vertical (V) sawtooth and parabola signals that are generated in the deflection circuit. Note the block diagram in Fig. 11-28.

Chroma encoding circuit

This circuit contains the color reproduction correction, gamma correction, chroma signal generator, and the chroma clip/suppression circuits. The chroma encoding block diagram is shown in Fig. 11-29.

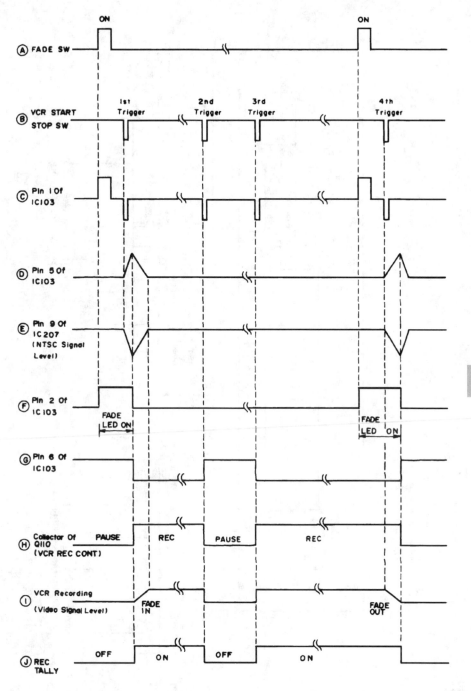

411

■ 11-27 *Fade timing chart waveform.*

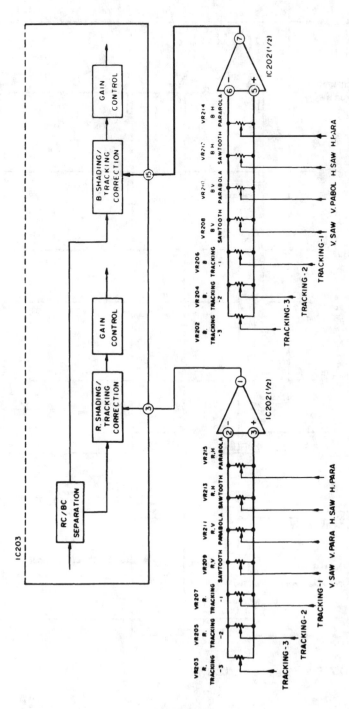

■ 11-28 Shading correction circuit.

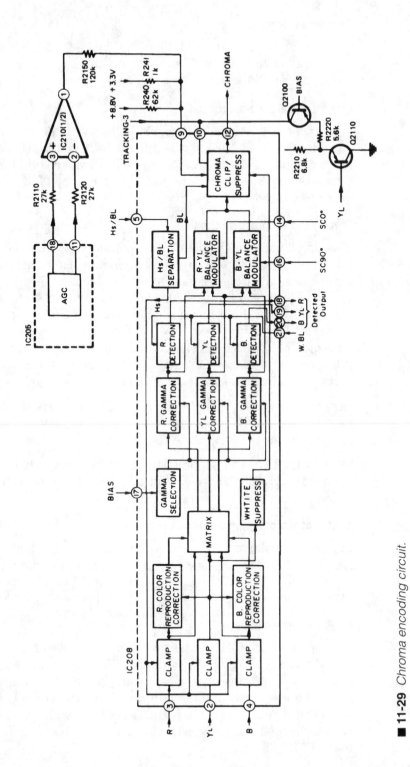

■ 11-29 Chroma encoding circuit.

Gamma correction circuit

This circuit receives the red, blue, and yellow signals from the color reproduction circuit. These signals, after gamma correction, are supplied to the chroma signal generator circuit.

Chroma signal generator circuit

This circuit matrixes the R, B, and YL signals with the color reproduction correction signal to generate the R-YL and B-YL color difference signals. It also modulates these baseband color difference signals on two subcarriers that differ in phase by 90 degrees to generate a chroma signal.

Chroma clip/suppression circuit

This circuit contains the high Y level chroma clip circuit and low-level chroma suppression circuit. This circuit clips the green color that results when an excessively bright object is shot with the camera. This is caused by the lack of a high current beam in the newvicon. When the beam of the newvicon is insufficient to discharge the target due to excessive light reflected from a brightly illuminated object, the red/blue signal component modulated and riding on top of the green signal is lost. Thus, the bright picture parts turn into an unnatural green color.

Hence, if this circuit is faulty, the bright picture parts would turn into an unnatural green color, and the chroma signal would appear on the blanking period of the NTSC signal.

The low-level chroma suppression circuit enhances color reproduction in a low luminance condition by improving the white balances.

NTSC signal processing circuit

In the circuit shown in Fig. 11-30, the chroma/luminance mixed signal, burst signal and the sync signal are mixed to produce the NTSC signal. The NTSC signal is applied directly to the portable VCR or a table model VCR via the power supply.

Playback sense circuit

This circuit (Fig. 11-31) senses the recording or playback mode of the portable VCR. When the portable VCR is set in the recording mode, this circuit sets the camera to supply the NTSC output signal to the viewfinder and the portable VCR for recording. When the portable VCR is set in the playback mode, this circuit shuts the camera function off and supplies the playback signal from the VCR to the EVF.

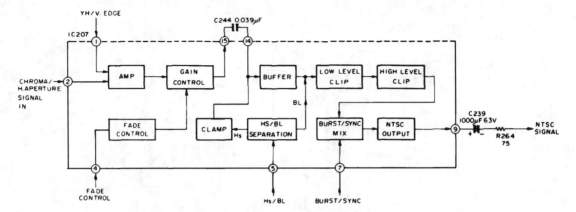

■ **11-30** *NTSC signal processing circuit.*

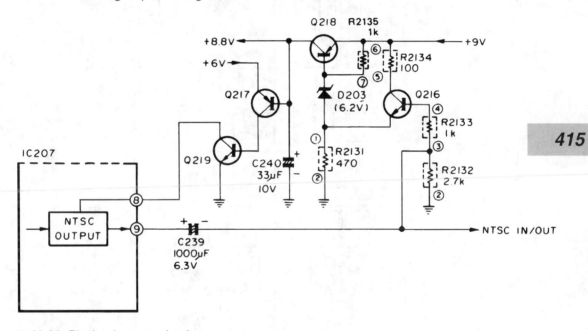

■ **11-31** *Playback sense circuit.*

Automatic white balance setting circuit

This circuit automatically sets the gain of the red and blue signals to match the gain of the yellow signal. Refer to Fig. 11-32 for this circuit diagram.

White balance indicator circuit

The circuit shown in Fig. 11-33 supplies a 15-Hz signal (1/4 pulse of Vs signal) to the indoor/white balance indicator D112. This LED will flash on and off if the white balance is improperly set.

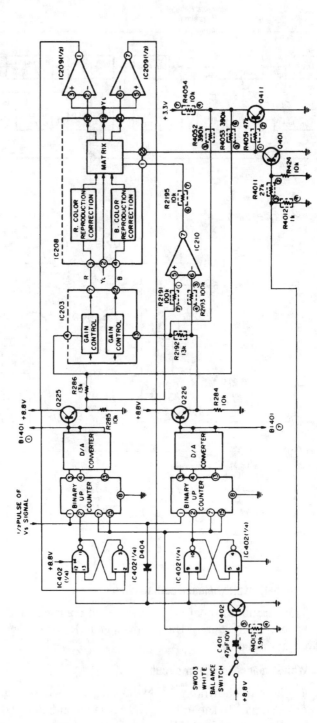

■ 11-32 *Automatic white balance setting circuit.*

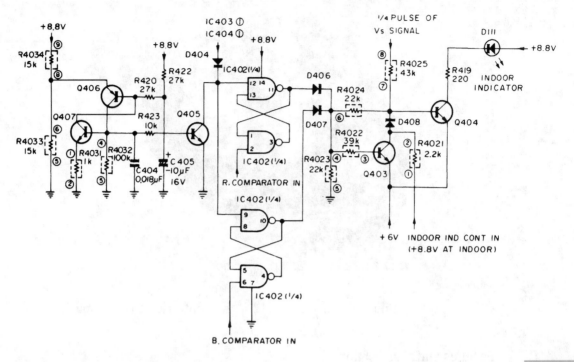

■ 11-33 *White balance indicator circuit.*

Zenith VM8300 8-mm camcorder[1]

The Zenith VM8300 is a new 8-mm camcorder in a horizontal format. This unit can be described as a binocular type. The camcorder has a sensitivity of four lux with a 1/3-inch CCD and 270,000 pixels. Horizontal resolution is 330 lines. A photo of this camcorder is shown in Fig. 11-34.

The main new feature of the VM8300 is "fuzzy logic" for the auto focus, auto iris, and auto white balance. Fuzzy logic is a relatively new mathematical approach intended to go beyond current methods of computation in which mathematical values of 0 and 1 are used to process data. Such a sharp boundary can cause abrupt changes in a conventional control system.

Fuzzy logic works by tuning the hard-edge world of binary control into "soft" grades. Values other than 0 or 1 are used for a fine evaluation of processed data. This method allows for a more flexible response to a given input. The output of a fuzzy system is smooth

---

[1]Information in this section is courtesy of Zenith Corp.

■ **11-34** *Zenith VM8300 camcorder.*

and continuous, which is ideal for the control of a continuously variable system.

Features and specifications

The Zenith VM8300 is a five-head camcorder (four video heads, one flying erase) with a sequential recording system; the drum diameter is 27 mm and the tape wrap is 292 degrees. The small diameter drum makes for a more compact tape mechanism.

This camcorder features FM, hi-fi, stereo, and audio recording capabilities with the built-in stereo microphone or through the input jacks.

Digital auto focus

The VM8300 has a full-range auto focus system for rapid focusing from 10 mm to infinity. This system digitally integrates the high band component contained in the video signal (performs multiplication and addition in every vertical period, hereafter referred to as evaluation value) and controls the focus motor by an 8-bit microprocessor to achieve maximum integrated value.

The change of AF evaluation value when the lens position is varied for a common object is shown in Fig. 11-35. In operation, the focus ring is controlled by the motor so that AF evaluation value is always increased. When AF evaluation value is decreased (the maximum value of the evaluation curve is identified), the focus motor operates in the reverse direction and the focus ring is returned

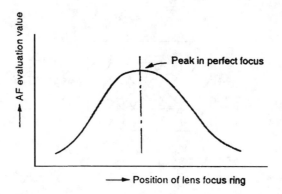

■ **11-35** *Initial operation at power turn ON.*

and stopped to a position where the maximum value of the evaluation curve is identified.

After that, according to the change of object information (the form of the curve of evaluation value is changed and the current position is not a summit of the curve), the microprocessor monitors the lens position so that it always searches a summit of the evaluation curve.

Initial operation at power on

When power is turned ON, the focus lens is once moved to infinity and returned to the center position. Then AF and other operations are started. (The position of lens focus is stored in the microprocessor by this operation.) If the focus motor or terminal switch is defective, AF, zoom, or high-speed shutter do not operate.

Operation instructions

IRSI signal output from IC916 is amplified by Q9302 after an increase in the luminance component by Q9301. The signal is divided into signals of more than 200 kHz, 600 kHz by Q9304 and input to pins 34 and 35 of IC931, the standard cell. These signals are selected together with the AE signal by the standard cell, then output to pin 50.

These time-shared multiplex signals are band-limited to 2.4 MHz by LPF T9301 via the buffer amplifier of Q9309, then peak held by Q9408 and Q9429. The signal is then converted into a digital signal by the A/D converter in the standard cell and integrated in every area as shown in Fig. 11-36.

Integrated AF evaluation value is fed to IC932, and AF operation is done based on AF evaluation value of focus zones 1 and 2. The se-

419

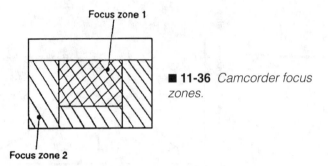

Focus zone 1

Focus zone 2

■ **11-36** *Camcorder focus zones.*

lection of these focus zones is determined by fuzzy logic of the microprocessor and controlled to be focused on the object center of the screen. The IRIS DRIVE signal is fed from Q9323 to pin 37 of the standard cell and changes AF operation according to the iris condition.

AF operation

When AF operation will not work, or does not move smoothly the following conditions are possible:

☐ Object outside focus zone. This system only obtains evaluation value calculation information within the focus zones and ignores objects outside the focus zone.

☐ Extremely low evaluation value. Poor S/N ratio under low illumination condition. No contrast on object.

☐ Sudden zooming operation. Objects are moving violently or too fast; severe panning or tilting of camera.

☐ Object within 1.1 mm, focus length cannot be focused structurally with telezooming.

Digital auto iris (AE)

Through the application of fuzzy logic the auto iris moves closer to the human eye. This lessens back lighting effects and makes for a better picture. The system is designed so that the iris will control counter fight or excessive light by controlling the AGC and gamma level based on the luminance level of each of six video areas.

The iris signal is fed from pin 17 of IC916 to pin 40 of IC931, the standard cell, through IC937 on the CA-1 board. The AE process is then accomplished in a certain time (IV) by a time-sharing system with the auto focus. At this time the signal is A/D converted and integrated into every area by the standard cell. This now becomes the amount (evaluation value) indicating the luminance level of each area. Microprocessor IC932 receives this evaluation value at the same time as the AF and judges the counter light or excessive light condition using fuzzy logic, based on characteristics that most objects are centered in the frame. This determines the compensation values of iris, AGC and gamma level output from pin 43 and 44 as PWM data.

The iris signal is amplified by IC937, detected, and input to pin 1 of IC938. It is then compared with the reference voltage input at pin 3 to drive the iris coil by the voltage difference. The PWM output at pins 43 and 44 of the microprocessor IC932 become the IRIS, REF, AGC, REF, and gamma control signals. When pin 1 of CN912 is LOW, the digital AE function is turned off, resulting in a standard condition where each compensation is not accomplished.

Auto white balance (AWB)

The color is compensated for more accuracy by applying fuzzy logic. A conclusion is derived from the color condition of the subject by analyzing color phase, distribution, and amount of high frequency component.

In the AWB circuit, the picture frame is divided into eight areas in the vertical and horizontal direction (total of 64 areas). The video information of each area is analyzed by a microprocessor, and the color is assumed and chroma compensation is preferred.

The B-Y and R-Y signals from pins 19 and 20 of IC913 are amplified by Q9103 and Q9105 and input to pins 45 and 43 of IC931 of the standard cell. The signals are clamped by a voltage at pin 44 and A/D converted in 64 areas, then input to the microprocessor. The microprocessor analyzes the brightness information, color level, hue, and area occupied by one color. The microprocessor then determines the control amount of the AWB circuit. The resultant control voltages are sent to pins 2 and 3 of IC932. The signals are input to pins 40 and 37 of IC913 as B and R GAIN control signals to control white balance.

Zenith VM6700 and VM6800 camcorders

Two new Zenith VHS-C camcorders, the VM6700 and VM6800, are similar in mechanical design to the VM6400. They can play back through a home deck with a cassette adaptor. These dual-speed camcorders provide 30 minutes of record time in the SP mode with a TC-30 tape or 90 minutes in the EP mode. A photo of the VM6700 is shown in Fig. 11-37 and a photo of VM6800 is shown in Fig. 11-38.

■ **11-37** *Zenith VM6700 camcorder.*

■ **11-38** *Zenith VM6800 camcorder.*

Both camcorders have a built-in gain up control that, when activated, allows shooting in lighting situations as low as one lux. That is about as much light as ten candles on a cake. The normal three-lux light sensitivity is excellent for low-light recording as well with less picture graininess than that produced with gain up.

Loading a cassette into the camcorder automatically places it in the RECORD/PAUSE mode. Auto focus, auto white balance, and auto iris are all programmed automatically. Manual focusing is also possible through the power focus controls.

The auto head cleaner automatically cleans the heads whenever a tape is loaded into the camcorder. The VM6800 steps up to an 8:1 power zoom lens that gives it equal zoom capability to the Zenith full-size models without adding to its size.

The fader control allows the VM6800 to create professional-looking scene transitions. You can "fade out" from a picture to black or "fade in" from black to a picture.

Full-size Zenith camcorder

The VM7060 is the latest of the full-size camcorders. This two-lux, four-head camcorder brings all of the advantages of immediate compatibility with fun-size decks. It will provide 160 minutes of record time in the SP mode. The two extra heads allow for playback of EP recorded tapes. A photo of this camcorder is shown in Fig. 11-39.

423

■ **11-39** *Zenith VM7060 standard-size camcorder.*

The major advantage to a fun-range, auto-focus system is that it can be focused quickly, accurately, and automatically on subjects from infinity to macro. It brings the subject right up to the lens without making any manual adjustments.

The high-speed shutter makes it possible to capture high-speed action on videotape. Pause and slow motion playback of the action is enhanced with even fast moving objects like golf clubs or baseballs frozen in time.

The flying erase head permits clean, noise-free edits from one scene to the next. The retake feature allows the user to search for the point on the tape where a new recording is to begin without leaving the record/pause mode. Playback will show a perfectly clean cut from the first recording to the second with no noise or "glitch" between scenes.

The fade control provides the option of "fading out" of one scene and "fading in" to the next scene rather than abruptly cutting from scene to scene. The fade control offers a smoother, more professional-looking transition.

The camcorder offers DATE and TIME insert recording with a touch of a button. The correct date and time are stored (even if the power is interrupted to the camcorder) by the lithium battery supplied with the VM7060. The display can be turned on or off any time by pressing the TIME/DATE button.

The built-in self timer automatically begins recording in ten seconds. An hour's worth of action can be condensed into a single minute with the time lapse recording feature. The camera can be programmed to record one second of video every 30 seconds, one, or two minutes. The manual time lapse mode can create animation recordings.

A built-in character generator, or titler, enables the user to display a title or caption along with the video. It can store two title pages that can be displayed and recorded at any time. Each title consists of one line of up to 12 characters. In addition to words, such symbols as musical notes, hearts, or happy faces can be used.

12

Case histories:
problems and solutions

THIS CHAPTER FEATURES SOME ACTUAL VCR CASE HISTORY problems and solutions for various brands of videocassette recorders. These troubles cover old and newer model machines. The chapter concludes with techniques for troubleshooting VCR servo systems.

Zenith JR9000

This section provides the playback and record checklists for verifying proper operation of this unit. Then a list of questions is included to help you compile symptoms of various faults, followed by a note of caution for servicing these units. The section concludes with case histories pertinent to the Zenith JR9000.

VCR playback checklist

This checklist shows you how to check a VCR with a test tape. Note the JR9000 setup in Fig. 12-1.

☐ Install alignment tape (Zenith part number 868-27).

☐ Push the cassette lift assembly down. The standby lamp should light up, and tape threading should be completed in three seconds or less.

☐ Depress the play button. The TV monitor screen should blank video and mute audio, followed by the presentation of picture and sound.

☐ Depress the stop button, then depress the fast-forward button. The VCR should stop at the end of the tape.

☐ Depress the rewind button. The VCR should stop at the end of the tape.

☐ Depress the eject button. The tape should unthread and the cassette lift assembly should pop up within five seconds. During the unthreading, the stand-by light should be on.

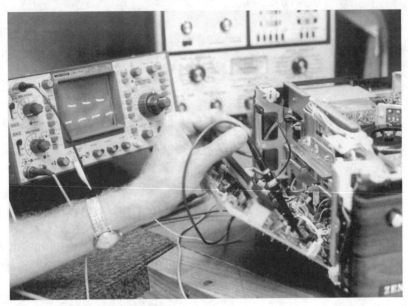

■ **12-1** *Older model Zenith VCR being checked out on my service bench.*

☐ Reinstall the tape, and locate the beginning of the one-hour playback portion. Depress the play button with the tape in the one-hour position. The VCR should blank the video and mute the audio, followed by a normal picture and sound.

☐ As the two-hour portion passes the heads, the video should blank and the audio mute. Next, the VCR should automatically change capstan speed, followed by a normal video and sound reproduction.

☐ Check the memory mode of operation.

VCR record checklist

☐ Insert a blank cassette.

☐ Select a program to be recorded.

☐ Record a portion in the one-hour mode.

☐ Switch to the two-hour mode and continue to record at least five counts on the counter.

☐ Play back the tape to determine which modes are normal. This procedure is not necessary each time, but it does help to isolate problems more accurately. These checklists should also be followed after completing repairs. The machine being serviced should be operated in both the one- and two-hour modes.

Another important consideration involves the functioning of the chroma and luminance systems. If no chroma is present in the playback of a recorded signal that is in color, then the chroma processing section of the VCR must be checked out.

Because the JR9000 VCR contains a combination of mechanical and electronic systems, another step is in the type of fault-mechanical, electronic, or both.

Compiling symptoms

☐ Is the record mode operating properly?

☐ Is the play mode operating properly?

☐ Is the one-hour mode of record/playback functioning correctly?

☐ Is the two-hour mode functioning correctly?

☐ If the recorded signal includes chroma information, does this chroma appear to be normal at the output?

☐ Is the malfunction a mechanical or an electronic problem?

Caution for bench servicing

When placing the VCR on its left side (the RS-L board side) with the cover removed, damage can occur to the P-7 board if it is done improperly. The P-7 board, which supplies the regulated +12 V to most sections of the recorder, is located on the left-front corner of the machine. Should *you* roll the machine from its normal horizontal position onto its left side, the P-7 board could crack. This crack has been known to open the conductive path that leads to the output pins of CN5002; this cuts off the +12-V output.

To avoid this damage, the machine should be *lifted* not *rolled*, from one position to another. Always take this precaution when servicing any VCR with the covers removed.

Zenith JR9000 case histories

This section describes actual problems and solutions encountered with this model. Some symptoms pertain only to this model, while others might be more universal. Each symptom is listed, followed by the corresponding service technique and actual cause of the problem, plus perhaps some background information for clarification.

Trouble symptom

The standby lamp did not light during the threading and unthreading (eject) operations. Other functions of the machine were normal.

Background information   The threading/unthreading operations are functions of the SY-2 board, as is the standby lamp circuit. On the SY-2 board, transistor Q9 is the device that operates the lamp. The lamp is in the collector circuit of Q9, with one side of the filament connected to 12 V (B+).

Service techniques   The standby lamp could fail to operate because of several conditions: loss of the 12 V; a defective lamp; a defective Q9 transistor. A voltmeter check of the collector of Q9 can determine which of these conditions exists. This voltage point was measured during the threading and unthreading modes. The reading indicated a high voltage, meaning that the B+ was present, and the lamp filament was good. A clip lead to ground from the collector lit the lamp.

The cause   Transistor Q9 was defective (open).

Trouble symptom

The eject function did not operate. The threading function was normal. The dc threading motor turned continuously as long as the unit was turned on.

Service technique   The clue here is the constant operation of the threading motor. The SY-2 board controls the threading and unthreading functions. The current through transistor Q10 is the current through the threading motor. If this motor is constantly running, Q10 is always conducting. The operation of Q10 is controlled by transistor Q3, which is only cut off during normal threading and unthreading. This causes a high voltage on its collector and on the base of Q10, turning on Q10 and operating the motor. Voltage measurement indicated that this was the case. After threading, Q3 should turn on and Q10 off.

The cause   Transistor Q3 was defective (open).

Trouble symptom

The VCR would not unthread. Depressing a function button put the recorder into auto-stop. The ac motor would not turn.

Service techniques The cassette must be in and the threading switches must be closed to deliver the 12 V (B+) to the proper circuits. The voltage level on pin 19 of IC4001 on the SY-2 board must be high, and pin 21 must be low. Transistor Q3 must be off, and Q10 must be on. Switch 4004 must be in the non-eject position. An obvious first measurement would be the +12 V beyond the cassette IN switch (normally closed with a cassette pushed down into the operating position). This measurement point is pin 1 of connector CN4001. The +12 V did not appear at this point but was found at pin 10 of the same connector. Pin 10 connects to the other side of the cassette IN switch.

The cause The mechanical movement of the cassette IN plunger did not actuate the cassette IN microswitch. A slight readjustment of the position of the microswitch solved this problem.

Trouble symptom

The VCR would not record or play in the two-hour mode. The recorder operated normally in the one-hour mode. A known good tape cassette, recorded in the two-hour mode on another VCR, played back properly on the defective machine. Thus, the actual symptom was that there was no operation of the two-hour record mode.

Background If you check the pre-emphasis route of the luminance signal during the two-hour record operation, note that it includes the outside circuitry between pins 21 and 6 of IC1001 on the YC-L board.

Following the luminance/video signal with a scope through these stages showed a proper waveform up to and including the base of Q36. The collector and emitter of this transistor showed no signal. Transistor Q36 was cut off because of a faulty bias.

The cause Transistor Q49, the thermal compensator in the emitter circuit of Q36, was open.

Trouble symptom

There was no chroma and/or luminance during playback of a good recorded tape. This indicated the fault was in the VCR record system. In addition, the E-E signal, without any button activation, was defective.

Service technique Block diagrams in service manuals usually show the paths of the luminance, chroma, and E-E signals from the input to the output of the VCR. It is always good practice to check

429

inputs and outputs of these blocks (in this case the YC-L board) first before looking inside the circuitry of the blocks.

A look at the composite video input signal at pin 6 (CN1001 connector) revealed an improper video signal. The source of this signal is the PJ-7 board, incorporating the video in/out jacks and the output of the IF-1 board. Next, these sources were checked.

The cause   The shielded leads going to and from the video in/out jack sat the rear of the VCR had been pinched under the rear cover mounting stud by the cover when it was installed. This resulted in a shorted coax cable and a loss of signal.

Trouble symptom

No playback was possible in any mode. A known good recorded cassette in the machine did not produce a picture on the TV screen. Also, the machine did not record properly. Thus, this fault was common to both the record and playback functions.

Service techniques   Most of the record and playback circuitry is located on the YC-L board. The first step is to check the dc power to the board, in this case +12 V. No dc voltage was present. Fuse F5001 on the P-7 board was open. An ohmmeter check indicated a short to ground for the 12-V supply, which was traced to the YC-L board. The YC-L board has two 12-V circuits. One is switched on all the time the recorder is on. The latter 12-V potential was the one that was shorted. Locating a short to ground for the 12-V B+ conductive path on a PC board can be difficult.

Some help is available on the YC-L board, however. Several jumper wires connect sections of the B + PC patterns together. Thus, unsoldering a jumper can isolate the sections of the B+ line for resistance checks. This technique centered around the IC5 area of the YC-L board.

The cause   There was a short between the primary (connected to ground) and the secondary (connected to the 12-V line) of transformer T13.

Trouble symptom

Playback of tapes recorded in the two-hour mode was not possible. All other functions were normal.

Service techniques   The problem was isolated to the de-emphasis process circuitry of the two-hour playback of the luminance signal. The luminance signal could be traced through the system up to the Q1029 switching transistor, but not beyond.

The cause   The switching transistor was defective.

Co-channel cable interference

Sometimes on cable TV systems, a beat can develop between a channel-3 or channel-4 signal on the cable and the corresponding output frequency of the VCR's RF modulator. Two conditions appear to be required for this co-channel interference to occur: an unusually high (80 mV or more) cable signal level; a mistermination of the cable input.

ac motor failures

Some ac motors have failed in the Zenith JR9000 units. The failures were traced to the thermal protective device within the motor, which was modified on later production model runs.

Audio/control head failures

This failure has occurred when the leads going to the head assembly broke. Two methods of connecting these leads have been used in production. Automatic lead connection sometimes did not leave any slack in the wires, causing breakage later from mechanical vibration. Manual connection of the head leads provides more slack and is used in all later production of these models.

Trouble symptom

The tracking control seemed to drift in and out of adjustment intermittently. With the case removed, the recorder operated perfectly. The problem looked the same as when the head drum brakes were unplugged from the CS board.

Service techniques   All scope waveforms appeared to be normal. AU adjustments were set properly. The recorder operated perfectly with the covers removed. When the recorder was turned on end to disassemble it (tuner end), the 3.15-amp fuse blew on the P-5 board. The short was traced to the CS board.

The cause   Component leads protruding from the foil side of the CS board were touching the metal shield on the AD-L board, causing a partial loss of the drum braking, and with the recorder on end, the additional pressure caused the B+ to short, as several other leads were grounded to the shield.

The UHF tuner bracket and frame were straightened. The rear of the UHF tuner had been pressed against the component side of the CS board, causing it to swing in toward the AD-L board.

431

Trouble symptom

There was no audio during playback of a tape recorded on this VCR. A known good tape recorded on another machine produced normal sound when played back on the defective VCR. Hence, the problem was in the audio record section of the VCR.

Service technique   The audio record system includes the audio heads, the AD-L board, the tuners and IF, the MC-4, and PJ-7 boards. The heads are not the problem, because the playback with a good tape was normal. During record, the IF audio output was normal. It was also okay at pin 4 of connector CN7010 and at the input to the AD-L board (pin 1 of CN3603). No audio signal was present at the AD-L board output (pin 1, CN3604) to the audio heads.

The cause   Signal tracing from the AD-L board input into the board quickly determined that transistor Q611 and Q612 were defective. They are the active devices in the first two stages of amplification during the record mode. Replacing these transistors solved the problem.

Trouble symptom

Snow was present on channel 4, but not on other channels.

Service technique   These symptoms indicate an RF problem. Thus, the function blocks that are part of the RF signal path were checked. This included antenna input, VHF splitter (inside the VCR), and the VHF tuner. Direct connection of the antenna lead into the TV set's antenna terminal gave a good picture.

The cause   The VHF splitter in the VCR machine was replaced. The best way to check the splitter is to bypass the RF signal around it.

Trouble symptom

On playback, the picture on the TV set screen broke up. This symptom occurred during the playback of a known good recorded videotape. A tape recorded on this VCR played back normally on a good machine. Thus, the fault would probably be in the VCR playback system.

Service technique   The RS-L board, the YC-L board, and the RF modulator are included in the playback system. The output of the RS-L board is at the center of this playback signal path. Checking with a scope on this output at CN2004 (pin 5) revealed an abnormal signal. A similar condition was present at pins 16 and 17 of

IC2001 on the RS-L board. The input signal to IC2001, at pins 23 and 24, was found to be normal.

The cause   A close visual examination of the board area near IC2001 showed a short between pins 17 and 18. Scraping the space between the board conductors leading to the pins cleared the short and restored the VCR to normal operation.

Trouble symptom

There is "wow" in the reproduced sound. Video is normal in the playback mode. A known good recorded tape gave the same symptoms. Thus, the fault is in the playback circuits.

Service technique   Characteristically, the "wow" condition of sound is directly related to the movement or the tracking of the tape. The tape movement is a mechanical operation, whereas the tracking is mostly an electronic operation.

The cause   Replacement of the CS board solved the problem, thus the electronics of the tracking were involved. In this particular case, Q4631 of the tracking chain circuits was defective. When defective, it gives a "wow" or "warbling" sound during playback.

Trouble symptom

The VCR would not operate or unthread the tape. The tape would not move.

Service technique   Tape movement for the various operating modes is mechanically controlled by the operation of the capstan and pinch roller. The capstan motor was not rotating. An ohmmeter test at the CM board indicated a shorted winding condition. The resistance between the orange and white leads measured 12 to 20 ohms instead of the usual 200 ohms.

The cause   The capstan motor was defective. A new motor solved the problem.

Trouble symptoms

There was buzz or hum in the sound when a local UHF channel (20) was recorded. This condition only occurred in one large area of a city. All the recorders used had the same problem. The following conditions were observed:

☐ Very strong channel 20 signal.

☐ Adjusting recorder AGC did not remove buzz.

☐ Using various TV sets did not eliminate buzz.

433

☐ Buzz was not from the power supply, but it appeared to be a video signal in the audio, because buzz level varied with picture contrast.

☐ Attenuating the channel 20 signal input removed the buzz in the E-E mode but it still appeared in the playback of the recorded tape.

☐ The attenuated channel 20 recording exhibited the buzz when played back on a different VCR.

Service technique Analysis of the composite video waveforms out of the recorder IF only proved that the signal was exceptionally strong and that the UHF channels were not controlled quite as effectively as VHF by the AGC circuits. No distortion or video was present in the sound. Sync level in the composite signal appeared to be about 28 percent for channel 20 as compared to 30 percent for other UHF signals. A two percent difference is not much change in amplitude with a service type scope, but at least there was some reduction in sync level of channel 20 as compared to the other stations.

Because the buzz was evident when a recorded tape was played in different machines, it was assumed that the buzz was recorded on the tape. However, the audio was clean at the audio output jack of the recorder.

Analysis of the signals and voltages fed to the RF modulator proved that an excessive composite video signal fed to the RF modulator was causing overmodulation of the picture carrier and resulting in video intrusion in the sound.

The cause Following the adjustment procedures from the modulator backward through the playback circuitry in search of the excessive video resulted in these findings.

☐ All levels were high, but apparently the recorder was able to handle the strong signal and diminished sync in the record mode.

☐ In playback, the recorder was again able to cope with the conditions, but signal levels gradually increased from the recommended levels until the RF modulator input equaled 2.0-V PP video.

☐ The most effective means of controlling the video signal proved to be the adjustment of VR21 (on the YC-L board) to the specified 1.3 V PP at the emitter of Q25. Original level was about 1.9 V PP before adjustment on channel 20. The original level was normal on any other channel. After adjustment to

434

1.3-V PP for UHF 20, other channels read as low as 1.0 V PP. The only noticeable change in the picture on other channels was a slight change in brightness. Originally, all the UHF channels seemed to play back slightly brighter than VHF channels, probably because of the stronger (local) signals.

Note: Do not attempt to compensate by adjusting VR24 (1 V PP at TP3) for the proper video signal to the modulator. It would not reduce the signal sufficiently, and it caused deterioration in the playback of the other channels.

The recorder is able to handle a nonstandard signal and record properly with this adjustment. All of the tapes now play well, with the video playback signal reduced.

Trouble symptom

After 15 minutes of playing time, lines appeared in the picture on the TV set. This effect was seen on all channels. Direct connection of the antenna leads to the TV set's tuner input terminals did not display any lines. Thus, the fault was in the recorder.

Service techniques   The location of this fault could be in the VCR RF input, the VHF splitter, the IF-1 board, or in the RF modulator. The RF modulator is the easiest unit to change, so this is a good place to start.

The cause   The RF modulator was changed and the lines in the picture cleared up. The RF modulator was defective.

Trouble symptom

On playback, no signal was visible on the TV screen (the raster was blanked). A known good tape was then played back on the faulty machine with the same results. Thus, it appeared that the fault lay in both record and the playback. An additional symptom was the absence of sound (audio muted).

Service procedure   The functional blocks that are common to both record and playback are as follows:

☐ Videotape.

☐ Audio head.

☐ Video heads.

☐ Control heads.

The chances of a defective videotape were remote. If the audio head had been open, there would have been some noise in the speaker instead of the muted sound condition.

435

If the video heads had been open, there would have been a visible TV raster, without video, instead of a blanked (black) raster.

The control head could have been open, because both record and playback were affected. An open control head would result in no control track on the tape during recording.

The cause　The clue here was the fact that both the audio was muted and the video was blanked out on playback. This condition occurs if the CS board circuitry does not sense a good, steady control track signal. Thus, it appeared that the control signal system should be checked first (both record and playback).

During the record operation, a scope check at pin 1 of CN2005 (RLS board) showed a good 30-Hz signal going to the control head. During playback, no recovered control track signal was present at this point or at pin 12 of IC502. The control head was then suspected to be faulty. It was discovered to be open. Replacement with a new audio/control/erase head assembly returned this VCR to normal operation.

Trouble symptom

The VCR would not complete the threading operation and the standby lamp remained lit. Hence, no other function could be activated. The threading ring did not complete its cycle. Electronic circuitry is part of the threading system, but the mechanical action might be easier to check first.

Service procedure　A careful visual check of all significant parts of the threading mechanism disclosed a broken tip on the threading arm assembly.

The cause　The arm assembly tip was replaced, and the threading operation then worked properly. The broken part of the threading arm assembly allowed the arm to fold in the wrong direction, preventing the completion of the threading cycle.

Trouble symptom

After being used for several hours, the JR9000 VCR would exhibit a tracking problem. Recordings made on a known good machine would play back improperly. Visible symptoms on the TV screen resembled symptoms that appear when the drum brake cable is disconnected from the CS board. The picture would alternately clear up and then break up as if the tracking control was being rotated back and forth. As the recorder continued to heat up, the symptoms became more severe.

Service procedure   The symptoms indicated a temperature warm-up situation, and the evident tracking problem pointed the way to the CS board (where the tracking control circuitry is).

The recorder was allowed to cool down to room temperature. It was then turned on, placed in the playback mode, and operated normally. Applying a heat lamp to portions of the CS board localized the defect in the area of IC4601, which was causing the tracking problem. Coolant applied to IC4601 restored the recorder to normal operation. Replacement of the chip solved the problem.

Trouble symptom

The cassette would not eject from the VCR.

The cause   A broken eject spring, part of the slide lever assembly, was the trouble.

Trouble symptom

Intermittent color playback on VCR.

The cause   An intermittent delay line was found on the YC-L PC board.

Trouble symptom

The tape cassette would not load into the machine.

The cause   One section of the functions switch was open. This switch can also have intermittent contact problems.

Trouble symptom

A situation of poor playback performance was noted when recording and playing back programs received over cable systems that delivered very short sync signals. On playback, the pictures showed white flaring, and a buzz was audible. A scope check at the TV receiver's video detector showed the white going to zero carrier level, indicating that the VCR modulator was being overmodulated. The emitter of Q25 on the YC-L board (JR) had a peak-to-peak signal of 2.2 V when the recorded cable program was played back. When the alignment tape was played back, the Q25 emitter signal read the specified 1.3 Vpeak-to-peak.

Service procedure   On the JR9000 VCR, two adjustments can be made to solve this problem, one for each record and playback. In the Text:record mode, while observing the E-E level, adjust (VR5) on the YC-L board slightly to remove the unwanted video/audio symptoms.

In the playback mode of a tape recorded on the VCR being serviced, adjust VR21 (YC-L board) while observing the signal on the TV screen. Adjust the control only enough to clear the buzz and sync compression symptoms.

Trouble symptom

Would not record. Capstan slows down during record mode and then stops.

The cause   Dirty contacts on switch S601 on the AD-L board.

Trouble symptom

Intermittent, slow tape movement on both record and playback modes. When tape is running too slowly, sync and picture breakup problems occur.

The cause   Intermittent open winding on the supply reel brake solenoid. Replace the brake solenoid.

Trouble symptom

No response from any of the control functions that relate to the SRP board, such as partial threading or unthreading, partial rewinding, or nonfunctioning of the counter memory circuit.

The cause   Defective IC1 (CX141) on the SRP board (Zenith part number 905103).

Trouble symptom

Cross modulation of a strong channel 3 on a much weaker channel 2 when a tape recorded on the machine was played back.

The cause   Too much difference in signal strengths of adjacent channels. Adjust AGC control located in the IF-4 board.

Trouble symptom

No eject function.

The cause   Spring leaf broken on the eject switch lever.

Trouble symptom

No eject function. dc threading motor continually turned. Machine went through threading cycle without tape cartridge installed.

The cause   Transistor Q3 on SY-2 board open.

Trouble symptom

No two-hour playback operation on the JR9000 machine.

The cause   Defective Q1029 transistor located on the YC-L board.

438

Trouble symptom

No chroma in playback operation.

The cause   Replace IC4 on the YC-L board.

Trouble symptom

No chroma on playback or record.

The cause   Zenith JR9000-T6 and IC2 on the YC-L board. Zenith KR9000-T6 and IC2 on the YC-2 board.

Trouble symptom

Sound on playback warbled, two-hour free speed would not lock in, and picture was flashing at about 2-Hz intervals intermittently.

The cause   Transistor Q4631 on the CS board defective.

Trouble symptom

No record on playback.

The cause   JR9000-Shorted T13 on the YC-L board. KR9000-Shorted T13 on the YC-2 board.

Trouble symptom

Playback and record buttons would not lock. Machine went immediately into auto-stop. The threading and unthreading cycles seem too fast.

The cause   Defective Q5501 transistor on P3 board. With this fault, the 12-V B+ will measure about 18 V.

Trouble symptom

Machine would not unthread. All control buttons put the machine into auto-stop. The ac motor would not turn.

The cause   Defective ac motor caused by shorted windings.

Trouble symptom

The VCR made poor recording. When played back, the recording on this tape caused the machine to go into mute (no video or audio) and the set has a blank raster.

The cause   Transistor Q501 is shorted on the RS-L board (for JR9000) or ARS board for (KR9000) causing loss of control track pulses.

Trouble symptom

No chroma on playback. Loss of 4.27-MHz signal.

439

The cause   Open transformer T13 on the YC-L board.

Trouble symptom

Intermittent loss of chroma on playback.

Servicing procedure   Touch-up ACK level adjustment on YC-L board.

Trouble symptom

Snowy picture.

Servicing procedure   Readjust the REC current and F-Chara adjustments on the ARS board.

Trouble symptom

Cassette lift assembly would bind up halfway down in position.

The cause   Tape-up sensor bracket pushed against cassette, causing binding. Cassette cartridge not pushed all the way into the lift assembly. Cassette guide assembly too far toward front of the machine.

Trouble symptom

Machine would not thread. Loss of dc (12 V) to the threading motor.

The cause   A broken P7 board feed-through connector was found. This caused a loss of the 12 Vdc to the motor.

Trouble symptom

Machine would not rewind the tape.

The cause   Defective ac motor (part no. 941-104).

Trouble symptom

Played back pre-recorded tapes normally, but no TV reception.

The cause   Fuse F5301 (0.5 amp) was open on the P-8 board. A short circuit was caused by the AD-L board being loose and not locked into the nylon clips. It was lying against the 18-V connection pin. To correct, clamp the ADL board into its nylon clips and replace the fuse.

Trouble symptom

No playback on pre-recorded (or other working VCR) or recorded (on faulty machine) videotapes.

The cause   Defective audio/control head assembly (part no. 949-102).

Trouble symptom

Picture breaks up and loses color on playback.

The cause   Defective playback/receive switch.

Trouble symptom

Picture breaks up in a symptom similar to that occurring when the control head is defective.

The cause   Defective IC501 chip on the RS-L board.

Trouble symptom

Noisy picture on playback.

The cause   AGC delay (in TV receiver) control is set incorrectly. Readjust the AGC delay control.

Trouble symptom

No tracking in the playback mode.

The cause   Defective IC601 (CX143A) and Q629 transistor located on the CS board.

Trouble symptom

Machine would not play back previously recorded tape. Tape recorded on defective machine played back on a good machine. Tapes played on the faulty machine were erased.

The cause   Inspection of the SRP board revealed a shorted diode (D4009). This allowed 10.2 V to be applied to the record circuit during playback. Replacement of this diode solved the playback problem.

Trouble symptom

Auto-stop would engage regardless of the function selected.

Service procedure   The waveform at TP-1 on the SRP board was not correct. Adjustment of VR4001 produced the proper wave form. The machine was aligned to specifications and now operates normally.

Trouble symptom

Fast-forward mode would not function. Tape would not eject.

Service procedure   The 12 V was not present. A check of the SRP board verified this. The power transformer did not supply ac to the board, but voltage was not found at the diode bridge.

441

The cause   Connector CN4005 (pins 1 and 2) were not making proper contact. The terminals were discolored and the connector was replaced.

Trouble symptom

In this case, the faulty machine would not record color but played back color from a prerecorded tape.

Service procedure   Pin 12 of IC2 (ACC on YC-A board) should measure 3.4 V during record and 0 V during playback. The voltage was 0 V in both modes.

The cause   Capacitor C39 (0.01 μF), from pin 12 of IC2 to ground was shorted. A replacement capacitor solved the problem.

Trouble symptom

Audio recorded by this VCR would have a warble (or wow) for two to four seconds after the pause lever was released. There was no evidence of picture breakup.

Service procedure   This machine had over 1,000 hours of operating time and visual inspection of the capstan during playback revealed possible binding.

The cause   The capstan and flywheel were disassembled and cleaned, and the shaft and bearings were lubricated. This solved the problem.

Trouble symptom

No chroma or occasionally weak chroma during playback. The black-and-white picture was normal.

Service procedure   It was found on the YC-L board that the playback ACK was 3.2 V and not the required 4 V. On the RS-L board, pin 12 of IC1 measured 2.7 V and should be 3.9 V. Note: Pin 12 connects to the mono/chroma switch inside the IC1 chip.

The cause   IC1 on the RS-L board was defective and replaced.

Trouble symptom

Intermittent blanking of picture during playback.

Service procedure   A check of the RS-L board playback circuitry revealed a variance in amplitude of the waveform at TP-505 (pin 9 of IC502 servo).

The cause   Connector CN2005 from the control head to the RS-L board was making intermittent contact. A replacement connector solved this problem.

442

Trouble symptom

When the tape was put in rewind mode, it sometimes slowed down or stopped.

Service procedure Check for evidence of wear or binding in tape pulleys or tape path.

Zenith VR8910W

This VCR utilizes a microprocessor system control to operate any of the machine's controls. Refer to Fig. 12-2 for the block diagram of this control system, which is located on the SS-9 board.

Before looking at any problems in this microprocessor control system, let's take a brief look at the record and playback circuit operation. See Fig. 12-3.

Playback operation

When the play button is pushed, pin 5 (PLAY IN) of IC501 (the microprocessor) changes over from a high to a low. By this operation, pin 26 (PLAY OUT) produces an output, and 12 V is output from Q502 and Q509.

At the same time, pin 15 (MUTE OUT) outputs a muting signal for about 2.8 seconds. Also, pin 32 (ANT SEL OUT) outputs, and a signal is produced to change over TV/VTR (the changeover electronic switch) in the antenna input section. The VTR is then output by Q508 and Q514.

Control output signals are also sent to other solenoids to complete the play operation. Motor control output is generated from pin 12 (CAPSTAN), pin 13 (DRUM), and pin 14 for the reel. These solenoids and motor control outputs are sent to the DR-1 substrate to drive each solenoid and motor.

Record operation

When the REC button is depressed, IC 501, pin 8 (REC IN), changes over from high to low. By this operation, pin 27 (REC OUT) outputs and REC 12 V is generated from the Q503 buffer transistor and Q510. Solenoid and motor control outputs are generated in exactly the same way as they are for the playback mode.

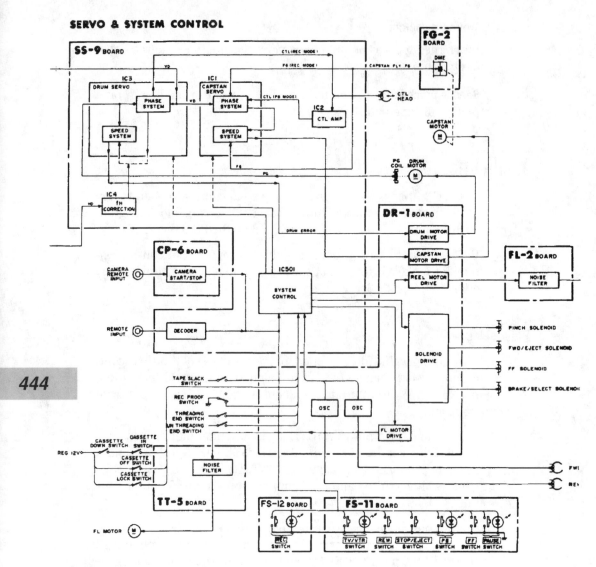

■ **12-2** *Block diagram of Zenith VR8910W VCR servo control system.*

Trouble symptom

The problem with this machine was that it would not play back a prerecorded tape. However, it would produce a picture in the fast-forward (FF) scan mode.

Also, when the machine was turned on, the red record LED was on. The VCR could record properly.

The cause   The problem was in the microprocessor SS-9 control board. The Q503 buffer record control transistor was showing

Q501 25A1175 Inital reset	Q528 25A1175 Inital reset		IC501 mPD553C-149 System control	Q502 2SC2785 Play out drive	Q505 2SC2785 Rew out drive	Q508 2SC2785 Annetenna selector	Q511 2SA1048 FF out switch
D504 RD7.5E-B1 Reset clear	D516 ISS119 FF mute	D515 ISS119 Reset DET		Q503 2SC2785 Rec out drive	Q506 2SC2785 Eject out drive	Q509 2SA1048 Play out switch	Q512 2SA1048 Rew out switch
				Q504 2SC2785 FF out drive	Q507 2SC2785 Pause out switch	Q510 2SB740 Rec out switch	Q513 2SA1048 Eject out switch

■ **12-3** *Transistor list.*

leakage. (The same problem could have developed in any of the other microprocessor control mode functions.)

Zenith VR9700 and VR9000 Beta machines

Trouble symptoms (system control)

You could have intermittent shutdown in forward functions (play, record, and fast-forward) or no play or fast-forward operation (when these buttons return to the stop mode immediately). There could also be an intermittent rewind action, or the unit shuts down before the end of the tape during rewind mode.

Service checks   The sensor circuit IC (see Fig. 12-4) contains an oscillator that has the tank circuit components (sensorText: coil) connected to pin 2 and a capacitor external to the IC between pin

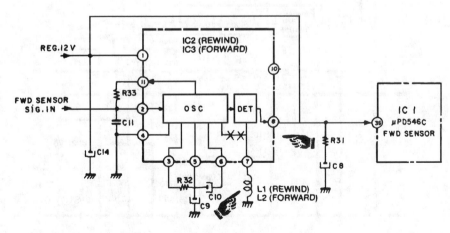

■ **12-4** *Block diagram of the Rewind/Forward Sensor circuit.*

2 and pin 4 (ground). The forward sensor oscillator in IC3 is on at all times, but the rewind sensor of IC2 is turned on only when the VCR is in the rewind mode. When the sensor foil at the end of the tape passes the sensor coil, the foil detunes the oscillator tank circuit, disabling the oscillator. This in turn biases the last stage in the IC (between pins 7 and 8) OFF, causing pin 8 to go high. This high causes IC1 to set up the stop mode. The coil at pin 7 of IC2 or IC3 is the dc return path for the base circuit of the stage between pins 7 and 8. When pin 7 is open, pin 8 goes high just the same as if the end of the tape had been sensed.

Zenith VR8500 VCR

Trouble symptom

The drum servo on this machine lost its lock as the VCR warmed up.

Service checks When the VCR was first put into operation, the drum free-speed control (RV9) located on the SS-9 board (see Fig. 12-5) could be adjusted for proper free speed. However, the drum servo would lose lock as the VCR warmed up. Readjustment of RV9 would not maintain servo lock. The voltage at test point TP2 should be adjusted for 5.2 V, but it would not go any higher than

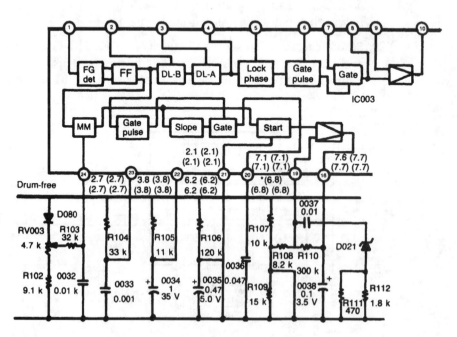

■ **12-5** *Drum Free Speed circuit found in Zenith Model VCR recorder.*

3 V when the machine warmed up. The voltage and waveform found at pin 22 of IC3 were not correct. When a circuit coolant was sprayed on capacitor C54 (connected to pin 22), the voltage at TP2 increased to near 10 V.

The cause   Capacitor C54 and R105 at pin 22 of IC3 are part of a slope (ramp) circuit. The output of this circuit is later sampled to determine the drum speed error voltage. Capacitor C54 was leaky and affected the adjustment range of the drum free-speed control RV9.

Capacitor C54 was replaced at pin 22 of IC3. The voltage at TP2 was then adjusted with control RV9 for a stable 5.2 V. A normal scope waveform then appeared at pin 22 of the IC3 chip.

Sony VCRs

Trouble symptom

The unthreading process requires the operation of the ac motor, which turns the take-up reel to pull the tape back into the cassette. The ac motor also rotates the video head disc assembly, developing the 30 PG pulses that are converted to the RF switching pulse. If this pulse is not present at pin 22 of IC4001 on the SY-2 board, the auto-stop function is activated. Hence, this machine would quickly shut down when a function button was pressed.

The cause   The ac motor was defective due to shorted windings.

Trouble symptom

The eject function would not operate as it does during threading. To perform the unthreading operation, the motor must rotate in the opposite direction. This reversal is normally accomplished when the threading switch moves to the eject position when the eject button is depressed. Investigation showed that the threading switch was not moving to the eject position.

The cause   The spring leaf on the eject switch lever was broken. This lever, connected mechanically to the eject button, activates the threading switch on the SY-2 board. Replace the spring leaf switch assembly.

Trouble symptom

No playback was possible in any mode. A tape attempted to be recorded in the machine did not play back in a known good machine.

Background information   An important clue to this problem is the ever-present blanking of the video. This condition occurs whenever the apparent control signal, as "seen" by the CS board, is not the normal 30-Hz waveform. In this machine, the automatic tape speed circuit on the CS board determines the operation of the blanking/mute signal. The input of this circuit comes from pin 14 of IC2502 (on the RS-L board). This input for normal operation is the playback control track signal. An oscilloscope connected to this point indicated that the CTL signal was not present.

The cause   The control head was defective (open). Thus, no CTL signal could be recorded or played back.

Trouble symptom

A cassette tape recorded on the VCR would not play back. Video blanking and audio mute are on continuously. Also, the recorded tape would not play back on a known good machine. Thus, the basic symptom indicates a record function fault.

Service technique   The symptoms are the same as in the problem above and so is the technique used to find the source of the problem—the lack of a recorded control track. In this case, the faulty component was not the control head. Scope checks revealed that the CTL signal was not reaching the head.

The cause   Transistor Q2501, the record CTL amplifier, was defective (shorted).

Trouble symptom

The luminance record and playback functions were normal. Chroma was not visible in the playback signal. A prerecorded videotape did not produce chroma when played back on another machine. Thus, the problem was no chroma on record or playback.

Service technique   In this machine, one circuit that is common to both the record and playback system is the 3.58-MHz crystal filter. This is the source of a signal that develops the ACK (automatic color-killer) voltage of +4 V when chroma is present in the composite video signal. Without this voltage, chroma in record and playback are not possible. Thus, this circuitry was checked and revealed a loss of the 3.58-CW signal.

The cause   Transformer T6, part of the 3.58-MHz filter, was defective (opening winding).

Trouble symptom

There was an intermittent shut-down of all functions in the VCR. The machine was thus intermittently in the auto-stop mode.

Service technique   There are several conditions that can cause activation of the auto-stop mode, for example, end of tape (during rewind or fast forward FF) or when the drum head assembly stops rotating. In this particular VCR, the symptoms were not related to the end-of-tape situation, and the slack sensor was not activating. Thus, attention was directed toward the drum assembly.

Visual observation indicated that the cylinder head drum was rotating steadily during the intermittent shutdown condition. When the cylinder head rotates, the two magnets on the bottom of the disc generate pulses as they pass over the PG coils on the fixed portion of the drum assembly. These pulses, coupled to the SY-2 board, permit normal tape and head movement. If the pulses are not going to the SY-2 board, the auto-stop function activates.

A scope check of the PG pulse path indicated that there was an intermittent signal being coupled to the SY-2 board. This discovery led to the PG pulse source. The B PG pulse was steady, but the A PG pulse was intermittent. The fixed coil from which this signal is developed was intermittently opening.

The cause   Replacement of the complete drum cylinder assembly was necessary to replace the defective coil. This replacement solved the problem.

449

Zenith KR9000

Trouble symptom

On these machines, the main drive belt could come off as soon as the motor begins to run.

The cause   The drum drive motor could be tilted. Use a washer under the rubber motor mounts to align the drive motor. The motor belt drive pulley might have to be changed. A newly designed pulley is now available to solve this problem.

Trouble symptom

The function buttons don't stay down. This problem might be intermittent or occur after the machine has warmed up. Look for this trouble on the SRP system control board.

The cause IC1 (905 to 103) could be defective. This chip is mounted on the SRP board, located on the front of the machine. Check the 200-kHz oscillators peak-to-peak voltage at TP-1 and TP-2.

If the machine shuts down during record, playback, or fast-forward, check TP-1 on the SRP board for 4.6 V PR Adjust or clean the VR-1 control if necessary. If it shuts down in the rewind mode, check TP-2 for 2.7 V PP. Adjust or clean control VR-2 if necessary. These controls adjust the levels of the 200-kHz oscillator.

Zenith VR9775 remote control VCR

This machine uses microprocessors to operate all of the system control circuits and modes.

Trouble symptom

The VCR would turn on and the remote control functions were operating correctly. The videocassette would load into the machine properly. However, when any function button was pressed (play, record, etc.), the VCR would run for about three seconds and then shut down. If all systems are not go in this machine, the microprocessor shuts it down to avoid any(more) damage. Let's first take a brief look at this systems control operation.

Background (system control operation) The system control operations are mostly performed by three microcomputer chips. These are CPU-L, CPU-2, and CPU-3, as shown in the block diagram of Fig. 12-6. All are located on the SS-10 board. Inputs for system control are produced by the function buttons on the front control panel, mechanical switches, malfunction detection signals from the servo block, the reel servo block, and input signals from the tuner timer block and camera. The CPUs make decisions based on the input signals and produce the correct outputs as per programs designed for them. The signals from the system control are sent to the servo, video, reel servo, audio blocks, etc. Almost all operations of this unit are controlled by the signals from the CPUS. The block diagram outlines the system control and input and output signals.

The VCR has two separate motors that drive the supply and take-up reels. For this reason, the system control has almost no mechanical sections. Various modes are turned on by electrical signals that are controlled by the servo and reel blocks based on instructions issued by the CPUS. These electrical signals are sent to the motors. Mechanical sections controlled are the cassette

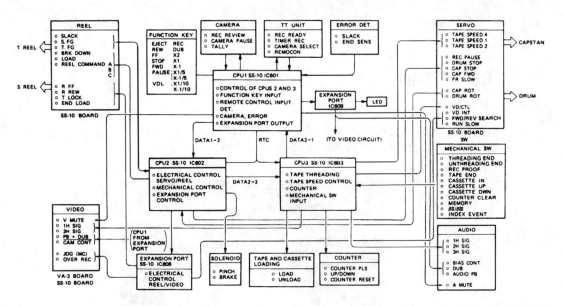

The following labels appear in the block diagram:

REEL (SS-10 BOARD)
- SLACK
- S. FG
- T. FG
- BRK DOWN
- LOAD
- REEL COMMAND A / B / C

T REEL
S REEL
- R FF
- R REW
- T LOCK
- END LOAD

FUNCTION KEY
- EJECT / REC
- REW / DUB
- FF / X2
- STOP / X1
- FWD / X-1
- PAUSE / X1/5
- / X-1/5
- VDL / X1/10
- / X-1/10

CAMERA
- REC REVIEW
- CAMERA PAUSE
- TALLY

TT UNIT
- REC READY
- TIMER REC
- CAMERA SELECT
- REMOCON

ERROR DET.
- SLACK
- END SENS

SERVO
- TAPE SPEED 4
- TAPE SPEED 1
- TAPE SPEED 2 → CAPSTAN
- REC PAUSE
- DRUM STOP
- CAP STOP
- CAP FWD
- FR SLOW
- CAP ROT
- DRUM ROT → DRUM
- VD/CTL
- VD INT
- FWD/REV SEARCH
- RUN SLOW

SS-10 BOARD

CPU1 SS-10 IC801
- CONTROL OF CPUS 2 AND 3
- FUNCTION KEY INPUT
- REMOTE CONTROL INPUT DET.
- CAMERA, ERROR
- EXPANSION PORT OUTPUT

EXPANSION PORT IC808 → LED

DATA1-2   RTC   DATA3-1   (TO VIDEO CIRCUIT)

CPU2 SS-10 IC802
- ELECTRICAL CONTROL SERVO/REEL
- MECHANICAL CONTROL
- EXPANSION PORT CONTROL

DATA2-3

CPU3 SS-10 IC803
- TAPE THREADING
- TAPE SPEED CONTROL
- COUNTER
- MECHANICAL SW INPUT

MECHANICAL SW (SW)
- THREADING END
- UNTHREADING END
- REC PROOF
- TAPE END
- CASSETTE IN
- CASSETTE UP
- CASSETTE DWN
- COUNTER CLEAR
- MEMORY
- INDEX EVENT

VIDEO
- V MUTE
- 1H SIG
- 3H SIG
- PB + DUB
- CAM CONT
- JOG (MC)
- OVER REC

CPU1 FROM EXPANSION PORT

VA-3 BOARD
SS-10 BOARD

EXPANSION PORT SS-10 IC808
- ELECTRICAL CONTROL REEL/VIDEO

SOLENOID
- PINCH
- BRAKE

TAPE AND CASSETTE LOADING
- LOAD
- UNLOAD

COUNTER
- COUNTER PLS
- UP/DOWN
- COUNTER RESET

AUDIO
- 1H SIG
- 2H SIG
- 3H SIG
- BIAS CONT
- DUB
- AUDIO PB
- A MUTE

■ **12-6** *Zenith VR9775 VCR block diagram of microprocessor control system with Input/Output signals.*

loading and threading, the brake solenoid for reels, and the pinch roller solenoid. These solenoids are of the self-latching type; once a pulse voltage (200 ms) is given, the solenoid remains in the same mode. No sensor has been provided for detection of tape slack, as revolutions of the reel motors are being controlled by reel motor servo circuits on the SS-10 board.

The cause   Therefore, when the microprocessor senses some problem, such as a jammed part or broken tape, it shuts the machine down. After a close inspection, the cylinder head was apparently jammed by a small sliver of metal. The microprocessor did not sense the PG pulses from the cylinder head (as it did not rotate) and thus it shut all systems down and prevented any other damage. After the sliver of metal was removed from the cylinder head, the VCR operated properly.

Panasonic PV1300

Trouble symptom

The cylinder heads would search intermittently. The picture would shrink and then go very snowy and have ghosts.

The cause   The problem was traced to an intermittent loss of the 30-Hz reference pulse from IC8004, the 3.58-oscillator/divider chip. Replacement of IC8004 returned the playback to normal operation.

RCA VBT200

Trouble symptom

Bars of snow appeared across the screen, as shown in Fig. 12-7. The trouble appeared to be a tape tension problem.

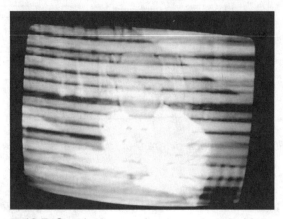

■ **12-7** *Streaks across the screen caused by a loading arm out of position.*

Service technique   Determine if both loading arms are going into the fully loaded position. If not, push them into place and note any difference. In this case, the problem cleared up.

The cause   The right loading arm was not going to the fully loaded position. Replace the loading arm assembly.

Panasonic PV3000

Trouble symptom

The machine would load but then immediately unload. It went into fast-forward and rewind modes, but it had no tape movement.

The cause   A fuse (F1001) for the +9-V regulator circuit was open. You must completely disassemble the machine and shields to replace the fuse.

Trouble symptom

The VCR would record and play back properly, except it would not play back video on the camera's viewfinder.

If the camera will record onto tape, the camera is okay because the camera video (out) and viewfinder video (in) are on the same line.

The cause   Testing indicated an open Q3020 video amplifier transistor. Q3020 amplifies video to the camera jack.

Trouble symptom

A loss of head-switching pulses caused horizontal lines to go across the monitor screen.

The cause   Troubleshooting with the scope indicated that the PG pulses were misshaped. These misshaped PG pulses were caused by a defective (open) C2502 capacitor (located on the PG amplifier board).

RCA VET 650

Trouble symptom

(SLP operation only.) The customer description of the symptom related to a tape speed selection problem with the special effects VCR. The symptom was described as "locked in SLP only, both in playback and record." Other speeds could be selected in the record mode, and programs recorded in SP or LP were played back at the SLP speed. Also, the stop LED remained lighted in all modes.

453

The cause   The most likely cause of the symptom described is a defective IC6007 (quad two-input AND gate) chip. Two of the AND gates in this chip are associated with circuitry that enables the one-second SLP operation whenever search forward or search reverse modes are selected. A failure in either of these AND gates (part of IC6007) can supply constant base bias to Q6061 resulting in continuous SLP operation.

RCA VET 250/450/650

Trouble symptom

One or two horizontal lines of interference appeared across the center of the picture of recorded material, whether played back on the same machine or a different one. The symptom was not evident when prerecorded programs (recorded on another machine) were played on the defective machine.

The cause   The symptom could be the result of buffer oscillator frequency drift in the defective VCR machine. If the buffer oscillator is not locked to incoming vertical sync, incorrect control track pulses will be recorded. During playback, these incorrect control track pulses cause head switching pulses (interference), usually near the center of the picture.

Service technique   Reset the buffer oscillator free run to a slightly slower frequency (35.8 ms rather than 35 ms). For models V-ET 250 and 450, adjust the buffer oscillator control R2015 (located on servo board) for a width of 35.8 ms (equal to 27.93 Hz). For model VET 650, adjust buffer oscillator control R6916 (located on the slow/still board) for a width of 35.8 ms (equal to 27.93 Hz).

RCA VDT150/VTE150

Trouble symptom

Will not record tape.

Probable cause   Defective buffer IC907. Also check safety tab switch.

Trouble symptom

Intermittent noise bands on playback (no cylinder lock in play).

Probable cause   Check for a defective R501 tracking control.

Trouble symptom

Poor recorded picture.

Probable cause   Check record amplifier IC205.

Trouble symptom

No picture on playback.

Probable cause   Switch transistors Q204, Q205, Q206, and Q207. Also, check video buffer transistor Q208.

Trouble symptom

Grainy picture and no color.

Probable cause   Video playback and no color. Check Video playback amplifier IC202.

Trouble symptoms

Trouble associated with upper D-D cylinder video heads. These problems might be snowy picture, poor-quality playback, no picture, or vertical jitter.

Probable cause   Check Upper D-D cylinder—it might need to be cleaned. Cylinder head might be worn or defective.

454

Trouble symptom

No picture on playback or E-E.

Probable cause Defective Q205 buffer transistor.

Trouble symptom

No video in record or E-E.

Probable cause Switch IC603 defective.

Trouble symptom

Snow on playback, no picture or no sound, no video information into TV set channel 3.

Probable cause Faulty RF converter.

Trouble symptom

Will not record.

Probable cause Video record amplifier IC201 defective.

Trouble symptom

Noise in color picture.

Probable cause Phase comparator buffer IC207 defective.

455

Trouble symptom

No chroma.

Probable cause Signal amplifier IC301.

Trouble symptom

Noise in video.

Probable cause Phase control (IC501) or capstan speed control (IC551) defective.

Trouble symptom

In play, sound is from the tape, but video is from off-the-air channel.

Probable cause Zener diode D209.

Trouble symptom

VCR problems associated with servo, such as capstan runs slow, audio wow, poor tracking, or no servo lock at any speed.

Probable cause Phase control (IC501) or capstan speed control (IC551) defective.

Trouble symptom

No operation of any type.

Probable cause Fuse F901 blown or a defective IC901 microprocessor chip.

Trouble symptom

Repeated F901 fuse failure.

Probable cause Motor drive IC902 or regulator transistor defective.

Trouble symptom

No camera pause function.

Probable cause Defective microprocessor IC902.

Trouble symptom

Problems associated with reel motor, such as take-up reel stops or does not turn, eats tapes, or no fast-forward or reverse.

Probable cause Defective reel motor.

Trouble symptom

Reel motor runs continuously.

Probable cause Motor drive IC902 or motor drive IC906 defective.

Trouble symptom No play, fast-forward, or rewind.

Probable cause Driver IC911 defective.

Trouble symptom No pause function.

Probable cause Microprocessor IC901 defective.

Trouble symptom

Intermittent D-D cylinder speed.

Probable cause Defective D-D cylinder motor.

Trouble symptom

No capstan operation in record mode. Playback mode operates properly.

Probable cause Phase match edit diode D668.

Trouble symptom

The machine is in the play mode at all times.

Probable cause Defective microprocessor IC903.

Trouble symptom

Hum or garbled audio.

Probable cause Faulty record kill relay RL401.

Trouble symptom

Buzz in audio on E-E and record. Does not record audio.

Probable cause Switch IC603 is defective.

Trouble symptom

Distorted audio.

Probable cause Play/record amplifier IC401 defective.

Late model four-head VCRs

Trouble symptom

The trouble appears to be a tracking problem, and a beat appears in the picture. If you see a clear band of the picture at the top of the TV screen with machine in the SP mode, this symptom is probably caused by a faulty contact on the relay in the video head circuit that selects the SP or EP play heads. The relay function is to short out the unused heads for either mode.

The cause Bad relay contacts can inhibit shorting function, allowing both heads to pick up signals. Thus, the VCR transmits a beat to the screen between the signal picked up by both SP and EP heads. The EP heads are mounted on the drum to make the contact with the tape after the SP heads so they don't pick up a signal until the SP heads have made part of their pass, which is the reason for the clear picture at the top of the screen. To solve this problem, replace the relay.

Forward-back tension

Measurement of the forward-back tension is one of the most critical measurements that you can undertake while servicing any VCR or camcorder. Failure to confirm that both the correct forward-back tension measurement exists and that the proper operation of the mechanism occurs can often lead to much time lost while troubleshooting.

In order to understand the importance of the back tension part of the service procedure, you must understand each of the various

457

functions of the mechanism and also what can occur when there is a malfunction.

Simply stated, the purpose of the back tension assembly is to ensure a stable back tension on the tape as it moves forward. This function can appear to be relatively unimportant at first; however, it is essential to video recording. Each of the rotating video heads must penetrate, to the proper depth, the primary plane of the oxide surface of the tape during the record and playback processes. Also, in order to ensure proper writing and reading of the video tracks, no undesired physical distortions of the videotape can be tolerated during either recording or playback. However, in practice, it is not possible to eliminate all physical distortions. The back tension lever assembly continuously stabilizes the tape motion as it is drawn out from the supply side of a cassette. By the action of this assembly, tension is kept constant on the tape, and physical distortions in the horizontal plane are kept to a minimum. Electronics in the VCR can be used to compensate for—and correct— the small variations in signals that can result from the remaining distortions.

Forward-back tension defects can be classified into three groups, each having unique and sometimes overlapping symptoms. The following points should always be checked when servicing any VCR problem.

Back tension too low

When the back tension is too low, the video heads can not penetrate the oxide plane with enough force. Low tension can cause insufficient magnetization of the tape during recording, resulting in a weaker signal being available during playback. A symptom that might appear to be caused by a playback preamp defect, or a worn head, might actually be caused by the tension being too low. Failure to check the tension first might cost lost time while troubleshooting circuits where no fault exists or, even worse, you might have replaced some major component (such as a video head) unnecessarily.

Back tension too high

When back tension is too high, the video head penetrates the oxide plane of the tape too deeply. This leads to premature wear of both the heads and the tape. Also, the tape is subjected to an increase in physical distortion, possibly resulting in irregular stretching of the tape, which can lead to many recorder malfunctions. Sometimes you might even hear an abnormal contact sound as the tape buffets against the surface of the heads. High tension can be the cause of various symptoms, including color dropout, inconsistent tracking, tight tracking control, audio distortion in hi-fi VCRs, excessive video head wear, and tape damage.

Inconsistent back tension

If back tension is unstable or inconsistent, there can be a variation of the symptoms in both the previous categories. Unstable back tension often is an intermittent condition—and usually a subtle one. For instance, this condition could cause you to seek intermittent color symptoms in the color circuits, when in fact the tension instability is causing excessive jitter in the signal. Remember that the chroma APC circuits can only compensate for a certain amount of change in phase.

It should now be evident that the back tension readings and mechanism operation must be checked carefully on every VCR to be serviced. Ideally, back tension should be checked at the beginning, center, and end of each reel of tape. This is useful information because it helps to identify the ability of a mechanism to respond to changing tape load conditions. Keep in mind that back tension is normally checked on a regular basis on all broadcast video recorders.

Forward-back tension can only be measured correctly by using special test instruments. One type of instrument measures the tension at any accessible point along the length of the tape. One such instrument is the Tentelometer. However, test instruments of this type are normally very expensive.

Two points should be noted when using either of these measuring devices. First, the desired tension readings should appear at the approximate midscale point on the measuring device indicator because this should provide optimum measurement accuracy. Second, the back-tension adjustment should typically provide a reading that is at the center, or slightly above the center, of the back-tension range specified in the service manual for the model being serviced.

Other VCRs: Problems and solutions

Zenith VRF250 tuner

Trouble symptom

Tuner audio distorts.

Service technique   When first checked, the tuner audio distortion was barely noticeable, but tape playback was good. After a short time the tuner audio became more distorted. As the VCR continued to warm up, the audio distortion increased. Checking the tuner and IF circuits caused the 4.5-MHz ceramic filter CF1 to be suspect. CF1 connects between both the video detector (pin 28) and the limiter (pin 13) on the IF board. It appeared that, as the temperature increased, the frequency response curve characteristics of ceramic filter CF1 shifted, resulting in the change of distortion.

The solution   Replacing the CF1 filter corrected the problem.

Zenith VR500 Mechacon/mechanism

Trouble symptom

Cassette will not eject and then goes into auto shutdown.

Service technique   A cassette was in the VCR when it was brought into the shop. This cassette could not be ejected when power was applied. The eject lamp was on, the pause lamp flashed, and the VCR entered the auto shutdown mode. The VCR top and two circuit board modules (02 video and 03 servo) were removed for access to the mechanism. Next, we checked to see if the slide lever on the worm gear shaft in the mode control motor (Fig. 12-8) bracket and worm gear assembly moved freely or would bind when the coil spring was compressed. Because the slide lever was binding, we replaced the mode control motor, bracket, and worm gear assembly. We also checked the relay lever arm assembly to see if it was bent or cracked at the metal posts. It was defective, so the assembly was replaced.

The solution   Replacing both of the assemblies resulted in normal operation.

Service hint   Replace both assemblies if either assembly is replaced. After these components have been replaced, apply 6 V to the mode control motor terminals and check that the unit loads and unloads without a tape being loaded. Once this has been confirmed, proceed with any other required repairs.

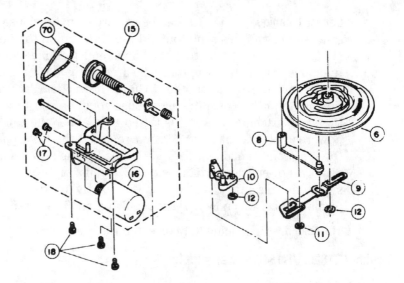

■ **12-8** *Mechanism for Zenith VR5000.*

Zenith VR1805 power supply/Mechacon

Trouble symptom

Enters rewind for a short time, then autoshut-off.

Service technique   When you plug in the VCR or press the power button the VCR enters the rewind mode for approximately 10 seconds, followed by autoshut-off. The clock display remained ON, but there was no SWD 5 V or SWR 12 V. Pin 15 of IC601 (CPU) on the 02 Mechacon board always had about 3.35 V, instead of 0 V. With the VCR power turned ON, pin 15 of IC601 was grounded. This restored the SWD 5 V and SWD 12 V. Everything appeared to be almost normal. However, when PLAY was pressed, the tape would wrap halfway around the video headdrum, and then unwrap.

Service hint   Watch the mode CTL motor. In this case, the motor would immediately enter the rewind mode for about 10 seconds and then enter the autoshut-off mode, causing the CPU on the (02) Mechacon Board to be suspected.

The solution   Replace IC601.

Zenith VR1810 power supply

Trouble symptom

Clock resets intermittently, especially in the timer mode.

Service technique   The VCR had been worked on previously for the same symptom. This symptom did not show up on the bench for about two weeks of setting the timer and working the various functions. Then the symptom appeared for good. As good luck would have it, we had another VCR of this model that was functioning normally. Thus, the assemblies could be substituted from the good machine. It was suspected that either the timer IC or power supply was at fault. We first replaced the front panel, but the symptoms remained. Then we subbed in the power supply assembly, power transformer, and (02) regulator boards, and the symptoms disappeared. The power supply checked out. The prime suspect was now IC1, the regulator, and it was replaced.

The solution   Replacing IC1 corrected the symptom.

Zenith VR1820, VR1810 power supply

Trouble symptom

Turns ON, but power LED is dead. Fast forward and rewind work, but will not play.

Service technique   All voltages were checked at the output of the (02) regulator board. All voltages checked okay except for the SWD 5 V, causing IC1 (the voltage regulator IC) to be suspected.

The solution   Normal operation returned after IC1 was replaced.

Zenith VR1810 power supply/mechanism

Trouble symptom

Will not operate at all. Blows fuses F2 and F3 when PLAY or RECORD are depressed.

Service technique   Before the fuses blew, the cassette loaded, the drum motor started, and the tape started to load. The fuses were replaced and all voltages in the power supply were checked. All voltages checked out normal. Because the play mode involves drum start, reel motor, and the mode/loading motor, any one of these could be binding or have some related defect. The cassette housing cover was removed. The end sensor was covered with tape, and CN1 and CN2 were disconnected from the (20) drum MDA board. Now the only motor that would operate, until loading was completed, was the mode motor. The fuses did not blow, but loading was slow (labored), and entered to stop mode before the loading was completed. The mode assembly was removed to check it not only for timing, but also for binding and lubrication. Because everything in the assembly appeared normal, the mode assembly was then reinstalled.

The solution   VCR now loads completely, CN1 and CN2 were reconnected, and the VCR operated normally without blowing fuses.

Zenith VR1810 power supply Mechacon

Trouble symptom

Intermittently shuts off or blows fuses F2 and F3.

Service technique   We checked the voltages to all the motors and found that the voltage to the reel motor was varying erratically. Then we checked the reel motor current and found the same indications, causing the reel motor to be suspected.

The solution   Replacing the reel motor restored normal operation.

Zenith VR1820 power supply

Trouble symptom

Intermittent video.

Service technique   The top cover was removed and the (03) main (A/V/S) board was raised. When video was lost, the video head drum would stop rotating and then start to reverse. It was then noticed that the video became intermittent when pressure was applied to the (03) Main (A/VIS) board. A close examination with a magnifying glass uncovered a poor solder connection at pin 7 of the ribbon cable J1. J1 is one of three ribbon cables that are connected between the (04) mechanism control board and the (03) main board.

Service hint   The ribbon cable J1, pin 7 supplies switched 5 V to the audio, video, and servo circuits on the (03) main and (43) pre/rec boards.

The solution   Resolder connections on ribbon cable J1.

Zenith VR1820 power supply/video

Trouble symptom

Erratic operation, intermittent failure of fuse F3.

Service technique   Traced the 5-V power supply circuits starting from the (02) regulator board. Eventually found a filter capacitor shorting two pins together in the (03) main video board circuit. The capacitor lead was shorting delay line DL102, pin 5 (+5 V) to pin 4 (ground).

463

The solution   Dressed the capacitor lead away from the delay line.

Zenith VR1820 power supply

Trouble symptom

No display.

Service technique   Checking the pins between the (01) power transformer board and the (02) regulator board showed a reading higher than normal (2.7 Vac at pins 4 and 5). This was also found at CN1, pins 7 and 14 on the (02) regulator board. Further checking showed an open foil connection at CN1 of the (05) motherboard between CN1, pin 7 and CN13, pin 1 or between CN1, pin 14 and CN13, pin 2.

The solution   Repairing the open foil paths corrected this condition.

Zenith VR1830 power supply

Trouble symptom

Channel indicator is on, but VCR will not load cassette.

Service technique   The first investigation was in the power supply circuit on the (02) regulator board. We measured the SWD 5 V, SWD and UNSWD 12 V, motor 12 V, and the UNREG 12 V. The UNSWD 12 V measured 10.22 V, while the motor 12 V measured 0 V. Troubleshooting this circuit located a shorted zener diode, D14.

Service hint   Referring to the schematic (Fig. 12-9), notice that the base (control) circuits of Q1 (motor 12 V) and Q2 (UNSWD 12 V) on the [19] power transistor board can be affected by zener diode D14 on the [02] regulator board.

The solution   Replacing the zener diode D14 restored normal VCR operation.

Zenith VR1805 Mechacon

Trouble symptom

Will not load cassette.

Service technique   First item checked was Q1 (the start sensor) on the [18] cassette housing board and Q1 (the end sensor) on the [13] end sensor board, both of which are located on the cassette housing assembly. We then inserted a light and, using a meter, checked to see if the sensors were conducting; they checked out. Then we checked voltages on connector CN606. These voltages

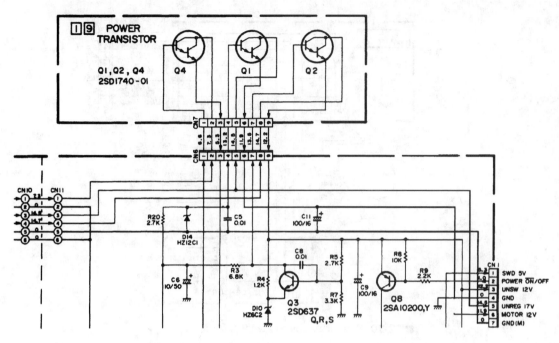

■ **12-9** *Power supply for Zenith VR1830 VCR.*

were also good. However, the 5.4 V was missing at lead 6 on the [131] end sensor board. The 5.4 V must be present at the emitter of Q1 (end sensor) when a cassette is inserted in the cassette housing assembly. The voltage reaches here through two leaf switches, CASS DET-1, and −2. After shorting pins 6 and 8 of the [13] end sensor board, we were able to activate and load a cassette. This indicated an open path in the leaf switch circuit. Notice circuit diagram in Fig. 12-10.

The solution   Leaf switch CASS DET-2 was found to have a mechanical malfunction. Replacing the switch restored the VCR to normal operation.

Zenith VR1820 Mechacon

Trouble symptom

While checking this VCR, the tape was found to be coming off the supply reel.

Service technique   This was happening because the capstan was running in the reverse direction. Checking the voltage on IC604 (capstan motor drive amplifier) showed that pin 8 had dropped to 0 V. Pin 8 carries the capstan drive signal from the servo circuit.

Other VCRs: Problems and solutions

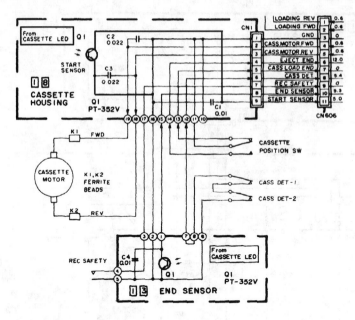

■ **12-10** *Sensor circuit in Zenith VR1805.*

Pin 3, the capstan motor (+) output from IC604 was HIGH, thus causing the capstan to run in a direction opposite to the desired tape movement direction.

The solution Replace IC604 and the symptom was corrected.

Zenith VR1825-1 Mechacon

Trouble symptom

Will not play. Tape caught in mechanism.

Service technique After removing the tape, all functions appeared to work normally except that in the play mode the take-up reel was not rotating. Voltage was not being supplied from IC603, pin 3 (reel motor control on the CTL board) to the reel motor. Input voltage is good on pin 6. Voltage on pin 2 drops from approximately 10 V to approximately 5 V when going from the stop mode to the play mode. Following the circuit back from pin 2 leads to the motor 12 V. The motor 12 V source was stable. Diode D606 showed a voltage drop of 1.9 V. The drop should be approximately 0.7 V.

The solution Replacing D606 returned the VCR to normal operation.

Zenith VR1820 servo

Trouble symptom

Tape damage because of intermittent capstan speed.

Service technique   A general check of the transport assembly revealed no irregularities. While testing the electronics of the servo system on the [03] main board, it was found that the capstan FG was not being amplified in the first stage of IC401. The waveform at pin 13 (the input to IC401) was found to be erratic. This symptom was traced to a defect in C404, the capacitor that couples the FG signal to IC401.

The solution   Replace capacitor C404.

Late model Zenith VCRs

The accompanying photos show the following Zenith models: VRJ410 (Fig. 12-11); VRJ210 (Fig. 12-12); VRJ220HF with hi-fi stereo (Fig. 12-13); four-head stereo VRJ420HF (Fig. 12-14).

■ **12-11** *Zenith VRJ410 VCR.*

■ **12-12** *Zenith model VRJ210 VCR.*

■ **12-13** *Zenith hi-fi stereo VRJ220HF VCR.*

■ **12-14** *Zenith VRJ420 HF VCR.*

Tips for servo troubleshooting

Servo problems fall in two slots: 1. A failure to lock in either mode; sometimes this failure can be intermittent. 2. A failure to lock at the proper time.

Symptoms for single servo VCRs

In playback, a servo failing to lock produces a picture that is normal for a few seconds, then is broken up by a noise band moving vertically through the picture. The noise band occurs rhythmically and is the clue to the technician that the playback servo is defective. Another symptom is horizontal breakup of the picture. The machine's record servo system might be good. This can be verified by making a recording and playing it back on a properly operating unit.

A defective record servo system produces a picture similar to that of a defective playback servo, but with a difference that can be easily noticed. The rhythmic noise will still be present, but an additional band will be moving vertically through the picture. Because of no servo control during record there will be no synchronization between the PG pulses and the vertical sync pulses. Therefore vertical sync was not recorded at the beginning of the track—its normal position—but was instead recorded randomly along the track. Because of this, head switching occurs anywhere along the track, creating the second band. This band is really the head switching transients that occur out of sync with the PG pulses.

Some troubleshooting tips

A fluorescent light can help identify servo problems in record or playback modes. Hold the light over the drum. Looking down at the drum and activating either record or playback, the parts on the drum will produce a strobe-like pattern that will be unstable until the servo locks up. As the servo locks up at 30 Hz, the strobe's pattern becomes stationary. If the pattern does not become station-

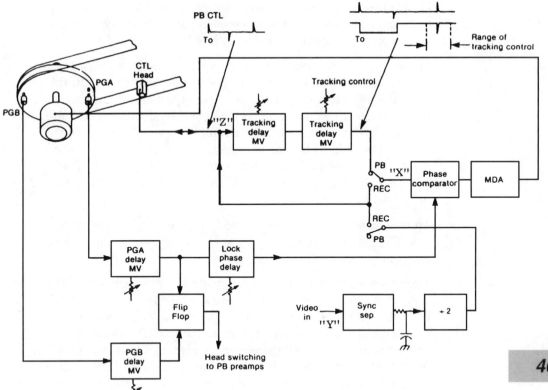

■ **12-15** *Basic servo troubleshooting blocks.*

ary, the servo is not locking. Troubleshoot the servo system at this time.

Several techniques are available to isolate problems to specific sections of the servo system of earlier models. For example, a unit that plays back prerecorded tapes properly, but has a servo problem when playing back its own recordings, indicates a definite servo problem. Referring to the block diagram in Fig. 12-15, the problem lies in the area between the sync separator and REC/PB switch. This is because all other blocks are used during the playback mode, which is operating properly.

A failure to lock in during playback, but recordings are good, points to a problem in the tracking delay circuits. All other blocks are good because they are used during record.

A loss of CTL pulses during playback triggers the video mute circuits, thereby producing a blank screen during playback. To troubleshoot the machine, the video mute circuit should be disabled. A picture might appear on the screen, but the rhythmic noise bands will be present, indicating the servo problem.

Should there be a servo problem in both the record and playback modes, circuits common to them are the likely suspects. In this case, the PG processing circuits, phase comparator, and motor drive circuits are common and should be investigated. The pulse from the PGA is used to develop a ramp for use in the phase comparator. Check for this ramp and the reference pulse (CTL in playback, Sync in record). If both are present, but are moving with respect to each other, then the comparator's output of the motor drive might be defective.

The tips given here can also be used in systems with capstan servos. The timing of the FG pulses, of course, would have to be checked for proper frequency.

The problems discussed so far are related to problems where the servo system does not lock up. There are other problems where it locks up, but at the incorrect time. These problems cause vertical sync pulses to be recorded too early or too late. The delay circuits between the PG coils and the phase comparator are the prime suspects for these problems. It is a good practice first to try to align these circuits according to the service manual's instructions. Any circuit that does not perform according to the service adjustment procedure is a suspect and should be examined.

The prime adjustments to be made are head switching, by adjusting PGA and PGB delays, lock phase delay, and the tracking delays. Tracking delays, PGA and PGB delays are adjusted during playback using the factor alignment tape. The lock phase delay is adjusted during record.

A contemporary servo circuit has most of its active servo circuits in a single IC chip. Defects in these circuits cause virtually the same symptoms to appear on the monitor, cuing the technician to the servo block. In this system both PG and FG pulses are used for controlling drum speed and phase, and capstan speed and phase.

Drum servo troubleshooting (Sony)

If the drum is not turning on a Sony unit, it will not shut down and the capstan will continue to pull tape to the reel. To check the drum motor, remove the drum error wire and connect an external 6.5 V to the terminal. The motor should turn. Failing to turn can indicate a defective drum motor.

If the motor turns, then the problem is either in the motor drive circuit or the drum/capstan servo IC. Adjust the external supply voltage to produce a 50% duty cycle (13.8 kHz) PWM signal at ICl/pin 34. The PWM at pin 37 will appear rhythmically on the scope's screen, but this is normal under this condition. The presence of PWM at both pins verifies the operation of IC1, and suspicion should then be focused on the LPF in IC5/pins 5 and 7, the amplifier in IC4/pins 12 to 14 and the drum motor drive (Q209 to Q214). Checking these circuits is easily done with voltage and component checks.

An absence of PWM signals at ICl/pins 34 and 37 prompts a scope check at pins 32, 33, and 40 for the presence or absence of PG pulses and leads the investigation toward the PG coils. Check connector CN18. It would be unlikely that all three PG coils will become defective simultaneously.

Drum servo lock problems (loss of horizontal sync and noise) are directly related to the internal operation of the IC. Check the drum phase PWM at pin 37. It should have about a 50% duty cycle (833 Hz) with a very slight jitter, indicating its corrective action. If the signal at pin 37 is incorrect, then adjustments of the drum free speed might correct the problem.

Also check the 3.58-MHz signal input at pin 28. It is used to develop internal reference VD for the phase detector. If all necessary signals and support voltages are present, then replace IC1.

Troubleshooting tile capstan servo

Should the capstan stop turning, the take-up reel will also stop turning. The reel sensors signal the system control microprocessor, which shuts down the unit. Using an external 5-V source at pin 4, with the wire on the connector disconnected, the capstan motor can be checked for proper operation. Verify the presence of 12 V at pin 2 and a good ground at pin 3. If the motor does not turn when 5 V is applied to pin 4, then it is probably defective.

Capstan problems can easily be identified by listening to the audio. Though the picture might be affected, wow and flutter will be dominant in the audio. This is a clue to the technician to troubleshoot the capstan servo.

Like the drum servo, the results obtained from checking the output PWM signals dictate the direction to follow to complete the repair. The procedure to troubleshoot the drum servo can also be used to troubleshoot the capstan servo. Troubleshooting involves

471

monitoring the timing and presence of the FG pulses, CTL pulses, and 3.58 MHz reference input. Capstan adjustments are made with the tracking controls and free speed adjustments.

Some VCR troubleshooting tips

A defective VCR might automatically point you toward the defective heads, servos, and system control problems that commonly cause VCR servicing headaches. In fact, technicians looking exclusively in these areas might overlook the silent but essential subassemblies of a VCR. These subassemblies are often assumed to be working, yet a simple defect can frustrate even the most experienced technicians if proper troubleshooting procedures are not followed.

The subassemblies in modem VCRs are not intended for component level repair, however. Most of them have to be purchased and replaced as a complete unit. The replacement cost of even one subassembly could exceed what a customer is willing to invest in the repair. And if you have to absorb the cost of one of these units later on, your costs go up.

You can be sure you are on the right track and eliminate unprofitable troubleshooting by proving the four common subassembly sections good or bad. Let's look at some quick troubleshooting techniques that use minimal testing.

Check the power supply

Without the correct power or B+ voltage supplied to the VCR circuits or other subassemblies, the VCR will not function properly. Technicians often forget this simple first step and go directly to the symptoms rather than to the cause of the defect. They eventually end up back at the power supply, but only after they have wasted a lot of time and profit.

Older VCRs use a simple power supply consisting of a transformer, diodes, regulators, and filtering components. Newer VCRs use switched mode power supplies. In either case, technicians have learned that a variable ac supply, isolated from the ac line, is a must for safe troubleshooting. The steps for checking the power supply are simple:

1. Adjust your ac supply to its lowest setting, preferably zero volts.

2. Turn on the VCR (you might have to repeat this as you perform step 3).

3. Watch the current as you slowly increase the voltage to the normal line voltage (stop as excess current is drawn).

4. Check a B+ supply voltages.

Excess current draw indicates a problem in the power supply or a circuit powered by the power supply. VCRs use a combination of analog and digital circuits. The power supply, either linear or switch mode, is called on to produce a wide variety of voltages, such as 5 Vdc for CMOS or TTL circuits, 12 Vdc at higher currents for motor driver chips, and voltages greater than 30 Vdc for tuning voltages. When testing the power supply, use an accurate digital voltmeter to verify the proper dc voltage levels and, just as important, check for noise. A quick and accurate way to measure dc voltage and ripple is with an oscilloscope that has digital PPV and DCV functions. Often when dealing with power supply defects, you only have a short time before components are damaged or the VCR shuts down.

Check the antenna (TV/VCR) switching circuits

If the VCR is powered OFF or is in the TV mode, a solid-state switching circuit connects the VCR ANTENNA IN terminal to the TO TV terminal. This circuit has to pass all VHF, UHF, and cable channel frequencies, from 54 MHz to over 900 MHz. Lightning can damage this circuitry, especially for VCRs connected to an outside antenna. A damaged switching circuit could result in no picture at all or a snowy picture on some channels and good reception on others.

Testing this type of circuit is simplified with an all-channel capable RF generator such as the Sencore VA62A universal video analyzer. Let's look at how it is done:

1. Connect a known-good TV to the TO TV connector on the VCR.

2. Connect the RF output of the VA62A to the ANT IN connector on the VCR.

3. With the VCR turned off, set the RF output level of the VA62A to 0 dB (1000 Uv). Virtually all TVs and VCRs should give a snow-free picture with this RF signal level.

4. Analyze the quality of the picture on several channels in all channel bands.

It is recommended to test both the VHF low band and VHF high band. On UHF check both the low and high ends (14 and 69). IF the TV is cable ready, make sure it is switched to normal instead of cable. (Note: Some TVs have mechanical switches and others are switched by using on-screen displays.

Cable channels 14 and above are different frequencies than broadcast channels 14 and above. To thoroughly check the VCR antenna switching circuit, the RF test generator must have cable channel frequencies.

A snowy picture or other problems during any of these tests indicates a problem in the antenna switching or its control circuit. With the proper test equipment, you can be sure from the beginning that you will not run into any surprises later.

Check the RF modulator

In the VCR mode, the antenna switching circuit routes the TV RF signal through demodulating circuits and then to the inputs of the RF modulator (see Fig. 12-16). The output is a modulated RF signal, switchable between channel 3 and 4.

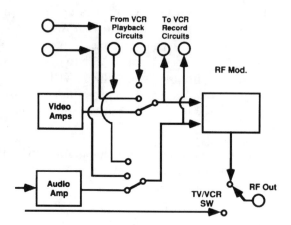

■ **12-16** *Without the proper testing methods, a defective RF modulator can lead you in the wrong direction.*

The RF modulator input can come from one of three possible sources: A/V input connectors, the VCR tuner/IF/video detector subassembly, or the VCR playback circuits. A quick check to see if you have a video input to the RF modulator is to check the signal at the video output jack. If you do not have a video signal at this point, you probably do not have a video input signal to the RF modulator either. In this case, you need to troubleshoot the playback or tuner circuits.

If you are not seeing a good picture on channel 3 or 4, you need to confirm that the modulator is working before you draw any conclusions. Unless you have a way to test the condition of the modulator, you could end up wasting time troubleshooting playback or other circuits.

Testing the RF modulator with the Sencore VA62A analyzer

1. Connect the VCR Ant Out jack to a TV (Ch 3 or 4).
2. Inject the video drive signal at the modulator input (usually 1 V-PP).
3. Analyze the quality of the picture.

A working modulator will produce a normal picture on the TV. If you do not see a marked improvement in the reproduced picture, make sure your VCR is in the VCR mode and the supply voltages are correct. You can even prove the condition of the modulator further by simply injecting channel 3 or 4 from the VA62A at the modulator's output. If a normal picture returns, you know the defect is within the modulator.

475

Testing the tuner/IF/detector circuits

Most VCRs have one or two "tin boxes" that contain the tuner/IF/detector sections. Your cost of $20 to $40 for these units is not unusual. One defective assembly could put you close to your customer's maximum limit for repairs. Fortunately, all three sections can be quickly tested using the Sencore VA62A video analyzer:

1. Inject the video IF (45.75 MHz) signal into the input of the IF section.
2. If the test pattern from the VA62A improves on the TV, the IF and detector sections are good.
3. If normal operation does not return, drive the output of the detector with the video drive signal (match the drive level with the circuit level).
4. If normal operation now returns, the problem lies between the IF input and the detector output.

With two simple signal injections, you have dynamically tested the condition of the IF and detector sections. If they test good, you need to test just one more section.

The last major subassembly that you should test is the tuner. It's an important and expensive assembly in the VCR and should be thoroughly analyzed on a variety of channels for proper operation. Some customers might use an outside antenna that receives VHF and UHF signals. Others will have cable and it might be different than your shop test cable. An off-the-air test signal is not very good. The only way to accurately analyze all tuners is with an all-channel adjustable level, RF generator such as the VA62A analyzer. You do not need to check all channels, but test two or three on each band for proper operation.

476

Glossary

ACC Automatic color control. Used to maintain an overall constant color signal level in the color circuits.

ACK Automatic color killer.

adjacent track The video track to the immediate left or right of the track of concern.

AFC Automatic frequency control. Used to phase-lock the color circuits to either the recording or playback color signal to achieve a stable color signal.

AFT Automatic fine tuning. A special circuit found in most recent TVs and VCRs that makes the local oscillator of the tuner follow the channel of concern in order to produce a stable IF frequency. In other words, if for any reason the TV station being received changes frequency, the AFT circuit automatically compensates so that no interference is visible on the screen. No manual fine tuning is necessary.

AGC Automatic gain control. Used to maintain an overall constant picture level in the luminance circuits.

APC Automatic phase control. Used to help phase-lock the color circuits either to the recording or playback color signal to achieve a stable color signal.

azimuth A term used to describe the left-to-right tilt of the gap of a recording head (viewed straight on).

balanced modulator A circuit designed to output the frequency sum or frequency difference of its two signals. Any special characteristics of one of the input signals are present in the output signal.

beats The unwanted signals produced when two original signals are allowed to be mixed together.

bipolar PG Pulse generator signals that have both positive and negative excursions.

burst A short-time occurrence (8 cycles to 10 cycles) of the 3.58-MHz subcarrier signal that appears right after horizontal sync but is centered on the blanking portion of the video waveform. Burst is used to keep the color oscillator of a TV receiver locked to the TV broadcast station.

candela   Unit of luminous intensity; equal to the luminous intensity of a surface of 1/600,000 square meter of a blackbody radiator at the temperature of freezing platinum under a pressure of 101,325 newtons per square meter (formerly candle; difference is negligible).

capstan   A small rotating metal dowel that, in conjunction with a pressure roller, drives the recording tape to ensure positive tape movement to the take-up reel.

chroma   The color portion of a video signal; the quality (hue, saturation) of a color.

chrominance   The color portion of a video signal; the difference between a reproduced color and a standard reference color.

clamp   The process of giving an ac signal a specific dc level.

complementary colors   Colors that lie on the opposite side of the white point in the chromaticity diagram from the primary colors, so that an additive mixture of the two, in appropriate proportions, can be made to yield an achromatic mixture.

control signal   A special signal recorded onto the videotape at the same time a video signal is being recorded. Used during playback as a reference of the servo circuits.

converted subcarrier   The process of frequency shifting the color 3.58-MHz subcarrier and its sidebands down to 629 kHz.

crosstalk   The name given to the unwanted signals obtained when a video head picks up information from an adjacent track.

C signal   The color portion of a video signal.

cue   To scan the playback picture at a faster-than-normal speed in the forward direction.

dark clip   After emphasis, the negative-going spikes (undershoot) of a video signal might be too large in amplitude for safe FM modulation. A dark clip circuit is used to cut off these spikes at an adjustable level.

delta factor   A term used to indicate that a playback signal has some jitter or wow and flutter. Delta factor, or a change in frequency, means that the color signal off the tape is not a stable frequency of 629 kHz, but rather a signal whose frequency at any instant is some small amount above or below 629 kHz.

densitometer An instrument that measures optical density by measuring the intensity of transmitted or reflected light; used to measure photographic density.

density The degree of opacity of a translucent material; the common logarithm of opacity.

deviation A term used to describe how far the FM carrier swings when it is modulated. In VHS the upper limit is 4.4 MHz.

dew detector A variable resistor whose resistance value depends on ambient humidity.

DDC Direct drive cylinder. As used in VHS, this means that the video heads are driven by a self-contained brushless dc motor using no belts or gears. DD cylinders produce pictures with better stability.

dihedral A term used to describe the relative position between the two video heads as they are mounted in the head cylinder. Perfect dihedral means that the tips of the heads are exactly 180 degrees apart.

DL Delay line.

dropout A momentary absence of FM or color signal due to uneven oxide or a coating of dust on the tape or video heads.

duty cycle In describing a rectangular waveform, the duty refers to the percentage of off-time and on-time for one complete cycle. A 50 percent duty cycle indicates there are equal periods of off-time and on-time for one cycle; this is a square wave.

E-E Electronics-to-electronics. The picture viewed on the TV screen when a recording is being made. This picture goes through some but not all of the recorder's circuits and is used to test the operation of said circuits.

emphasis The process of boosting the level of the high-frequency portions of the video signal.

EQ A shortened form of equalization, used in the audio record and playback circuits.

equal-energy source A light source for which the time rate of emission or energy per unit of wavelength is constant throughout the visible spectrum.

FG Frequency generator, used in the servo circuits.

field One-half of a television picture. A field consists of 262.5 horizontal scanning lines across the picture tube. Two fields are necessary to complete a fully scanned TV picture (frame). First, one field is scanned on the picture tube screen, starting at the top of the tube with line one, and ending at the bottom with line 262.5.

479

Then, the next field begins at the top of the tube again with line 262.5 and ends at the bottom with line 525. The lines of the second field lie in between the lines of the first field. This property of falling in between lines is called interlacing. The two sweeps of the picture tube, or two fields, make up one complete TV picture or frame. Frame repetition is 30 Hz, therefore field repetition is 60 Hz.

FL An abbreviation for filter.

flagwaving Term used to describe a TV set's ability to accept unstable playback pictures from a videotape recorder. All home VTRs have some degree of playback instability before the active picture is scanned. This can cause a bending or flapping from side to side of the top inch or so of the screen. This movement is called *flagwaving*.

F-number (F-stop) The ratio of the focal length of a lens to the maximum effective lens aperture or iris diameter; a lens of F/3.5 has a focal length 3.5 times its effective diameter. Indicates setting of lens aperture or iris; light passed by a lens varies inversely with the square of the F-number.

footcandle U.S. unit of illuminance; equal to one lumen per square foot.

footlambert U.S. unit of luminance; equal to one lumen per square foot; luminance produced by one footcandle being tristimulus values (X, Y, and Z recommended by CIE).

FM signal The luminance portion of the video signal used to control the frequency of an astable multivibrator. The output of this multivibrator is a frequency-modulated (FM) signal, shifting from 3.4 MHz to 4.4 MHz (pulse sidebands).

frame One complete picture. For more details, refer to field.

gate A circuit that delivers an output only when a specific combination of inputs is present. For use in analog or digital applications.

guard band This is the space between video tracks on the videotape when in the SP mode. Guard bands contain no information.

Hall effect IC An external magnetic field causes current to flow in this type of device.

HD Horizontal drive signal.

head cylinder A cylindrical piece of metal that houses the video heads. The tips of the heads protrude slightly from the surface of the cylinder so they can scan the tape as the cylinder spins.

head switching   The action of turning off the video head that is not in contact with the videotape. For example, a particular video head turns off 30 times per second.

head-switching pulse   The signal that is applied to the head amplifier to perform head switching. This is a square wave of 30 Hz with a 50 percent duty cycle.

helical   Describes a general type of VCR in which the tape wraps around the video head cylinder in the shape of a three-dimensional spiral, or helix. The video tracks are recorded as a series of slanted lines.

illuminance   Light incident upon a surface; luminous flux density (illumination).

interchangeability   A term used to describe how well a particular VCR can play back a tape recorded on another VCR of the same type.

interlacing   The property of how scan lines of two TV fields lie between each other. Refer to field.

interleaving   A term used to indicate that the harmonics of the chrominance signal lie between the harmonics of the luminance portion of the video signal as it is viewed on a spectrum analyzer. This indicates that the color information of a video signal does not interfere with, although it is broadcast at the same time as, the luminance information. Signals that have this interleaving property are not readily seen on a TV screen because of their virtual cancellation characteristics.

jitter   The name given for the effect on the playback picture if a VCR has too much wow and flutter. The picture appears to have a rapid shaking movement.

kelvin   SI unit of temperature; 0 degree K equals absolute zero. Used to specify temperature of blackbody radiator for color temperature.

lumen   Unit of luminous flux; equal to the flux through a unit solid angle from a uniform point source of one candela, or to the flux on a unit surface all points of which are at unit distance from a uniform point source of one candela (e.g. one candela at one foot from a one square foot surface).

luminance   1. The portion of video signal that contains the sync and black-and-white signal information. 2. Light emitted or scattered from a surface; formerly called brightness. (reflective screen luminance = illuminance X surface luminous reflectance (screen gain).

481

lux   SI unit of illuminance; equal to one lumen per square meter.

mired   A unit of measurement of the reciprocal of color temperature; equal to the reciprocal of color temperature in kelvins times 1 million. Used to specify filter correction of color temperature. Derived from Micro-REciprocal-Degree (mired value of light source + mired shift of filter = new mired value).

MMV   Abbreviation for monostable multivibrator. Usually, it's an IC device that gives a logic high or logic low output with a variable duration upon receipt of an input pulse or transition.

monochromatic color   Referring to a negligibly small region of the spectrum involving only one wavelength.

neutral density filter   An optical filter that reduces the intensity of light without appreciably changing its color.

nonlinear emphasis   Similar to regular emphasis with the difference that smaller level, high-frequency portions of the signal are given more of a boost than higher level, high-frequency portions.

NTSC   Abbreviation for National Television Systems Committee. These four letters identify the U.S. color television standard.

opacity   The light flux incident upon a medium divided by the light flux transmitted by it (also called filter factor when dealing with ND filters; inverse of transmission %).

PG   A pulse generator used in VCR servo circuits.

phot   SI unit of illuminance; equal to one lumen per square centimeter.

photometry   The measurement of light visible to the human eye.

primary colors   Three colors that can be combined in various proportions to produce any other color.

Q   A term used to describe the graphic response of a filter or tuned amplifier.

radiometry   The measurement of light within the total optical spectrum.

review   To scan the playback picture at a faster-than-normal speed in the reverse direction.

rotary chroma   The name of the process used in VHS to change the phase of the chrominance signal at a rate of 15,734 Hz (same as the TV horizontal sync frequency).

rotary transformer   A device used to magnetically couple RF signals to and from the spinning video heads, thus eliminating the need for brushes.

sample and hold   A process used in a comparator circuit where the value of a particular signal is measured at a specific moment in time and is stored for later use.

search   To scan the playback picture at a faster-than-normal speed in either the forward or reverse direction.

servo   Short for servo mechanism. This is an electromechanical device whose mechanical operation (for instance, motor speed) is constantly being measured and regulated so that it closely matches or follows an external reference.

skew   Another term for tension error. Skew is actually the change of size or shape of the video tracks on the tape from the time of recording to the time of playback. This can occur as a result of poor tension regulation by the VCR or by ambient conditions that affect the tape.

subcarrier   The name of the 3.58-MHz continuous wave signal used to carry color information.

SS   Slow and still picture modes.

tension error   Refer to skew.

time base stability   A term describing how closely the playback video signal from a VCR matches an external reference video signal in regard to sync timing, rather than picture content.

tracking   The action of the spinning video heads during playback when they accurately track across the video RF information laid down during recording. Good tracking indicates that the heads are positioning themselves correctly and are picking up a strong RF signal. Poor tracking indicates that the heads are off track and picking up low-level RF signals and noise.

tristimulos values   The magnitudes of three standard stumuli needed to match a given sample of light (X, Y, and Z recommended by CIE).

uniform chromaticity scale (UCS)   Chromaticity diagram formulated in 1976 similar to 1931 CIE except; equal distances on the diagram represent approximately equal perceived color differences and to the equal energy point.

VCO   Voltage-controlled oscillator. An oscillator whose frequency of oscillation is governed by an external voltage.

483

video head   The electromagnet used to develop magnetic flux that will put RF information on the tape. In VHS systems, two video heads are mounted in a rotating cylinder around which the videotape is wrapped. As the cylinder spins, each video head is allowed to alternately scan the tape.

video track   One strip of RF information laid down during recording onto a tape as the video head scans across the tape.

VHS   Abbreviation for the video home system type of VCR recording.

VTR or VCR   Videotape recorder or video cassette recorder.

V-V   Video-to-video, or the actual playback picture produced from a tape during playback.

VXO   Voltage-controlled crystal (x) oscillator. Similar to VCO except that a quartz crystal is used as a reference and can be varied.

white clip   After emphasis, the positive-going spikes (overshoot) of the video signal might be too large for safe FM modulation. A white-clip circuit is used to cut these spikes off, at an adjustable level.

XTAL   Abbreviation for crystal.

Y signal   The black-and-white portion of a video signal containing black-and-white information and sync.

Index

Illustrations are indicated in **boldface**.

487

488

489

490

491

493

About the author

Bob Goodman has been in the electronics service field for over 47 years, four of which were served in the U.S. Air Force (Korea) as a Radio-Radar Electronics Specialist. He also attended several electronics schools at Keesler A.F.B., Mississippi.

Goodman is a graduate of Capitol Radio Engineering Institute, Washington, DC, in the field of television engineering technology. He also completed two solid-state seminars (Motorola) at Louisiana Tech University and other electronics schools and courses.

Goodman holds a first-class Radio-Telephone License issued by the FCC and an Amateur Radio License (WA5KXH). Goodman is an engineering technician certified by the National Society of Professional Engineers and is a CET, Certified Electronics Technician.

Goodman has conducted advanced electronic troubleshooting seminars at many state and national electronic conventions and electronic vocational trade schools.

Goodman developed and wrote the first color TV service manual for TAB Books in 1968. Since 1968, he has written over 47 technical books for TAB/McGraw-Hill. The author's *Maintaining & Repairing VCRs* (first edition) has been translated into Chinese and published in Beijing, China, where he has sold over 100,000 copies. The second edition of this book was translated and published in China in 1990.